Advanced Physics

Advanced Physics

S. M. Geddes

MACMILLAN

First published 1981
Reprinted 1983, 1985

Published by
MACMILLAN EDUCATION LTD
Houndmills, Basingstoke, Hampshire RG21 2XS
and London
Companies and representatives
throughout the world

Printed in Hong Kong

British Library Cataloguing in Publication Data
Geddes, S M
Advanced physics.
1. Physics
I. Title
530 QC21.2
ISBN 0-333-27063-0

Contents

Wave Theory

Extra Topics

Preface

This book has been written for the GCE A-level candidate, aiming for a pass in the examination but probably not intending to proceed further in pure physics. Medical students and other non-mathematicians have been particularly in the author's mind, and the use of advanced mathematics has been deliberately kept to a minimum.

The subject is presented in a somewhat unconventional order, the result of more than fifteen years' teaching in bilateral and comprehensive schools.

There are ten units of work, designed to fit comfortably into the five terms of a normal sixth-form course and allow time for revision at the end. The easier parts of the various topics are covered first, so that in the second year they can be revised while being extended. Because of this sequence of development, the chapters should not be taken out of order. Previous knowledge of the subject to O-level (grade C) is assumed.

The simple numerical exercises set in the course of the text should be attempted when reached, as an aid to concentration. A selection of questions from recent A-level examinations is given at the end of each chapter, including both multiple-choice and longer questions, so that the requirements of the major examining boards are clearly illustrated.

Answers to all numerical problems are supplied at the end of the book.

S.M.G.

Acknowledgements

The author acknowledges gratefully the courtesy of the North East Surrey College of Technology in allowing access to their facilities of library and laboratory while this book was being prepared.

Thanks are also due to the following individuals: Mr C. Luce, Chief Technician, N.E.S.C.O.T., for help and advice with regard to many of the experimental techniques described here; Dr P. B. Coates, of the National Physical Laboratory, for up-to-date information concerning temperature measurements, and about the current-balance; Mr N. Geddes, Epsom College, and Mr S. J. Griffiths, formerly of Aberystwyth University, for supplying the answers to the exercises.

The author and publishers wish to thank the following Examining Boards for their co-operation:

The Associated Examining Board [AEB]
Joint Matriculation Board [JMB]
Oxford Delegacy of Local Examinations [O]
Oxford and Cambridge Schools Examination Board [OC]
University of Cambridge Local Examinations Syndicate [C]
University of London [L]

The author and publishers wish to acknowledge the following photograph sources:

J J Lloyd Instruments Ltd – Fig. 21
Dr A J F Metherell, University of Cambridge, Dept of Physics – Figs. 140, 168, 179, 183, 188, 192, 196
Science Museum, London – Fig. 70

Cover photograph – Royal Aircraft Establishment, Farnborough

The publishers have made every effort to trace the copyright holders of all illustrations but where they have failed to do so they will be pleased to make the necessary arrangements at the first opportunity.

Heat I

1 Measurement of temperature

Scales of temperature

Temperatures are now measured by scientists on the *thermodynamic* or Kelvin scale, using the symbol 'K'. 0 K, the zero of the thermodynamic scale, is *absolute zero*, which is defined from Charles' law (Chap. 30). However, temperatures are still often given in degrees Celsius because of their familiarity.

Many 'fixed points' have been defined for this scale of temperature, to cover the very large range now in use. The basic fixed point, in addition to absolute zero, is the *triple point* of water (defined later in Chap. 32) which has been given the value of 273.16 K so that the traditional ice-point and steam-point will remain at 0° C and 100° C for practical purposes.

The scale between these two basic fixed points is divided into 273.16 equal parts which are both kelvins and degrees Celsius, and it can be extended above the triple point indefinitely.

Thermometers

Any property of a substance which changes with temperature can be used as the basis of a thermometer. There are two main types of thermometer, those based on thermal expansion and those making use of electrical properties.

Thermal expansion

Most substances expand when they are heated, and the *expansivity* of a substance is defined as

$$\frac{\text{expansion}}{\text{(original size)} \times \text{(change in temperature)}}.$$

This expression can refer to changes of length, area or volume, and so the expansivity is called linear, superficial or cubic expansivity as appropriate.

Some values of the expansivity of common materials are given in Table 1.

Table 1

Substance	Expansivity
Aluminium	$2.3 \times 10^{-5}\ K^{-1}$ (linear)
Copper	1.7×10^{-5}
Iron	1.2×10^{-5}
Platinum	0.85×10^{-5}
Glass	0.8×10^{-5}
Pyrex glass	0.3×10^{-5}
Alcohol (at about 20°C)	$12 \times 10^{-4}\ K^{-1}$ (cubic)
Glycerol ,,	4.7×10^{-4}
Water ,,	2×10^{-4}
Mercury ,,	1.8×10^{-4}
All gases: cubic expansivity $= \frac{1}{273} = 3.66 \times 10^{-3}\ K^{-1}$ relative to their volume at 0°C	

Since all these values are small, it is permissible to take the cubic expansivity of a substance as 3 × linear expansivity if required, and its superficial expansivity as 2 × linear expansivity.

Bimetallic thermometers

As Table 1 shows, metals vary in their expansivity; this explains the bending of a compound bar when it is heated. This bending is the basis of many everyday thermometers and thermostatic devices. They are robust and cheap, direct-reading and compact, but of poor accuracy as the metals show fatigue after a while.
Range: no lower limit outside the laboratory, and up to the melting point of the materials used.

Liquid-in-glass thermometers

The values given in Table 1 for the expansivities of liquids are quoted for one temperature only, as a liquid expands much more rapidly when nearing its boiling-point. However, over suitable small ranges the expansivity of a particular liquid may be considered to be sufficiently constant for its use in a thermometer (this is never true in the case of water).
Range: Alcohol – 100° C → 50° C (roughly)
Mercury – 30° C → 300° C (roughly).

In use, these thermometers must always be given enough time for the heat to penetrate through the glass into the liquid. Ideally, the whole length of the stem occupied by the thermometric liquid should be inside the apparatus. Liquid-in-glass thermometers are compact, easy to use, direct-reading and cheap. Accuracy can be to about 0.1° C in ordinary use, but this lessens as the thermometer ages because of the gradual reduction of the strains set up in the glass during manufacture.

Gas thermometers

When a gas is heated, both its volume and its pressure increase. If one of these two factors is controlled, the other varies proportionately with the absolute or Kelvin temperature of the gas. A thermometer which attempted to measure the change of volume of a gas as it was heated would not be practical since the expansivity of gases is so high; the working volume of a thermometer must always be small so that it does not interfere with the conditions in which it is being used.

The constant-volume gas thermometer measures the change of *pressure* produced when a gas is heated. Experiment has shown that the coefficient of change of pressure is equal to the expansivity of a pure gas, and that it appears to be constant at temperatures above the liquefaction point of the gas.

The form of the apparatus commonly used in school laboratories is shown in Fig. 1. The gas is contained in a glass bulb from which a bent capillary tube leads to a mercury manometer. The constant volume which the gas is to occupy ends at a mark A which is opposite the zero of a vertical millimetre scale. The second arm of the manometer may be moved vertically against this scale. When the apparatus has been correctly adjusted at the temperature to be measured, the pressure of the gas can be calculated by adding the difference between the mercury levels at A and B to the value of the external atmospheric pressure acting at B.

The linear relation between pressure and Kelvin temperature means that both accurate interpolation and extrapolation are possible with relation to the calibration points. Readings from this thermometer form the *standard gas scale*, which is the practical way in which temperatures defined on the thermodynamic scale may be realised. The apparatus used for this purpose at the National Physical Laboratory is similar in principle to the simple form just described, although there the gas (helium) is contained in a one-litre gold-plated copper cylindrical vessel, and the arrangements for adjusting its volume and measuring its pressure are more sophisticated.

The gas thermometer is used only for standardising other thermometers and not for any practical applications, since it is bulky and its use takes up much time.

EXERCISES

Take the appropriate values for the expansivity of the various substances from Table 1 as required.

1 A bar of aluminium, 1.000 m long at 20° C, is heated to 100° C. What will its new length be?

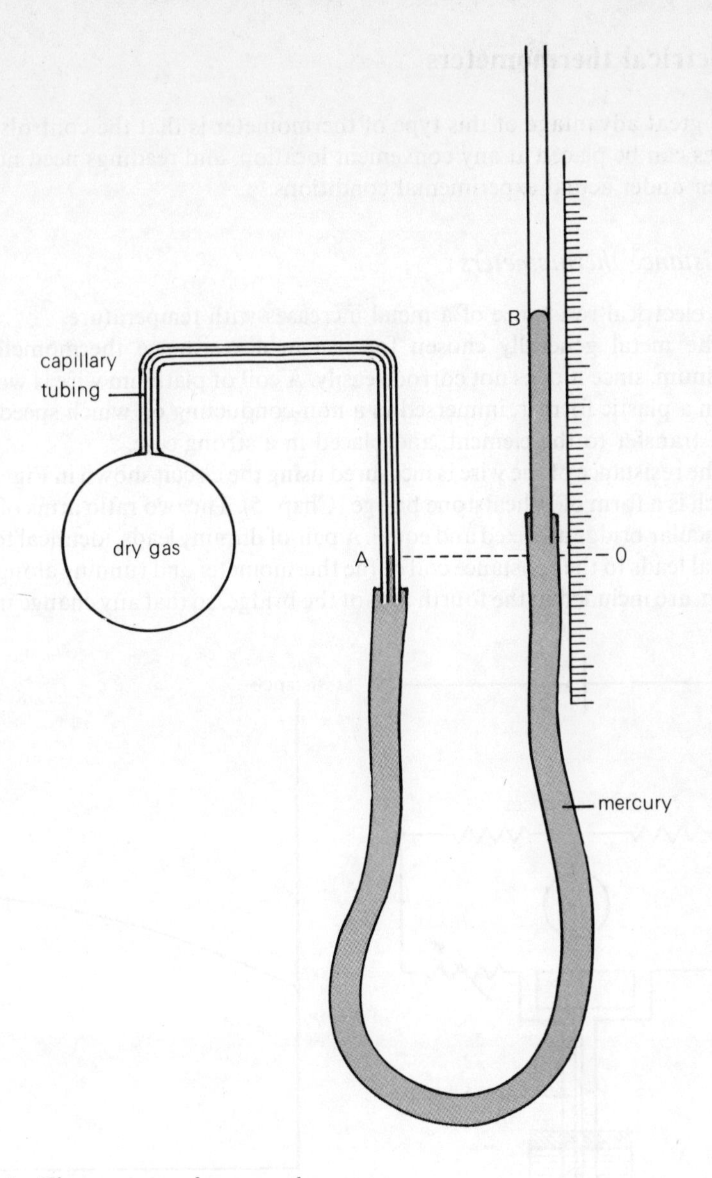

Fig. 1 The constant-volume gas thermometer

2 A specific gravity bottle made of glass has a capacity of 40 ml at 10° C.
(i) What will its capacity be at 80° C?
(ii) If it is filled with glycerol at that temperature, find the volume of the air space that will exist in the bottle when it and its contents have cooled down to 10° C again.

3 Referring to the apparatus shown in Fig. 1, when this is correctly adjusted at 0° C, the mercury at B stands 20 mm higher than at A. If the atmospheric pressure is equal to a height of 780 mm of mercury, calculate the new position of B when the apparatus is used at a temperature of 100° C.

Electrical thermometers

The great advantage of this type of thermometer is that the controls and scales can be placed at any convenient location, and readings need not be taken under actual experimental conditions.

Resistance thermometers

The electrical resistance of a metal increases with temperature.

The metal generally chosen for use in a resistance thermometer is platinum, since it does not corrode easily. A coil of platinum wire is wound upon a plastic former, immersed in a non-conducting oil which speeds up heat transfer to the element, and placed in a strong case.

The resistance of the wire is measured using the circuit shown in Fig. 2(a), which is a form of Wheatstone bridge (Chap. 5). The two ratio arms of this particular bridge are fixed and equal. A pair of dummy leads, identical to the actual leads to the resistance coil of the thermometer and running alongside them, are included in the fourth arm of the bridge, so that any change in the

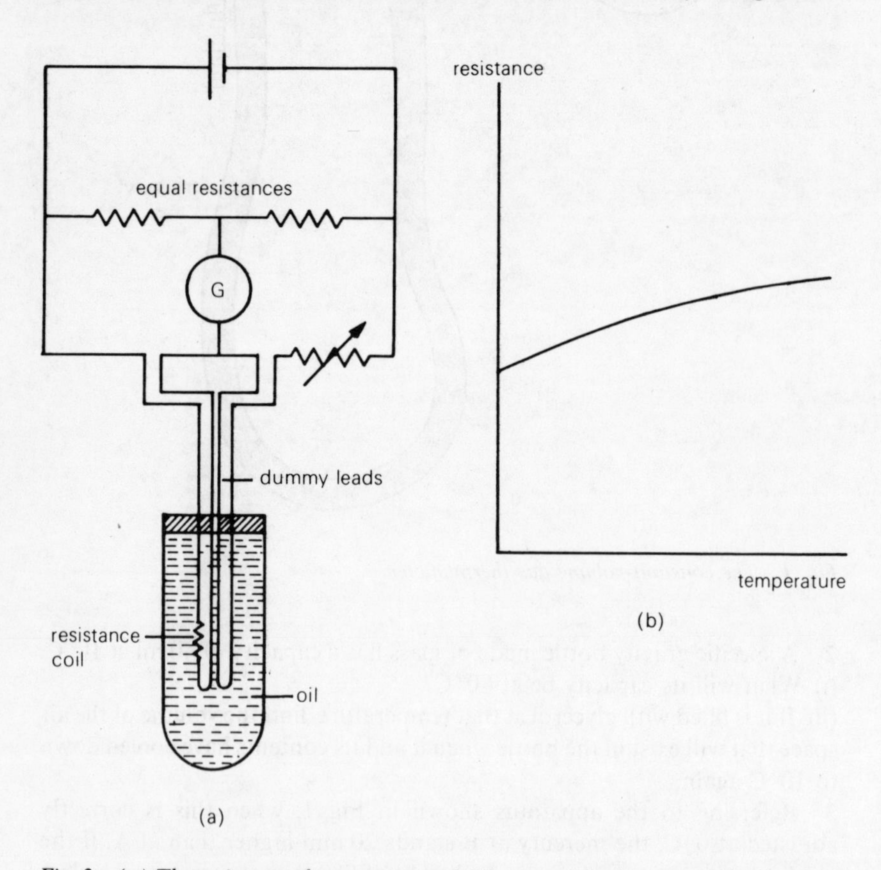

Fig. 2 (a) The resistance thermometer and Wheatstone bridge circuit (b) Graph of temperature and resistance

resistance of the leads themselves can be eliminated. The variable resistance also in the fourth arm is adjusted until the bridge is balanced, when it must be equal to the resistance of the coil.

The graph of resistance against temperature, Fig. 2(b), is a curve, and can only be considered as approximating to a straight line over a small range (certainly not more than one hundred degrees Celsius). Over larger ranges, a calibration curve must be used, which can be obtained by comparing the readings of the resistance thermometer with those of a gas thermometer under identical conditions.

The platinum resistance thermometer has been accepted as the standard for temperature measurements up to the fixed point of 630° C (the freezing point of antimony). It is extremely accurate but rather slow-reading because of its large heat capacity.

Thermocouples

If two different metals are joined together in an electrical circuit, and one junction is kept at a higher temperature than the other, a small e.m.f. (a few mV) is produced and will drive a current around the circuit. This is known as the *thermoelectric* or *Seebeck effect*. The current can be measured using a sensitive milliammeter or galvanometer, or the e.m.f. can be measured directly using a special form of potentiometer circuit.

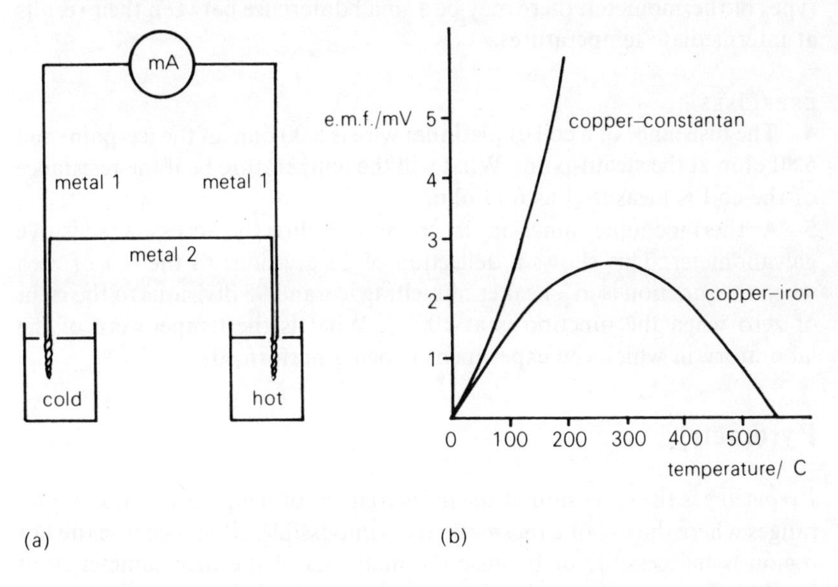

Fig. 3 (a) Circuit for use with a thermocouple (b) Graph of temperature and e.m.f.

Figure 3 shows the basic circuitry and the resulting graphs for two different combinations of metals, the cold junction being kept at 0° C and the hot junction at the temperatures given.

Both graphs are in fact curves. Since the curves are roughly parabolic, it is possible for the same e.m.f. to be produced at two different temperatures. Hence, if the thermocouple is to be used as a thermometer, its working range can be only up to the turning-point of the curve, which of course differs for different pairs of metals.

The combination of platinum with an alloy of 90 % platinum and 10 % rhodium forms a thermocouple which is the standard thermometer over the range 630° C to 1063° C (the freezing-point of gold). It is accurate, and very convenient in use since the junction of the couple occupies a very small space and can measure the temperature almost at a point; there is no time-lag in its response to changes of temperature.

General law of thermometry

If R_t is the reading of a thermometer at temperature t, then

$$t = \frac{R_t - R_0}{R_{100} - R_0} \times 100 \quad °C.$$

The value of R_t may be in any convenient experimental unit, as the units will cancel out on substituting figures into the expression.

This equation assumes that the graph of R and t is a straight line at least over the range 0–100°C. Since this is only approximately true for many types of thermometer, there may be a small difference between their results at intermediate temperatures.

EXERCISES

4 The resistance of a coil of platinum wire is 5.00 ohm at the ice-point and 6.80 ohm at the steam-point. What will the temperature be if the resistance of the coil is measured as 6.11 ohm?

5 A thermocouple junction is connected directly across a sensitive galvanometer. This shows a deflection of 23 divisions to the left of zero when the junction is in a beaker of melting ice, and 98 divisions to the right of zero when the junction is at 100° C. What is the temperature of the laboratory in which the experiment is being performed?

Pyrometry

Pyrometry is the extension of the measurement of temperature into higher ranges where the use of a thermometer is impossible, either because the hot region is inaccessible or because the materials of the thermometer itself would become molten under those conditions. Techniques must be based on measurements of thermal radiation.

The simplest method is to observe the colour of the hot body. An experienced furnace-man can tell the temperature of his furnace from the colour of the light coming from it. This principle has been developed scientifically in the *disappearing-filament optical pyrometer*.

The observer looks at the hot body through a small telescope. A filament of resistance wire crosses the field of view within the telescope, and a current can be passed through the wire from batteries stored in the handle of the instrument, so that the wire becomes hot. The observer adjusts the current by turning the knob of a small rheostat, until the colour of the heated wire matches that of the hot body as nearly as possible, i.e. the filament 'disappears' against its background. An ammeter incorporated in the instrument gives the temperature of the hot body directly, having previously been calibrated against a standard pyrometric lamp. Commercial pyrometers work generally over a range of about 700°C to 3000°C.

A disadvantage of this type of instrument is the personal factor involved in matching the colours. It may also be dangerous to the eye to view an intensely hot body directly. More accurate pyrometric measurements are based on photoelectric effects (Chap. 11).

Examination questions 1

1 Describe how you would use a metre bridge to investigate the variation in resistance of a length of copper wire with temperature as measured with a mercury thermometer. How would you use the results of your experiment to determine room temperature on the copper resistance temperature scale?

A heating coil is immersed in a copper calorimeter containing water at 0° C. The calorimeter is well lagged and heat losses are negligible. The initial rate of rise of temperature is $5.0\ \mathrm{K\,s^{-1}}$. After an interval of time, during which the voltage applied to the heater is kept constant, the temperature rose to 30.0° C and the rate of rise of temperature at this point became $4.5\ \mathrm{K\,s^{-1}}$. Given that the resistance R_θ at temperature θ° C is related to the resistance R_0 at 0° C by the formula $R_\theta = R_0(1+\alpha\theta)$, calculate the value of α. [AEB June 1979]

2 An advantage of the platinum resistance thermometer is that

A it may be used to measure rapidly changing temperatures.

B it has a linear scale, because the resistance of a piece of platinum varies directly as thermodynamic temperature.

C it may be used to measure steady temperatures with very high accuracy.

D it absorbs energy from its surroundings very slowly so that it does not disturb the condition of the body under test when placed in contact with it.

E it is the only type of thermometer that can measure accurately temperatures over 3000 K. [C]

3 (a) Explain how a temperature scale is defined.

(b) Discuss the relative merits of (i) a mercury-in-glass thermometer, (ii) a platinum resistance thermometer, (iii) a thermocouple, for measuring the temperature of an oven which is maintained at about 300° C. [JMB]

4 The value of the property X of a certain substance is given by

$$X_t = X_0 + 0.50t + (2.0 \times 10^{-4})t^2$$

where t is the temperature in degrees Celsius measured on a gas thermometer scale. What would be the Celsius temperature defined by this property X which corresponds to a temperature of 50°C on this gas thermometer scale? [L]

5 Explain the physical principles underlying one electrical method of measuring temperature that can be used in a school laboratory. Assuming that you possess a mercury thermometer correctly graduated in terms of the Celsius temperature scale, describe how you would calibrate the electrical thermometer for the temperature range 0–300° C and how you would use it to measure temperature. How would you attempt to calibrate the electrical thermometer if no such mercury thermometer were available?

Discuss the advantages and disadvantages of the instrument you describe as a means of measuring temperature. Assuming that it is used for temperatures up to 600° C, discuss the order of accuracy you would expect it to give at the higher temperatures outside its range of calibration. [O]

2 Calorimetry

The basic problem of accurate calorimetry (the measurement of quantities of heat) is that of avoiding unwanted heat losses, so that the measurements made of mass, specific heat capacity and temperature change provide all the information necessary to use in the equation:

$$\text{heat gained} = \text{heat lost.}$$

Determination of the specific heat capacity of a metal

Mechanical method

Figure 4 shows one of many possible forms of apparatus designed for this experiment. The general principle is that a measurable quantity of heat is supplied through friction to a known mass of the metal, and the resulting rise in temperature is observed.

In the form shown in Fig. 4, a stout belt of fabric is wrapped around a metal cylinder. The belt is kept taut by a load hanging from one end, while the other end is secured to a dynamometer (a spring balance) hanging from a rigid support. The cylinder is rotated steadily within this belt, either by hand or by a low-power motor. The frictional force produced can be taken as equal to the difference between the weight of the hanging mass and the reading of the dynamometer.

A thermometer is placed in a hole along the axis of the cylinder. The bulb of the thermometer should be wrapped in thin metal foil and covered with a little oil, so that it makes good thermal contact with the cylinder.

During the experiment, the cylinder is rotated through a predetermined number of revolutions so that the mechanical work done by the frictional force can be calculated. This calculation involves knowing the circumference of the cylinder, and its diameter should be measured using vernier callipers.

A rough run of the experiment will be needed to check that the temperature rise produced is sufficient to be measured accurately. The thermometer should be a sensitive one, reading over about 10–30° C in 0.05° C divisions. Its heat capacity is usually stated by the manufacturer. The load attached to the belt can be altered at this stage if necessary.

The cylinder must be weighed if its mass is not known. In the apparatus illustrated, the longitudinal holes in the cylinder serve to reduce its mass and so increase the temperature rise.

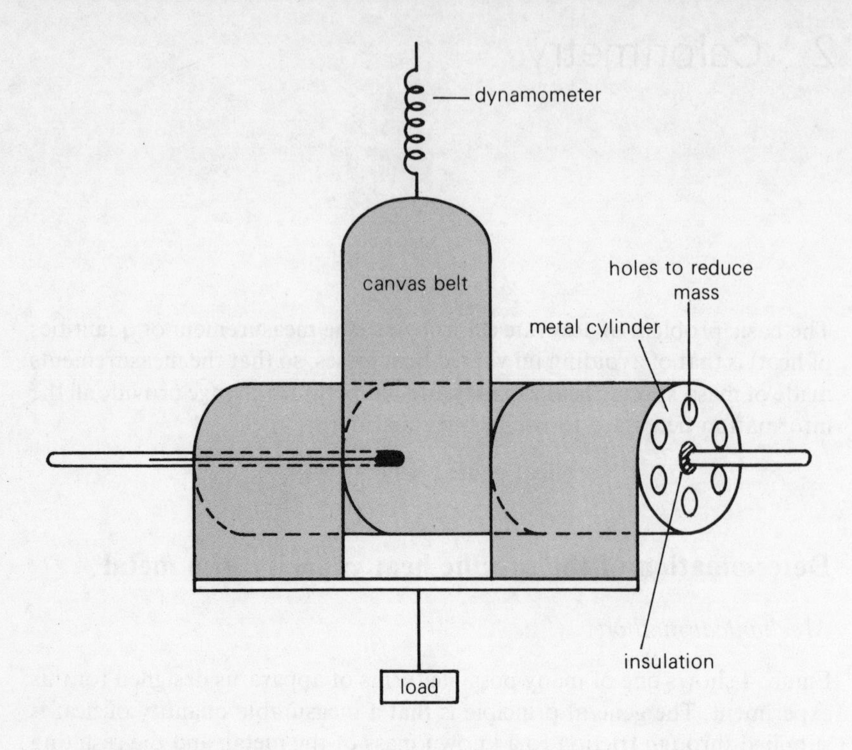

Fig. 4 Mechanical method for determination of specific heat capacity

Loss of heat to the surrounding air is minimised by using a shield supplied with the apparatus, fitting around the cylinder but with slots through which the ends of the friction band can pass. An insulating layer around the axle of the cylinder reduces heat loss by conduction. Since in practice the temperature rise is only a few degrees, if the experiment is carried out speedily, heat losses can be ignored.

The accuracy of this method cannot, in any case, be expected to be high.

Electrical method

The metal used for this experiment should be in the form shown in Fig. 5, a block drilled with two cylindrical holes, one at the centre to contain a small immersion heater and the other to hold a thermometer. A little oil is placed in the holes to improve thermal contact.

A suitable heater is available at 25 W. This delivers a steady supply of heat to the metal, and the steady temperature rise produced is observed by taking a series of readings of the thermometer at noted times. These readings should be graphed to check that the rise is indeed steady, i.e. the graph should be a straight line. The supply of electrical energy to the heater must be monitored throughout the experiment by the use of both ammeter and

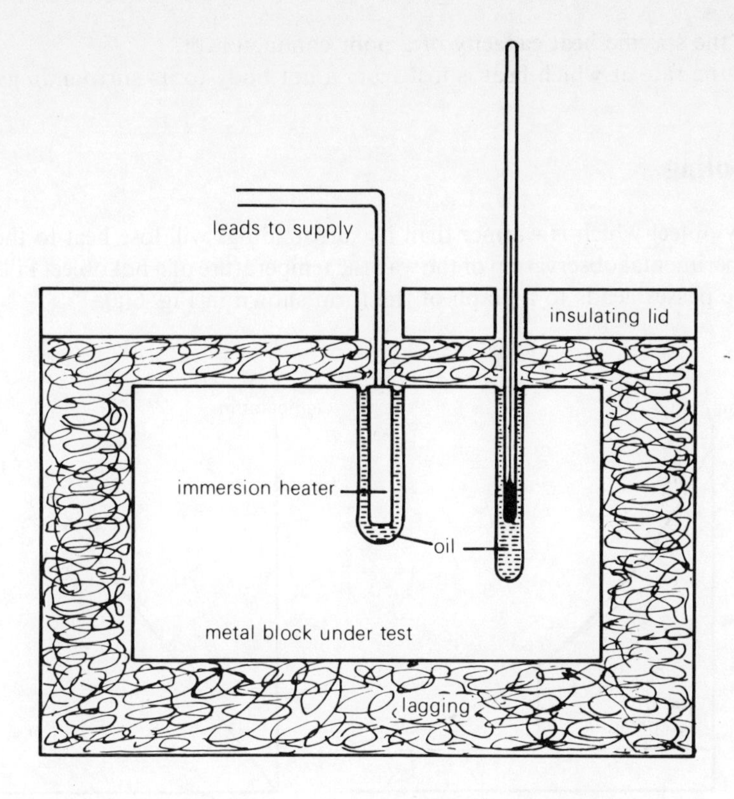

Fig. 5 Electrical method for determination of specific heat capacity

voltmeter, and a rheostat should be included in the circuit to help to keep the current constant.

The mass of the metal block must be determined, and the heat capacity of the thermometer should be included in the calculations if it is known.

In its simplest form, this experiment ignores heat losses into the lagging, which is excusable if the readings are taken fairly quickly and the temperature rise is small. For greater accuracy a cooling correction should be applied to the results, by carrying out a separate experiment to measure the actual rate at which heat is lost from the block under these conditions. How this is done is described in detail in connection with Lees' disc, in Chapter 3.

Experiments using non-metals

If the substances to be used in experiments are not good conductors of heat, then the heat exchange process may take some appreciable time, and the losses to the surroundings cannot be ignored.

The methods discussed in the following sections may be applied to the experimental determination of

(a) the specific heat capacity of a poor conductor, or
(b) the rate at which heat is lost from a hot body to its surroundings.

Cooling

Any object which is warmer than its surroundings will lose heat to them. Experimental observation of the way the temperature of a hot object falls as time passes, leads to a graph of the form shown in Fig. 6(a).

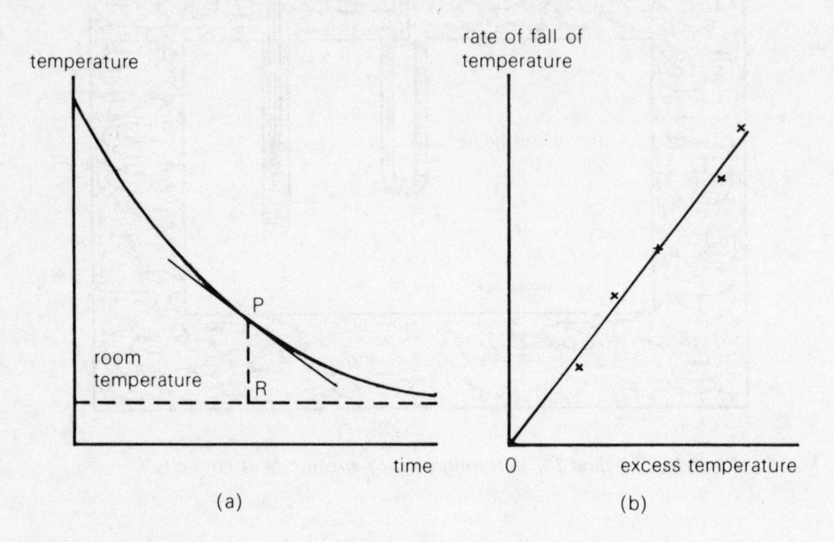

Fig. 6 Graphs to illustrate the cooling of a hot body

This is an exponential curve, showing that **the rate of fall of temperature is proportional to its excess above room temperature**. This is *Newton's law of cooling*. The straight-line graph shown in Fig. 6(b) is obtained from the first by drawing tangents to the curve at suitable points; the slope of the tangent at P is then plotted against the excess temperature PR.

The two graphs obtained are, of course, applicable only to the particular system and conditions used in the experiment. Other experimental arrangements will yield similar graphs, and by comparing these graphs, values for ratios of specific heat capacities, etc. can be obtained.

Cooling corrections in calorimetry

Suppose a large hot object is dropped into cold water, and the temperature of the water is then taken at intervals.

The results obtained will be of the form shown in Fig. 7. Over the portion AM, heat is flowing from the hot body into the water. At M, the two have reached the same temperature, and from then on, the system is cooling back to room temperature.

Loss of heat to the surroundings will also have taken place during the time AX, with the result that the measured maximum temperature M will not be as high as it should be.

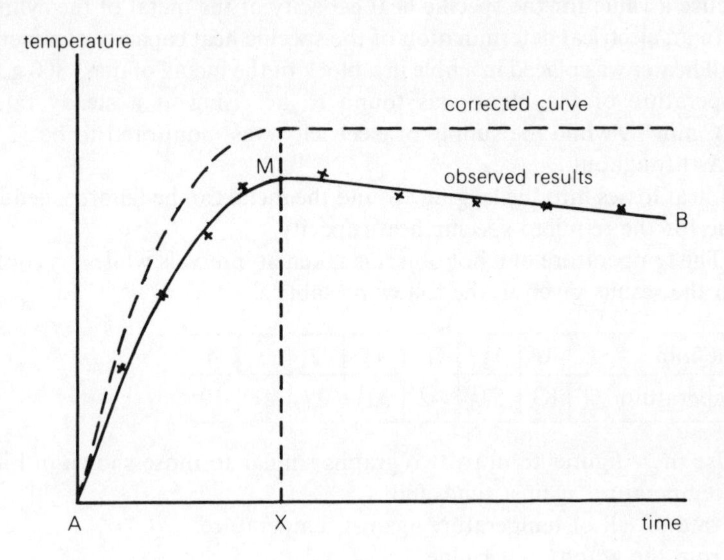

Fig. 7 Temperature–time graph for a method-of-mixtures experiment

There is a simple method of finding the necessary correction which may be used under certain experimental circumstances. This is when the rate of loss of heat to the surroundings is low, and the second part of the graph, from M onwards, is approximately a straight line of low gradient.

Calculation of the slope of this line gives the rate of fall of temperature of the system, x degrees per second (say). This result is then applied to a succession of points along the curve, so that one second after A, the temperature should have been x degrees higher; after two seconds, it should have been $2x$ degrees higher, and so on. Application of this correction over the whole graph will give it the shape shown by the dotted curve in Fig. 7, which is as would be expected if there were no external heat losses.

This simple method of correction is permissible if the time AX is not too long and the slope of MB is shallow. In other circumstances, or for high accuracy, rather more complicated calculations are required to give a valid figure for the cooling correction to be applied.

EXERCISES

1 In an experimental run using the apparatus shown in Fig. 4, the following results were obtained:

Diameter of cylinder = 100.0 mm
Mass of cylinder = 0.45 kg (The thermometer used was a thermocouple, whose heat capacity can be ignored.)
Load = 1.5 kg
Dynamometer reading during run = 4.5 N
Temperature rise after 100 revolutions = 1.95° C.

Deduce a value for the specific heat capacity of the metal of the cylinder.

2 In an electrical determination of the specific heat capacity of a metal, a small heater was placed in a hole in a block of the metal, of mass 800 g. The temperature of the block was found to be rising at a steady rate of 2.5°C min^{-1}, while the supply of electricity was monitored to be 12.0 V, 2.4 A throughout.

If heat losses into the lagging around the metal can be ignored, deduce a value for the required specific heat capacity.

3 The temperature of a hot object is taken at intervals while it is cooling, with the results given in the following table.

Time/min	0	$\frac{1}{2}$	1	$1\frac{1}{2}$	2	3	4
Temperature/°C	83	$57\frac{1}{2}$	42	$32\frac{1}{2}$	27	21	19

Use these figures to draw two graphs similar to those shown in Fig. 6:
(a) temperature against time, and
(b) rate of fall of temperature against temperature.

From the graphs, determine
(i) the temperature of the surroundings, and
(ii) the rate at which the object's temperature is falling when it is 30°C above that of the surroundings.

4 In a method-of-mixtures experiment, a hot object is dropped into some water, and the temperature of the water is taken every half-minute afterwards, until the maximum is clearly passed. The results are given in the following table.

Time/min	$\frac{1}{2}$	1	$1\frac{1}{2}$	2	$2\frac{1}{2}$	3	$3\frac{1}{2}$	4	$4\frac{1}{2}$	5	$5\frac{1}{2}$	6
Temperature/°C	35.2	37.6	38.2	37.7	36.8	36.0	35.4	34.8	34.0	33.2	32.5	32.0

Use these figures to draw a graph similar to the one shown in Fig. 7, and deduce the corrected shape of the graph if no heat had been lost from the water while the readings were being taken.

What is the corrected maximum temperature of the mixture?

Continuous-flow methods

The apparatus shown in Fig. 8 is a laboratory version of a design due to Callendar and Barnes in 1899, used for determination of the *specific heat capacity* of a liquid.

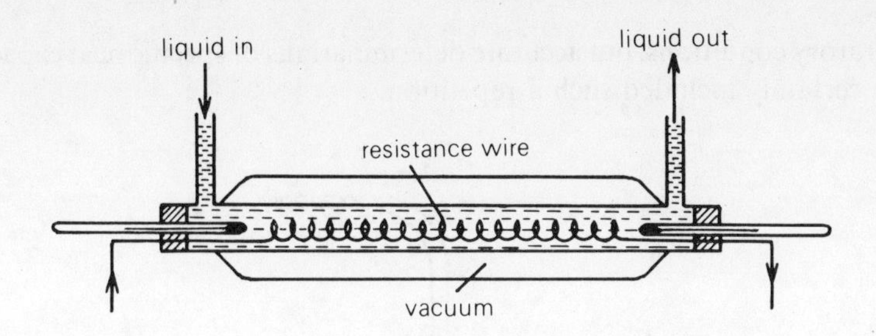

Fig. 8 Continuous-flow method for determination of specific heat capacity of a liquid

The liquid flows slowly and steadily past a spiral of resistance wire which carries a current. The heat developed in the resistance is carried away by the liquid, and eventually a steady state is reached in which all parts of the apparatus remain at constant temperature. From then on, *no heat is being absorbed by the walls of the vessel, nor by the thermometers, and their heat capacities need not be involved in the calculations.* This is one great advantage of continuous-flow experiments.

The glass tube of the commercial apparatus is about 0.3 m long and 20 mm in diameter. It is surrounded by a vacuum jacket to act as lagging. A pair of suitable mercury thermometers is supplied to fit the rubber bungs at each end of the tube, and measure the temperatures of the incoming and outgoing liquid. These thermometers should be calibrated in 0.05°C divisions.

The liquid enters and leaves the apparatus through side tubes, and must be supplied from a 'constant-head' so that its rate of flow is uniform. This rate must be determined at some convenient stage of the experiment, by collecting the outflow in a measuring cylinder over a measured period of time. If the liquid is not water, its density must be known, or found.

A suitable electricity supply will have been quoted by the manufacturers of the apparatus, probably about 12 V, 2.5 A. This should be monitored during the run of the experiment, using both ammeter and voltmeter and controlling the current by the use of a rheostat.

While the experiment is in progress, readings of the two thermometers should be noted, about every five or ten minutes, in order to be sure that it reaches a steady state. This may easily take half an hour or so. The last pair of readings will be used in the calculation. The second advantage of the continuous-flow method is that *there is no urgency over taking these temperature readings.*

It is possible to allow for the small amount of heat which escapes through the walls into the surroundings. This can be done by re-running the experiment with a slightly different head of liquid, and adjusting the supply of electricity until the thermometer readings are the same as before. In this case, the rate of heat loss to the surroundings will be unchanged, and this unknown term in the equations can be eliminated by combining the results of the two experimental runs. There is not time to do this under normal laboratory conditions, but accurate determinations of specific heat capacity have certainly included such a repetition.

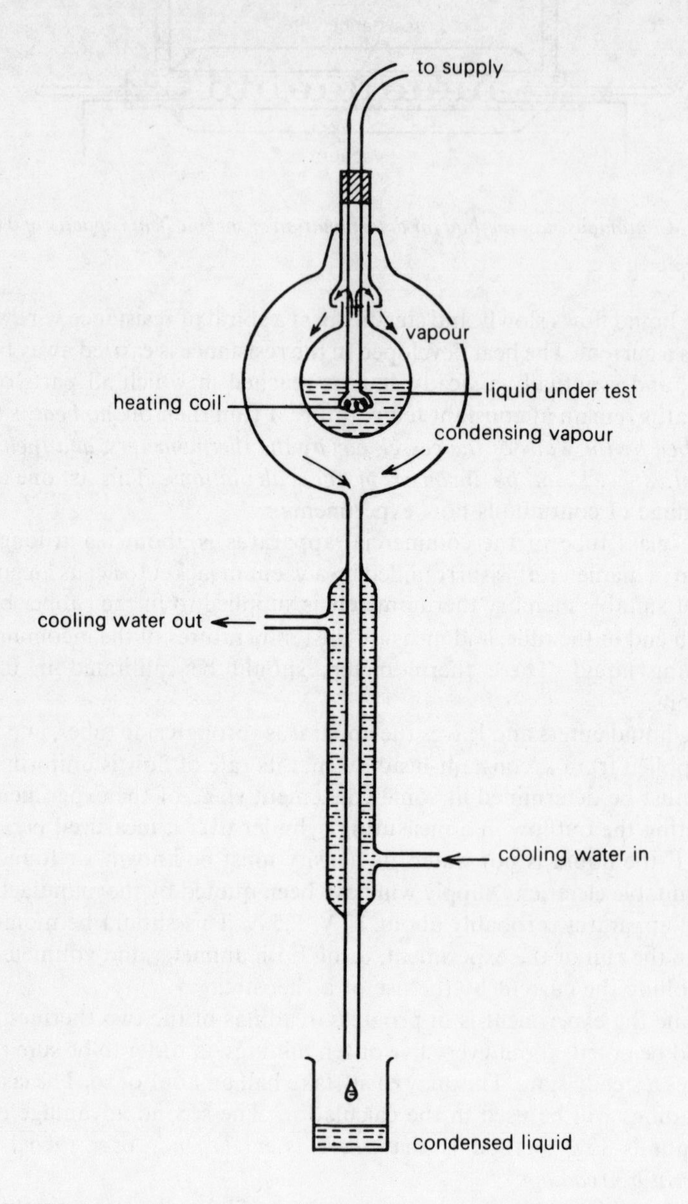

Fig. 9 Continuous-flow method for determination of specific latent heat of vaporisation of a liquid

Figure 9 shows a specially designed piece of glass apparatus used in the determination of *specific latent heats of vaporisation* of liquids. About 100 ml of the liquid under test is placed in the inner vessel, where it is raised to boiling-point by a small electrical immersion heater; a typical rating for this heater could be 24 V, 36 W.

The vapour from the liquid escapes through holes at the top of the inner flask, and then passes downwards through the outer vessel to a tube leading out at the bottom. This is connected to a Liebig condenser as shown, where the vapour is condensed so that it can be collected. When the experiment is running steadily, the rate of condensation of the vapour is found by collecting the drops of outflowing liquid during a measured time as described before, while the supply of electrical energy is determined from ammeter and voltmeter readings, monitored and controlled as before.

Although the glass walls of the apparatus will be at constant temperature and their heat capacities need not be included in the calculation, allowance must be made for the heat losses *through* the outer wall in this experiment, since no lagging is used. However, the experiment reaches its steady state much more quickly than in the previous case, and it is not difficult to re-run it using a different value of current. Since the supply of electrical energy has been altered, a different amount of liquid will be vaporised per second, yet the temperatures obtaining in the various parts of the apparatus in the steady state will be the same as during the first run. The heat loss through the outer wall will therefore be the same in both runs, and may be eliminated by combining the two sets of results.

EXERCISES

In an experiment using a version of Callendar and Barnes' apparatus, the following results were obtained:

steady readings of thermometers,
liquid entering = 18.8°C
liquid leaving = 21.1°C
volume of outflowing liquid collected in 300 s = 900 ml
density of liquid = $1000\ \text{kg m}^{-3}$
average voltmeter reading = 12.0 V
average ammeter reading = 2.6 A

5 Ignoring heat losses from the apparatus, deduce a value for the specific heat capacity of the liquid.

6 To allow for heat losses, the experiment was re-run with the following different results:

Volume of outflowing liquid collected in 300 s = 600 ml
Average voltmeter reading = 12.1 V
Average ammeter reading = 1.8 A

Deduce (i) an amended value for the specific heat capacity of the liquid, (ii) the rate at which heat was escaping through the walls of the apparatus during the experiments.

Examination questions 2

1 Explain what is meant by *specific latent heat of fusion.*

A mass of tin, melting-point approximately 230°C, is heated to a temperature of 300°C and allowed to cool to 100°C. Describe how, using a thermocouple to measure the temperature, you would obtain a cooling curve over the given range. Sketch a typical cooling curve. Discuss the factors that determine (i) the time it takes for the tin to solidify completely, and (ii) the slopes of the curve at different points.

0.005 kg of water, initially at 20°C, is poured into a large Dewar vessel containing liquid nitrogen at its boiling-point (– 196°C). Calculate the mass of nitrogen vaporised.

(Specific latent heat of vaporisation of nitrogen = 2.50×10^5 J kg^{-1}
specific latent heat of fusion of ice = 3.34×10^5 J kg^{-1}
specific heat capacity of water = 4.2×10^3 J kg^{-1} K^{-1}
mean specific heat capacity of ice = 2.2×10^3 J kg^{-1} K^{-1})

[AEB, Nov. 1974]

2 Explain what is meant by the *internal energy* of a system and its *heat capacity*. Show how they are related.

Summarise the advantages of electrical methods for the determination of specific heat capacities.

A thermally insulated crystal of mass 0.028 kg is heated electrically at a constant rate of 3.0 W. The following readings of the temperature T of the crystal were taken at various times t:

T/K	179	196	207	210	210	210	215	230	243	256	264
t/s	0	200	400	560	640	740	800	1000	1200	1400	1520

(i) Draw a graph of T against t.

(ii) Use the graph to find the specific heat capacity of the crystal at a temperature of 250 K.

(iii) Sketch a graph showing the way in which the specific heat capacity of the crystal varies with temperature. (The calculation of *values* of the specific heat capacity, apart from that at 250 K, is *not* required.) [C]

3 (a) In terms of the kinetic theory of matter explain why energy must be supplied to a liquid in order to vaporise it.

(b) Describe, with the aid of a labelled diagram, an electrical method for the determination of the specific latent heat of vaporisation of a liquid. Explain how the result is derived from the readings taken.

(c) When a piece of ice of mass 6.00×10^{-4} kg at a temperature of 272 K is dropped into liquid nitrogen boiling at 77 K in a vacuum flask, 8.00×10^{-4} m^3 of nitrogen, measured at 294 K and 0.75 m of mercury pressure, are produced. Calculate the mean specific heat capacity of ice between 272 K and 77 K. Assume that the specific latent heat of vaporisation of the nitrogen is 2.13×10^5 J kg^{-1} and that the density of nitrogen at s.t.p. is 1.25 kg m^{-3} [JMB]

4 (a) Describe and give the theory of a mechanical method for finding the specific heat capacity of *either* water *or* copper. Explain how corrections for loss of heat might be made.

Explain briefly how one could make use of the value so obtained to find the specific heat capacity of the other material.

(b) An electric kettle with a 2.0 kW heating element has a heat capacity of 400 $J\,K^{-1}$. 1.0 kg of water at 20°C is placed in the kettle. The kettle is switched on and it is found that 13 minutes later the mass of water in it is 0.50 kg. Ignoring heat losses, calculate a value for the specific latent heat of water.

(Specific heat capacity of water = $4.2 \times 10^3\ J\,kg^{-1}\,K^{-1}$.) [L]

5 Which one of the following statements about the continuous-flow method for determining the specific heat capacity of a liquid is *incorrect*?

A Temperatures are steady and therefore may be accurately measured.

B The thermal capacity of the apparatus is irrelevant.

C The effect of heat losses is small and simple to correct for.

D The determination can be made over any chosen temperature range.

E For a given power input the temperature rise is independent of the flow rate. [O]

3 Thermal conduction

The type of energy called *heat* is, on the atomic scale, kinetic energy of the individual atoms. The kinetic theory of matter gives simple, acceptable explanations of expansion and change of state, and also of the way in which heat travels through a substance by conduction, the atoms in turn vibrating more vigorously and passing on some of this extra energy to their neighbours.

Substances which are good conductors of heat tend to be those with a simple lattice structure while the poor conductors have larger, more complicated molecules, often man-made or organic. The process of conduction of heat through metals in particular is very similar to that of the conduction of electricity (Chap. 4).

Variation of temperature along a conductor

If a metal bar is heated at one end, and its temperature taken at intervals along its length, the results will be of the form shown in Fig. 10(a). Heat is

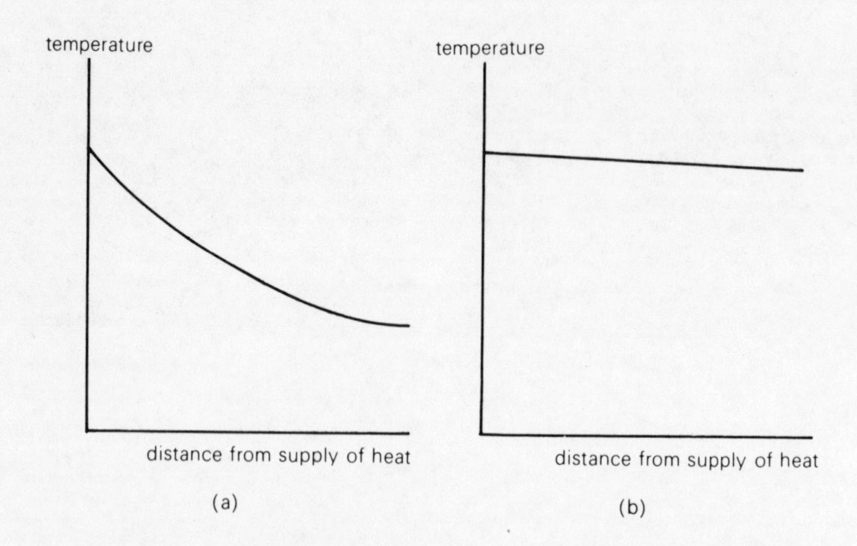

Fig. 10 Graphs of temperature variation along a metal bar heated at one end, (a) unlagged and (b) lagged

not only travelling along the bar by conduction, but it is also being lost to the surroundings by convection and radiation.

If the bar is well lagged at the sides, by a thick layer of insulating material, the temperature fall will be more nearly as shown in Fig. 10(b). In this case, the whole of the heat is travelling along the bar by conduction. Experimental work on thermal conductivity needs to be carried out under these conditions.

Conductivity

The thermal conductivity of a material is defined as **the amount of heat energy (in joules) conducted per second between two opposing faces of a 1 metre cube of the material, when a 1°C temperature difference exists between the faces.**

This wording is very similar to the definition of electrical resistivity though in fact, thermal conductivity actually corresponds to a factor called electrical *conductance* which equals 1/resistivity.

For a conducting bar of length l and area of cross-section A, perfectly lagged at the sides, this definition gives

$$\text{heat conducted through bar per second} = \frac{(\text{thermal conductivity}) \times (\text{area of cross-section}) \times (\text{temperature difference between ends})}{\text{length of bar}}$$

or, in symbols,

$$W = \frac{kA(t_1 - t_2)}{l} \qquad (1)$$

where t_1 and t_2 are the temperatures of the ends of the bar and k is the thermal conductivity.

Some values of conductivity are given in Table 2.

Table 2

Material	Thermal conductivity/W m^{-1} K^{-1}
Copper	403
Aluminium	236
Iron	84
Lead	36
Glass	1.1
Cardboard	0.21
Asbestos	0.11
Cotton wool	0.025

Measurement of conductivity of a good conductor

Figure 11 shows the form of the apparatus commonly used, called Searle's bar. The bar is made of the material to be tested and is about 300 mm long and 40 mm in diameter. One end of the bar can be heated, by either an electric coil or a steam pipe, and cold water can be passed through a coil soldered around the other end. Two holes A and B, drilled in the bar, are to hold the bulbs of mercury thermometers usually supplied with the apparatus. The whole apparatus is packed with a thick layer of lagging and enclosed in a stout wooden box so that only the tubes leading in, and the stems of the thermometers, project outside.

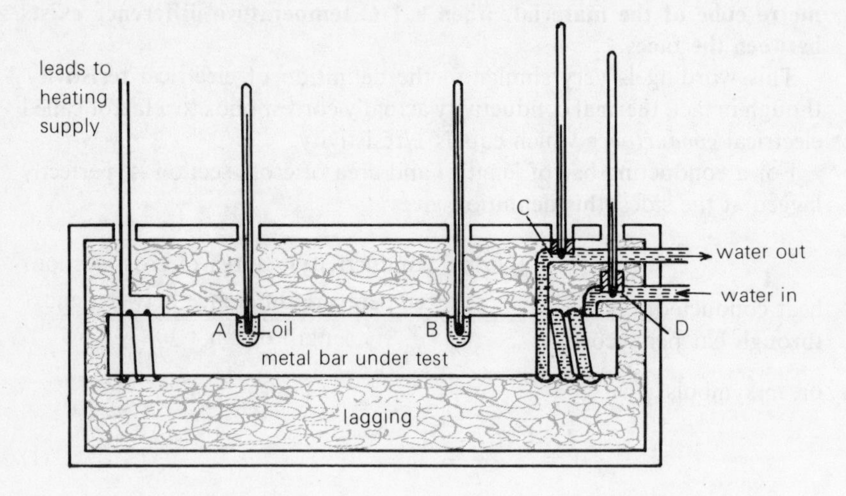

Fig. 11 Searle's bar

The heat is turned on, and a slow stream of cold water is passed into the cooling coil, entering at D so that it leaves, slightly warmer, at C which is towards the hot end of the bar.

A little oil is placed in the holes A and B, to ensure good thermal contact with the bulbs of the thermometers. Readings of the thermometers are taken at intervals until the apparatus is seen to have reached a steady state (this will take half an hour or so).

The heat flow through the bar is determined by measuring the rise in temperature it produces in the cooling water as it flows round the bar. Two more thermometers are placed through side tubes at C and D. The readings of these thermometers should be recorded along with the others. The rate of flow of the water must be measured at some stage during the experiment, by collecting a sample in a measuring cylinder during a measured time.

Heat passing out of the bar per second

$$= W$$

$$= \frac{\text{(mass of water collected)} \times \text{(specific heat capacity of water)} \times \text{(difference in thermometer readings C and D)}}{\text{time of collection of water}}.$$

The distance between the centres of the holes at A and B, and the diameter of the bar, should both be measured using vernier callipers.

Equation (1) can now be applied to the system.

Searle's method is not suitable for poor conductors, since (a) the temperature differences would be smaller and so less accurately measurable, and (b) the heat lost into the lagging would be comparable with that transferred through the material under experiment.

Either of the two following experiments is to be preferred.

Measurement of conductivity of a poor conductor

Lees' disc

The material to be tested must be available in the form of a flat disc, generally about 60 mm diameter and 2–3 mm thick. It is used, as in a sandwich, between two thick copper discs, and the system hangs freely in the air supported by three strings, as shown in Fig. 12.

The upper side of the system is heated steadily, either electrically or by steam, and the heat is conducted down through the three discs A, B and C, and is lost into the air. The process is allowed to continue until it reaches a steady state, which will probably take at least an hour. This is monitored by

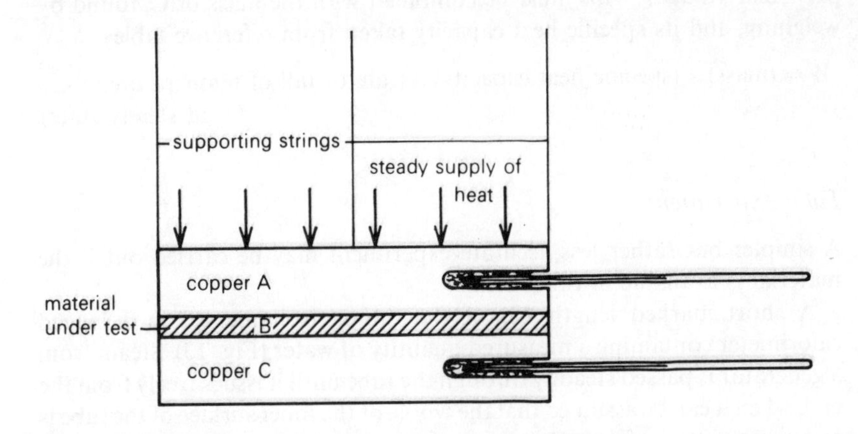

Fig. 12 Lees' disc

observing the readings of two thermometers inserted horizontally into special holes in the copper discs (containing a little Vaseline to ensure good thermal contact).

In the steady state,

$$\text{heat passing through the test disc B per second}$$
$$= W$$
$$= \frac{(\text{thermal conductivity of B}) \times (\text{area of disc}) \times (\text{difference in thermometer readings})}{\text{thickness of disc}}.$$

The measurements of the disc should be taken using vernier callipers for the diameter and a micrometer screw gauge for the thickness, taking care not to squeeze the disc.

The value of W, the quantity of heat passing through the disc per second, is found by a separate experiment carried out on the copper disc C. It must be assumed that no heat is lost to the air through the sides of the test disc, so that W is also the quantity of heat lost per second from disc C to the air. The truth of this assumption will depend on the thickness of the test disc.

Discs A and B are removed from the system leaving disc C supported as before. The heating element is used to raise the temperature of C well above the temperatures involved in the first part of the experiment. The heater is then removed and a thick layer of felt lagging placed instead over the top surface of C. The temperature of C is observed at intervals as it cools, and the results plotted to give graphs as shown in Fig. 6 (p. 14) in the preceding chapter.

Suppose that during the first part of the experiment, the steady reading of the thermometer in disc C was t_c. The rate of heat loss W, at this temperature, will be the same in the two experiments since the same conditions have been specifically maintained. The straight-line graph derived from the cooling curve will give the rate of fall of temperature at the particular value t_c. This must be combined with the mass of C, found by weighing, and its specific heat capacity taken from reference tables.

$$W = (\text{mass}) \times (\text{specific heat capacity}) \times (\text{rate of fall of temperature at steady state}).$$

Tube experiment

A simpler but rather less accurate experiment may be carried out if the material is available in tube form.

A short marked length AB of the tubing is immersed in a lagged calorimeter containing a measured quantity of water (Fig. 13). Steam from a generator is passed steadily through the tube until it issues freely from the end, when it can be assumed that the whole of the inner surface of the tube is at a temperature of 100°C.

The temperature of the water in the calorimeter will be rising slowly, as

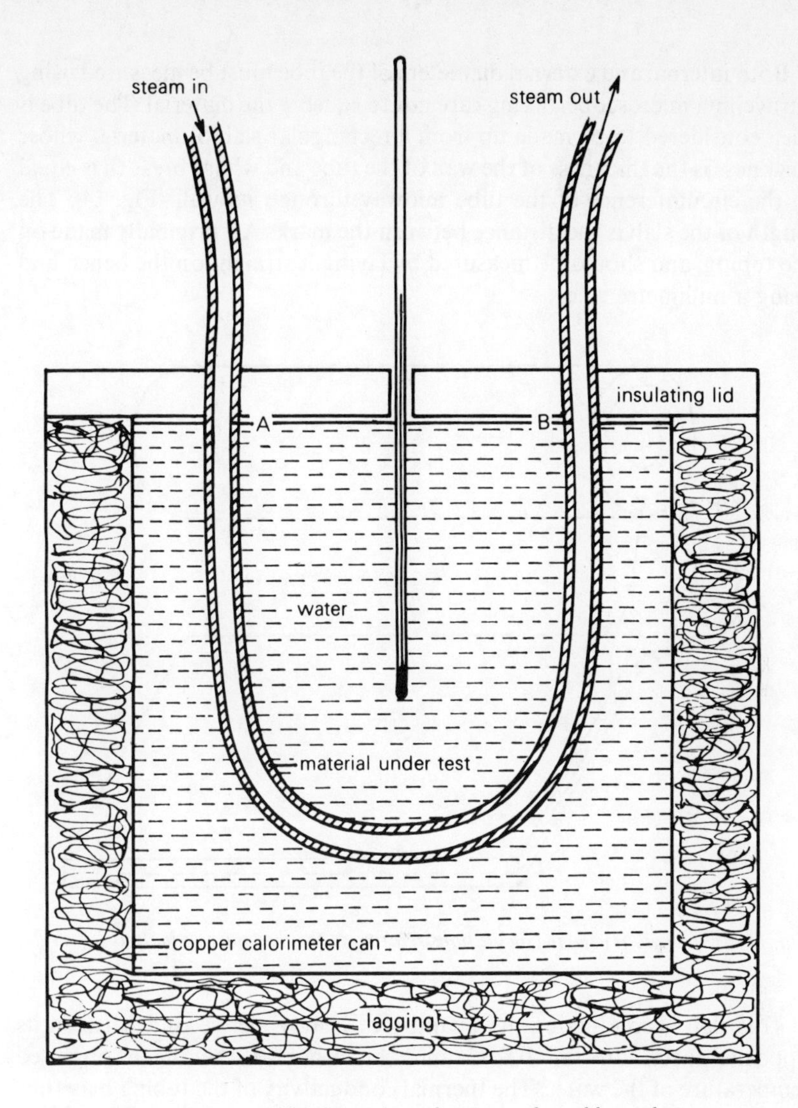

Fig. 13 Determination of the thermal conductivity of a rubber tube

heat is conducted into it through the tube. The water must be stirred frequently, and the rate of rise of temperature is found by taking readings from the thermometer at timed intervals. It is best to start the experiment with the water a few degrees below room temperature and allow it to rise an equal amount above, so that any heat loss through the lagging will effectively be cancelled out.

The inner can of the calorimeter is weighed and its water equivalent is calculated. The rate at which heat is passing into the water, W, can then be calculated from the readings taken.

W = (mass of water + water equivalent of calorimeter) × (specific heat capacity of water) × (mean rate of rise of temperature).

Both internal and external diameters of the tube must be measured using a travelling microscope, taking care not to squeeze the material. The tube is then considered to be made up from a rectangular slab of material whose thickness is the thickness of the wall of the tube and whose breadth is equal to the circumference of the tube midway through its wall (Fig. 14). The length of the slab is the distance between the marks AB originally made on the tubing, and should be measured by laying it straight on the bench and using a millimetre rule.

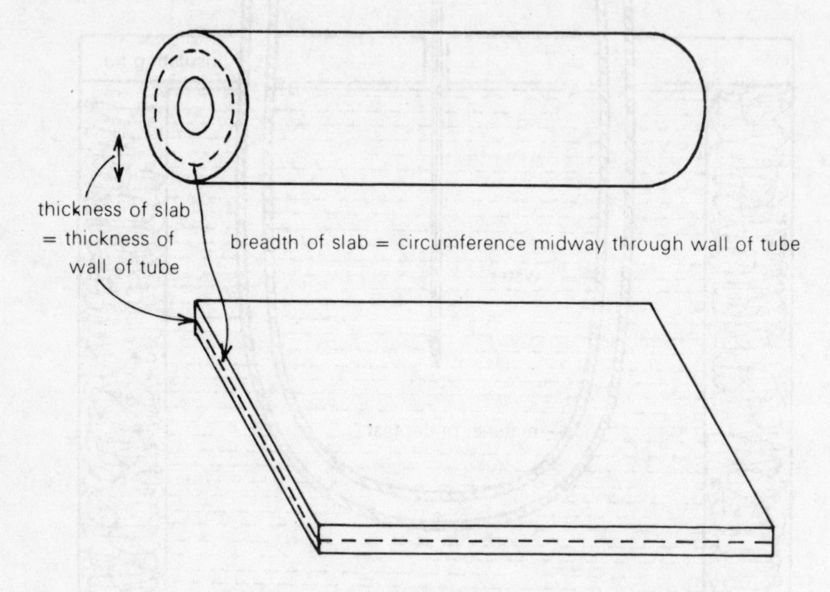

Fig. 14 Dimensions used in calculations relating to the thermal conductivity of a tube

This slab is conducting the quantity of heat W per second between its opposite faces, one at 100°C and the other at the mean observed temperature of the water. The thermal conductivity of the tubing may thus be calculated.

EXERCISES

Take any necessary values of thermal conductivity from Table 2.

1 Calculate the quantity of heat conducted per second through a block of copper measuring 0.20 m × 0.05 m × 0.05 m if a temperature difference of 10°C exists between the two square faces, and the four long faces are perfectly lagged.

2 Find the thermal conductivity of felt if 2.9 joule/s pass through a circular layer 5 mm thick and 60 mm diameter whose opposite faces differ in temperature by 30°C.

3 At what rate is heat being lost through a 1 m × 0.5 m window pane from a room kept at 22°C on a day when the temperature outside is 2°C? Assume that the glass is 3 mm thick.

4 A steam pipe of diameter 20 mm is lagged by a layer of asbestos material 10 mm thick. Calculate how much heat escapes per hour through a 1 m length of the pipe if the temperature at the outside of the lagging is 30°C.

Heat insulation

The topic of heat insulation is of great practical importance these days when so much emphasis is being placed on saving fuel. In the domestic field, there are the applications of double-glazing, loft lagging, and cavity wall insulation using injected foam filling.

Results for use in everyday life have also been obtained as a spin-off from the space programme, from the two problems of first keeping the astronauts warm in space and subsequently preventing burn-up as the capsule re-enters the atmosphere. New insulating materials have been developed in answer to these problems, and are now used in connection with fire-proofing, heat insulation, protective clothing, etc.

All problems in this field require individual solutions. To see the various factors involved, consider the question of the lagging of a hot water pipe.

Heat from the hot water will pass through the lagging and into the air, and a steady state will be reached in which the rate of heat loss from the hot water is constant. The temperature of the outer surface of the lagging will be above room temperature, and heat will be given off from it by both convection and radiation. If the lagging is thick, its total surface area will be large, and the total heat lost in this way may be quite large even if the lagging does not appear to get very warm. On the other hand, if the hot water pipe itself is narrow, its surface area will be comparably small, and heat losses by convection and radiation from the *unlagged* pipe might actually be no greater than those from the lagging just considered.

Calculations using experimental data show that a critical diameter of about 10 mm exists; for piping smaller than this, no saving of heat is achieved by lagging.

Examination questions 3

1 (a) Describe an experiment to measure the thermal conductivity of a poor conductor (for example, glass in the form of a thin disc) and explain how the result is calculated from the experimental observations.
(b) A room has a single window measuring 2 m × 3 m and having glass 3 mm thick. Assuming that the internal and external surfaces are maintained at 20° C and 0° C respectively, calculate the rate of loss of heat through the glass. Comment on your answer. Explain how such losses are reduced by double glazing. (Thermal conductivity of glass = 1.1 $W m^{-1} K^{-1}$.) [AEB, Nov. 1978]

2 In an experiment to measure the thermal conductivity of a good conductor, the specimen is normally made long in order to

A make end corrections unnecessary.
B ensure a parallel flow of heat along the bar.
C obtain an adequate temperature difference between the ends of the bar.
D correct accurately for heat loss from the sides of the bar.
E obtain a steady-state condition. [C]

3 (a) Explain what is meant by *temperature gradient.*
(b) The ends of a perfectly lagged bar of uniform cross-section are maintained at steady temperatures θ_1 and θ_2. Sketch a graph showing how the temperature varies along the bar and account for its shape.
(c) Describe an experimental arrangement which attempts to fulfil the conditions given in (b) and outline the measurements you would make in order to determine the thermal conductivity of the material of the bar. Show how to calculate the thermal conductivity from your measurements.
(d) An iron pan containing water boiling steadily at 100°C stands on a hot-plate and heat conducted through the base of the pan evaporates 0.090 kg of water per minute. If the base of the pan has an area of 0.04 m^2 and a uniform thickness of 2.0×10^{-3} m, calculate the surface temperature of the under-side of the pan.
(Thermal conductivity of iron = 66 $W m^{-1} K^{-1}$,
specific latent heat of vaporisation of water = 2.2×10^6 $J kg^{-1}$.)
[JMB]

4 Sketch graphs to illustrate the temperature distribution along a metal bar heated at one end when the bar is (a) lagged, and (b) unlagged. In each case assume that temperature equilibrium has been reached. Explain the difference between the two graphs.

By considering the relative increase in surface area explain why asbestos lagging of thickness 20 mm will be more effective in reducing the *total* heat losses from a copper pipe carrying steam at 100° C if the pipe has a diameter of about 60 mm than if the pipe has a much smaller diameter.

A window pane consists of a sheet of glass of area 2.0 m^2 and thickness 5.0 mm. If the surface temperatures are maintained at 0°C and 20°C calculate the rate of flow of heat through the pane assuming a steady state is maintained. The window is now double-glazed by adding a similar sheet of glass so that a layer of air 10 mm thick is trapped between the two panes. Assuming that the air is still, calculate the ratio of the rate of flow of heat through the window in the first case to that in the second. Why, in practice, would the ratio be much smaller than this?
(Conductivity of glass = 0.80 $W m^{-1} K^{-1}$,
conductivity of air = 0.025 $W m^{-1} K^{-1}$.) [L]

Electricity and Magnetism

4 Resistance

Current flow in conductors

The flow of an electric current in a conductor is due to the movement of negatively charged electrons.

Most conductors belong to the set of elements known as *metals*, having atoms which are very easily ionised. Their outermost shells contain only one or two electrons, and in the solid state most of these become detached from their parent atoms and drift as an 'electron gas' between the positive ions of the lattice (Fig. 15).

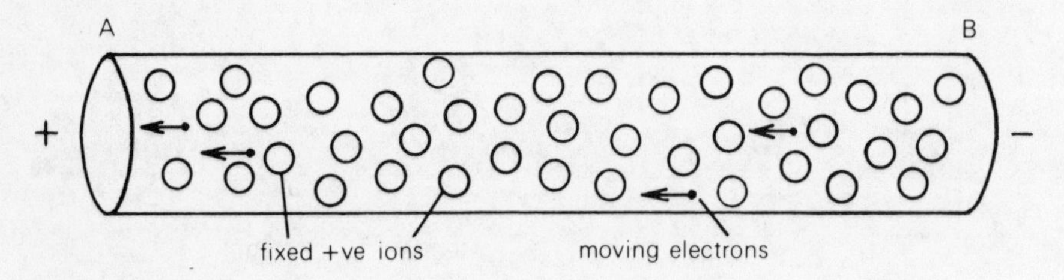

Fig. 15 The electron gas in a conductor carrying a current

Applying a potential difference between two points on the conductor will influence the electrons to move in one particular direction, thus producing an electric current.

The opposition to the motion of the electrons offered by the ions of the lattice is seen as electrical *resistance*.

Resistance is measured in ohms, where **1 ohm is defined as the resistance of a conductor such that a current of 1 amp flows through it when a potential difference of 1 volt is applied between its ends.**

Ohm's law

This is an experimental conclusion. The law states that **the current through a conductor is proportional to the potential difference between its ends, provided that the temperature remains constant.**

It is generally used in the familiar form

$$\text{current, } I = \frac{\text{potential difference, } V}{\text{resistance, } R}$$

Production of heat

As the electrons move between the ions of the lattice, there are frequent collisions in which some of the kinetic energy of the electrons is transferred into vibrational energy of the 'fixed' ions. The kinetic theory correlates an increase in energy of vibration of the atoms of a solid with a rise in temperature, and the conductor heats up.

The rate of production of heat in a conductor is given by:

power = current × potential difference,

or in the usual symbols,

$$P(\text{watt}) = I(\text{amp})V(\text{volt}).$$

If Ohm's law holds, this relation may also be expressed in the forms

$$P = I^2R$$

and

$$P = \frac{V^2}{R}.$$

The amount of heat energy developed in a given time is

heat = current × potential difference × time.

Change of resistance with temperature

If a metal gets hot, it expands. On the atomic scale, this is because its atoms vibrate more vigorously in their positions in the lattice. This will make it more difficult for the electrons to pass in between them when an electric current flows, so the resistance of the conductor increases.

In general, the resistance R_t of a conductor at temperature t°C changes according to the equation

$$R_t = R_0(1 + \alpha t + \beta t^2),$$

where α and β are constants for the particular metal, called the *temperature coefficients* of resistance.

Over small ranges of temperature, this equation may be approximated to

$$R_t = R_0(1 + \alpha t)$$

since β is very small compared to α.

The graph sketched in Fig. 2(b) (p. 6) is an illustration of these results, which were there used with the resistance thermometer.

Other substances, particularly some semiconductors, may show a decrease of resistance if the temperature rises. Alloys generally show very little change of resistance with temperature.

Drift velocity of the electrons

Referring back to Fig. 15, suppose a particular electron moves from A to B in 1 second, under the influence of the applied potential difference.

The total current flowing will be given by (total number of free electrons in cylindrical volume between A and B) × (charge on one electron), since current is equal to the amount of charge passing a point in 1 second.

Appropriate numerical values can be given to these terms, depending on the density and valency of the particular metal. For a current of 1 A flowing through a copper wire of the thickness commonly used for a connecting lead in the laboratory, it turns out that the electrons forming the current are moving only at about 5 mm s^{-1}.

This is a very slow drift, and the result emphasises (a) how very abundant 'free' electrons must be, in copper, to add up to a reasonable quantity of charge passing per second, and (b) that this 'drift' of the individual electron is quite different from the almost instantaneous establishment of a current throughout a circuit, which takes effect with the speed of light.

Resistivity

The resistance of a wire will
(a) increase with the length l,
(b) decrease for larger areas of cross-section A, and
(c) depend on the material of which it is made.

If the *resistivity* of a particular material is ρ, then the resistance of a wire is given by

$$\frac{\text{resistivity} \times \text{length}}{\text{area of cross-section}} = \frac{\rho l}{A}.$$

Resistivity is defined as the resistance of a 1 metre cube of the material. It is measured in units of ohm m. The resistivities of some common metals and alloys are given in Table 3.

Metals of low resistivity are required as connecting leads in circuits. For heating elements, a fairly high resistance is needed, and the alloy nichrome is used. The filaments of electric light bulbs need to withstand high temperatures without melting, and tungsten is used for these.

Manganin wire is generally used to make up standard resistors, since its resistivity changes little with temperature. A resistance box comprises a set

Table 3

Material		Resistivity at 0° C/Ω m
Elements:	Silver	1.47×10^{-8}
	Copper	1.55×10^{-8}
	Gold	2.05×10^{-8}
	Aluminium	2.50×10^{-8}
	Tungsten	4.9×10^{-8}
	Iron	8.9×10^{-8}
Alloys:	Manganin	41.5×10^{-8}
	Constantan	49×10^{-8}
	Nichrome	107×10^{-8}

ot coils of insulated wire wound on plastic bobbins, the length and thickness of the wire varying so that a range of different resistances is supplied.

High resistances, for use in electronics, are made from a mixture of powdered carbon and cement, set into the shape of a cylinder which is then painted with coloured bands to indicate its approximate value. Carbon, although not a metal, is a fairly good conductor of electricity.

EXERCISES

1 A potential difference of 4 V is applied across the ends of a 10 Ω resistor. Calculate (i) the current through the resistor, (ii) the power developed in it, (iii) the quantity of heat given out in 20 s.

2 A platinum wire, whose resistance at 0° C is 8.0 Ω, is raised to a temperature of 100°C. Find its new resistance if

(i) the average temperature coefficient of resistance of platinum over this range is $4 \times 10^{-3}\ K^{-1}$,

(ii) the two temperature coefficients, α and β, are taken as 4.02×10^{-3} and -6.0×10^{-7} respectively.

3 If the charge of a single electron is 1.6×10^{-19} coulomb, calculate

(i) how many electrons per second pass any point in a wire carrying a current of 3.2 A,

(ii) how fast these electrons must be moving if the cross-sectional area of the wire is 1 mm^2 and there are 10^{28} free electrons per m^3.

4 Using the table of resistivities shown above, calculate the resistance of an iron wire 0.80 m long and 0.15 mm in diameter.

Current flow in liquids

A great many liquids are decomposed when an electric current passes through them. This process, known as *electrolysis*, will not be dealt with in this book as it is more the concern of the chemist than the physicist.

It is possible to measure the resistance of a cylinder of liquid using a

Wheatstone's bridge circuit, but alternating current must be used to avoid any electrolytic effect, and headphones or a cathode-ray oscilloscope are then needed as detectors of the balance-point instead of a galvanometer.

The inverse of resistivity, called *conductivity*, is generally quoted for a liquid. It naturally depends upon concentration in the case of a solution.

Measurement of the internal resistance of a cell

The internal resistance of an electrochemical cell depends on its dimensions, the nature of the electrodes and electrolyte, and the age of the cell. Dry cells in good condition have a resistance of a fraction of an ohm, but this increases to one or two ohms as the cell runs down.

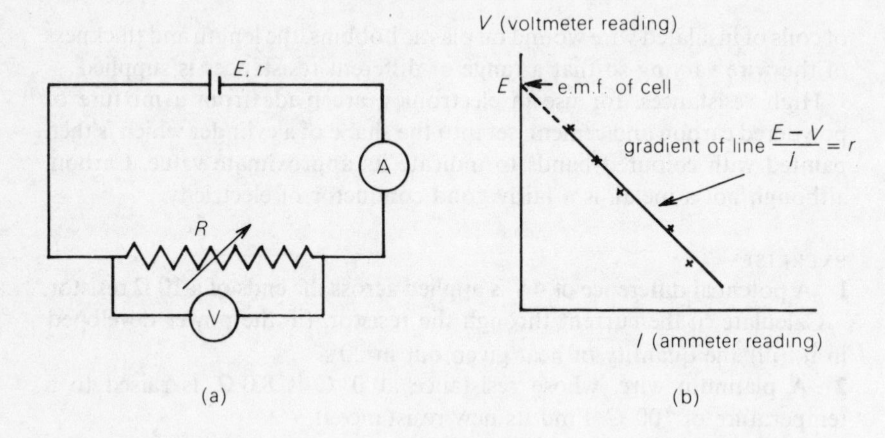

Fig. 16 (a) Circuit used to measure the internal resistance and e.m.f. of a cell (b) Graph of ammeter and voltmeter readings

The circuit shown in Fig. 16(a) provides a simple method for finding the internal resistance of a cell. The variable resistance should be a wire-wound rheostat. Suppose its resistance is R, and the internal resistance of the cell is r while its e.m.f. is E. The ammeter will show the current I through the circuit and the voltmeter will show the potential difference V across the rheostat.

Applying Ohm's Law to the different parts of the circuit,

for the rheostat $\quad I = \dfrac{V}{R}$

for the whole circuit $\quad I = \dfrac{E}{(R+r)}$

and for the cell $\quad I = \dfrac{(E-V)}{r}. \quad (1)$

The term $(E - V)$ is called the 'lost volts', meaning the part of the e.m.f. which drives the current through the cell itself and so is not available for use in the external circuit.

The rheostat should be adjusted to give different values for I, and V should be noted each time. The graph of I and V is then plotted. This will be a straight line of the form shown in Fig. 16(b).

Rewriting equation (1),

$$Ir = E - V$$

or

$$V = E - Ir.$$

Comparing this with the standard form of the equation of a straight line,

$$y = mx + c,$$

it can be seen that the gradient of the graph line gives r, the internal resistance of the cell.

The e.m.f. of the cell may be determined at the same time, since it is equal to the value of V at the point where the line of the graph (produced) cuts the axis.

A more accurate variation of this experiment is to use a single standard resistance as R and measure V using a potentiometer circuit. The cell is subsequently used on its own with the potentiometer so that its e.m.f. is directly measured. From these values for R, V and E, r may be calculated.

Combination of resistors

In series

Resistors connected as shown in Fig. 17 are said to be *in series*. The same current I flows through each resistor in turn.

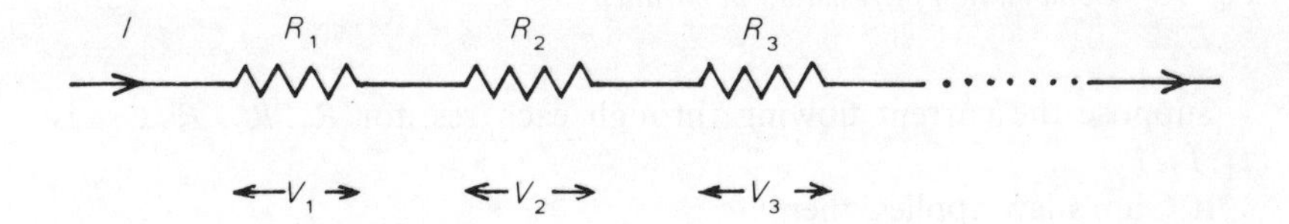

Fig. 17 Combination of resistors in series

Suppose the potential difference (p.d.) set up across each resistor $R_1, R_2, R_3 \ldots$ is $V_1, V_2, V_3 \ldots$.

If Ohm's law applies, then p.d. = current × resistance

i.e.

$$V_1 = IR_1, V_2 = IR_2, V_3 = IR_3 \ldots.$$

If the set of resistors is equivalent to a single resistor R, then the potential difference V set up across this when a current I flows through it must be equal to the sum $(V_1 + V_2 + V_3 + \ldots)$.

Substituting for $V_1, V_2, V_3 \ldots,$

$$V = IR_1 + IR_2 + IR_3 + \ldots$$
$$= I(R_1 + R_2 + R_3 + \ldots).$$

But for the single resistor, $V = IR$.

Hence $$\mathbf{R = R_1 + R_2 + R_3 + \ldots}$$

In parallel

Resistors connected as shown in Fig. 18 are said to be *in parallel*. The current I entering the network divides among the resistors, but the same potential difference V exists across each.

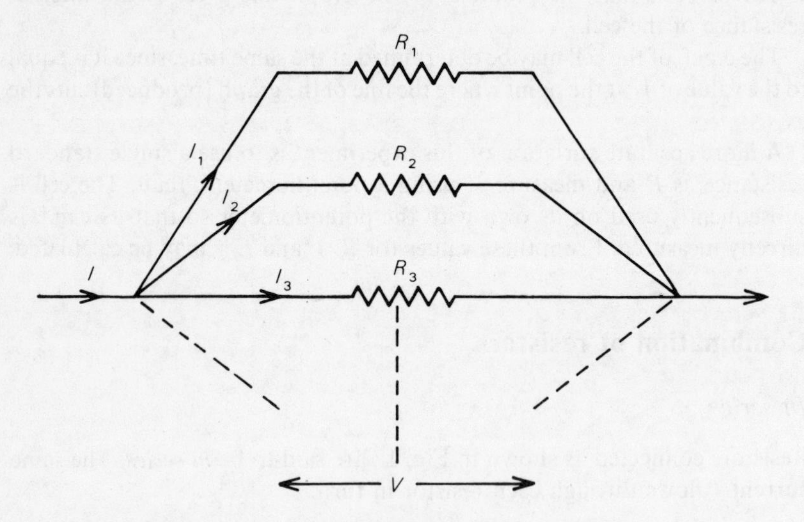

Fig. 18 Combination of resistors in parallel

Suppose the current flowing through each resistor $R_1, R_2, R_3 \ldots$ is $I_1, I_2, I_3 \ldots$.

If Ohm's law applies, then

$$\text{current} = \frac{\text{p.d.}}{\text{resistance}}$$

i.e. $$I_1 = \frac{V}{R_1}, I_2 = \frac{V}{R_2}, I_3 = \frac{V}{R_3} \ldots$$

Hence $$I = I_1 + I_2 + I_3 \ldots$$
$$= \frac{V}{R_1} + \frac{V}{R_2} + \frac{V}{R_3} \ldots$$
$$= V\left[\frac{1}{R_1} + \frac{1}{R_2} + \frac{1}{R_3} \ldots\right].$$

But for the single equivalent resistor R,

$$I = \frac{V}{R}$$

Hence
$$\frac{1}{R} = \frac{1}{R_1} + \frac{1}{R_2} + \frac{1}{R_3} + \ldots .$$

Kirchhoff's laws

The arguments just given are illustrations of the applications of Kirchhoff's laws, which are used to calculate the currents through the components of more complicated networks.

Law 1. For any junction of the network, the sum of the currents entering the junction is equal to the sum of the currents leaving it.

Expressed mathematically, this is $\Sigma I = 0$.

Law 2. For any closed loop of the network, the algebraic sum of the e.m.f.'s around the loop is equal to the sum of the products of the algebraic current and the resistance for each component of the loop.

Expressed mathematically, this is $\Sigma E = \Sigma IR$.

Applying these laws to different parts of the network as often as necessary leads to a set of simultaneous equations for all the various currents. These equations are most easily solved by matrix methods.

EXERCISES

5 A current of 0.14 A is observed to flow when a cell of e.m.f. 1.5 V is connected across a 10 Ω resistor. What is the internal resistance of the cell? Calculate the 'lost volts' under these operating conditions.

6 Find the single resistor equivalent to the combination of four resistors of values 1 Ω, 2 Ω, 4 Ω and 10 Ω, connected (i) in series (ii) in parallel.

If a potential difference of 2.0 V is applied across the combination, calculate the current through each resistor in each case.

Examination questions 4

1 A resistor of resistance 1.0 kΩ has a thermal capacity of 5.0 J K^{-1}. A p.d. of 4.0 V is applied across it for 120 s. If the resistor is thermally insulated, the final rise in temperature is

A 8.0×10^{-4} K B 3.2×10^{-3} K C 9.6×10^{-2} K

D 0.38 K E 9.6 K [C]

2 A bulb is used in a torch which is powered by two identical cells in series each of e.m.f. 1.5 V. The bulb then dissipates power at the rate of 625 mW and the p.d. across the bulb is 2.5 V. Calculate (i) the internal resistance of each cell and (ii) the energy dissipated in each cell in one minute.

[JMB]

3 A heating coil is to be made, from nichrome wire, which will operate on a

12 V supply and will have a power of 36 W when immersed in water at 373 K. The wire available has an area of cross-section of 0.10 mm^2. What length of wire will be required?

(Resistivity of nichrome at 273 K = $1.08 \times 10^{-6}\ \Omega$ m,
temperature coefficient of resistivity of nichrome = $8.0 \times 10^{-5}\ \text{K}^{-1}$
[L]

4 The resistivity of a metal such as copper increases with increasing temperature because

A the conduction electrons make more frequent collisions with each other.

B the conduction electrons make more frequent collisions with the atoms of the metal.

C the metal expands and offers more resistance to the flow of electrons.

D it is more difficult to cause electrons to leave their parent atoms.

E free positive ions in the metal undergo more collisions with neutral atoms.
[O]

5 Explain why the electrical resistance of a uniform conductor is proportional to its length and inversely proportional to its area of cross-section.

Give a qualitative explanation for the electrical resistance of metals and use it to account for (a) the increase in the resistance of metals with increasing temperature, and (b) the generation of heat when current flows in a resistor.

Two resistors, with resistances R_1 and R_2 at 0°C and whose temperature coefficients are α_1 and α_2 respectively, are connected in series. Derive an expression for the effective temperature coefficient of resistance of the combination.
[OC]

5 D.C. circuits

The Wheatstone bridge

This is a *theoretical* circuit which is the basis of several applied forms in both d.c. and a.c. work. It is designed for the accurate measurement of an unknown resistance.

The circuit consists of four resistors, R_1, R_2, R_3 and R_4, connected in a loop as shown in Fig. 19. A low current enters the loop at A and leaves again at B; this current must be sufficiently low to avoid causing any temperature change in the resistors.

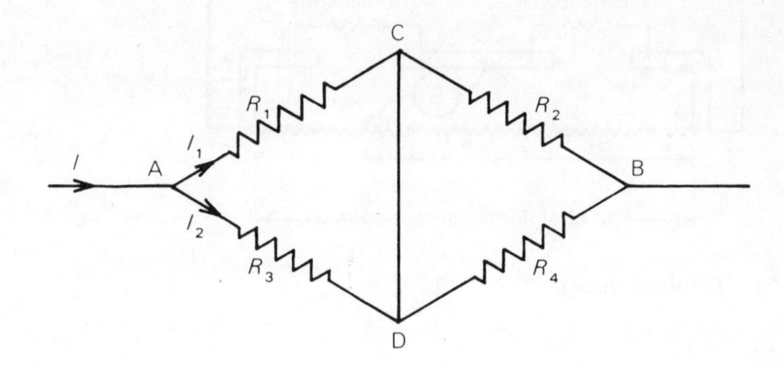

Fig. 19 Wheatstone bridge circuit

Suppose the incoming current I divides into I_1 through AC and I_2 through AD. *If no current flows through the connection CD*, then the current through R_2 will also be I_1 and that through R_4 will be I_2.

Since p.d. = current × resistance (Ohm's law),

for R_1, p.d. between A and C = I_1R_1

and for R_3, p.d. between A and D = I_2R_3.

But since no current flows along CD, C and D must be at the same potential,

i.e. p.d. between A and C = p.d. between A and D

$$I_1R_1 = I_2R_3. \quad (1)$$

Similarly, for R_2 and R_4,

$$I_1R_2 = I_2R_4. \quad (2)$$

Dividing (1) by (2),

$$\frac{R_1}{R_2} = \frac{R_3}{R_4}. \tag{3}$$

This equation holds only if no current flows through CD, and in practical forms of the circuit a sensitive galvanometer is connected between C and D. If the resistances are correctly adjusted so that the galvanometer shows no deflection, then the bridge is said to be *balanced.*

The metre bridge

Figure 20 shows this practical form of the Wheatstone bridge circuit, commonly used in school laboratories to measure an unknown resistance.

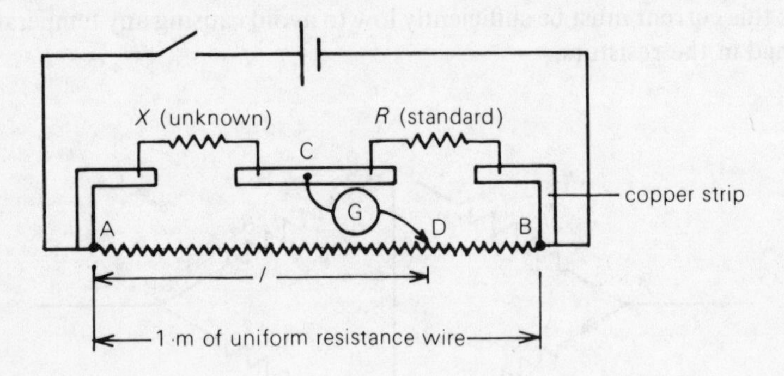

Fig. 20 The metre bridge

The two resistances R_3 and R_4 are the two parts of a uniform resistance wire AB which is one metre long. The wire is mounted alongside a metre rule on a base-board which also carries three pieces of stout copper or brass strip to form connections with the rest of the circuit. The unknown resistance X is connected across one gap between these strips, and a standard resistance R is placed across the other gap. A low voltage supply is connected across the wire AB.

A screw terminal, in the centre of the strip between X and R, is the point C of the Wheatstone bridge circuit, and the point D is a jockey, a sharpened brass knife-edge which can be placed in good electrical contact at any point along AB. A galvanometer is connected between C and D.

After the current has been switched on, the jockey is touched lightly to the wire, first near A and then near B. The galvanometer should deflect in opposite directions for these two positions, showing that a balance-point exists. The jockey is then tapped at various points along the wire until balance is reached, the galvanometer showing no deflection. It should be possible to locate the balance-point within about a millimetre on each side.

If the length of wire AD = l mm, then DB = $(1000 - l)$ mm and equation (3) can be applied to give

$$\frac{X}{R} = \frac{l}{(1000 - l)}$$

or

$$X = \frac{l}{(1000 - l)} R.$$

This expression has most accuracy if the balance-point is near the middle of the wire, and a rough run of the experiment will show if this is so. Various values of standard resistance should be available so that the most suitable can be used as R.

There are two possible sources of trouble in this experiment:

(a) Contact *resistances* at the terminals, and at the points A and B where the bridge wire should be soldered to the connecting strip. Any error due to these can be checked by interchanging X and R, when the balance-point should be found at the same distance l but measured from B instead of A.

(b) Contact *potentials* at the same places due to corrosion of the metals with age. These can be investigated by reversing the voltage supply, which should make no difference to the balance-point.

Four independent values of X can be obtained in this way for every value of R used. The results may be averaged if they do not differ too greatly, otherwise the board should be stripped and the terminals cleaned.

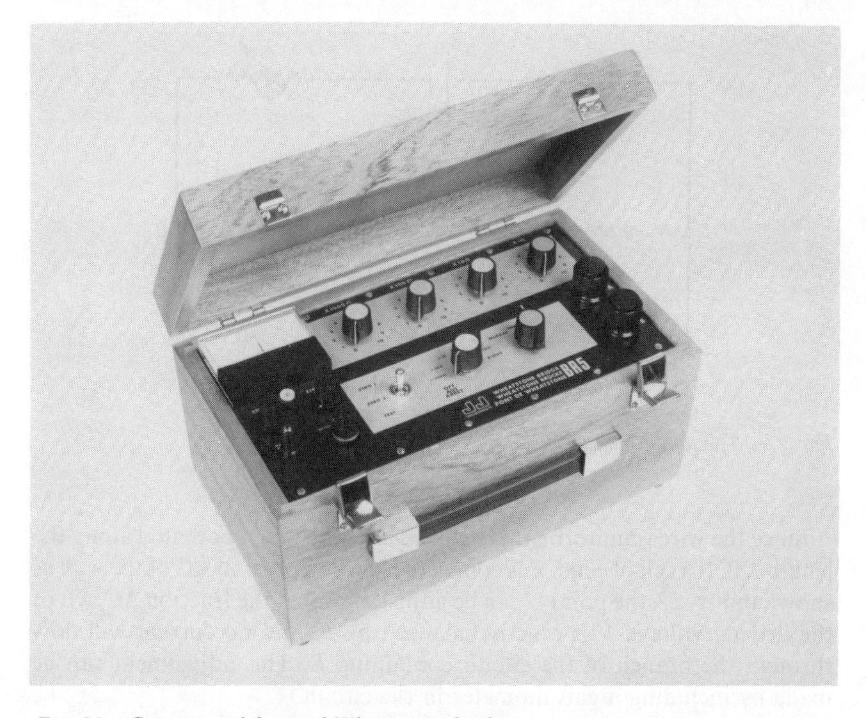

Fig. 21 Commercial form of Wheatstone bridge

A better and more robust form of the bridge circuit can be bought complete (Fig. 21). This incorporates standard wire-wound resistance coils for all four arms of the bridge, and the values of the resistances can be altered by rotating appropriate knobs.

EXERCISES

1 The galvanometer in a metre bridge circuit shows no deflection when the jockey is placed on the 600 mm mark, using a standard resistance of 4 Ω across the left-hand gap of the bridge. What is the value of the resistance across the right-hand gap?

2 A metre bridge circuit involving a standard resistance of 3 Ω is used to measure an unknown resistance thought to be about 3.7 Ω. Where is the balance-point expected to be?

The potentiometer

This circuit is *not* a form of the Wheatstone bridge in spite of a superficial resemblance. As its name implies, it measures potential differences (voltages).

The simplest form of potentiometer consists of a uniform length of resistance wire AB, 1 m long, mounted on a board alongside a millimetre scale. A current from a low-voltage source V is passed through the wire (Fig. 22).

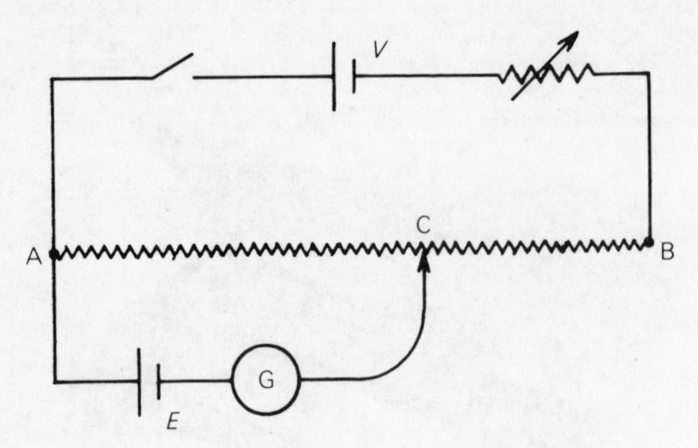

Fig. 22 The potentiometer

Since the wire is uniform, there will be a steady fall in potential along the length AB. If a cell of e.m.f. E is connected across a portion AC of the wire as shown in Fig. 22, the point C can be adjusted so that the fraction AC/AB of the driving voltage V is exactly balanced by E, and no current will flow through the branch of the circuit containing E. This adjustment can be made by including a galvanometer in the circuit.

Balance can only occur if the two cells are connected to the terminal at A in the same sense. The existence of a balance-point should be checked (as for the metre bridge) by touching a jockey lightly to the wire first at A and then at B, to see that this gives galvanometer deflections in opposite directions.

If E is greater than V, no balance-point can exist, and V must be increased. A rheostat may be used to cut down the p.d. across the wire if E is small, so that the balance-point will occur near the centre of AB.

The circuit is used to compare two e.m.f.s E_1 and E_2 by using them in turn in the position E, and finding the two balance-points C_1 and C_2. Then,

$$\frac{E_1}{E_2} = \frac{AC_1}{AC_2}.$$

This gives an accurate comparison of the actual e.m.f.s E_1 and E_2 as these cells are not supplying any current at balance, and are virtually on 'open circuit'.

Provided that V is steady, there should be no error introduced even if some contact potentials exist. To check this, the standard procedure is to balance the circuit using E_1, then to replace it with E_2, and finally to use E_1 again. There should be no significant change in the position of the jockey at C_1 on repeating the experiment in this way.

The results may be further checked if all the cells are reversed and the experiment then repeated. The balance points C_1 and C_2 should be the same as before.

For higher accuracy, a longer potentiometer wire may be used. This can be obtained in helix form, wound on a bobbin, with a tapping point controlled by the rotation of a knob.

Measurement of an e.m.f. using the potentiometer

Since no current is drawn from the cell E, it is permissible to use a *standard cell* in this position and so obtain a highly accurate value for any e.m.f. compared with it.

A standard cell, such as the Weston cadmium cell, is made up using pure chemicals in prescribed proportions, and its e.m.f. can be calculated from the chemical data. This e.m.f. will be maintained as long as the chemicals do not react, i.e. as long as no current is drawn from the cell, which is the case if the potentiometer circuit is used correctly.

Comparison of resistances using the potentiometer

Two resistances R_1 and R_2 may be compared by passing a steady current I through the two resistors connected in series (Fig. 23). This will set up potential differences of $E_1 = IR_1$ and $E_2 = IR_2$ across the resistors, and these may be connected in turn into the potentiometer circuit.

For low resistances, this method is preferable to the use of the metre bridge for two reasons:

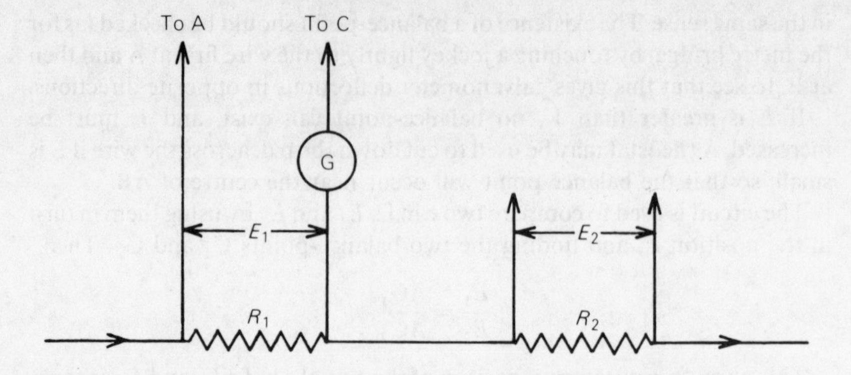

Fig. 23 Comparison of two resistances using the potentiometer

(a) the current I may be kept low, so that the resistors will not heat up;

(b) the resistances of the leads do not affect the experiment as no current flows through them.

Measurement of current using the potentiometer

If the current to be measured can be passed through a standard resistance, the potential difference set up across this resistance can be measured on the potentiometer, using a standard cell as described earlier. This method can be employed using different currents in order to check the calibration of an ammeter.

The potential divider

An adaptation of the potentiometer circuit is sometimes used to cut down a supply voltage which is too high for its particular purpose.

As shown in Fig. 24, the supply voltage is connected across two high resistances R_1 and R_2 in series, and connections taken for use from across R_2 alone. By applying Ohm's law, a simple calculation shows that the voltage for use across R_2 is the fraction $R_2/(R_1 + R_2)$ of the main supply V.

Such a circuit should be used only if the currents to be drawn are low (as with semiconductors), for otherwise much energy is wasted as heat in the resistors.

EXERCISES

3 A potentiometer circuit is balanced with the tapping-point on the 650 mm mark when a Leclanché cell is being tested, and the balance-point changes to the 470 mm mark when the Leclanché cell is replaced by a Daniel cell. Calculate the ratio of the e.m.f.s of the two cells.

4 The p.d. set up across the ends of a 0.100 Ω resistor carrying a certain

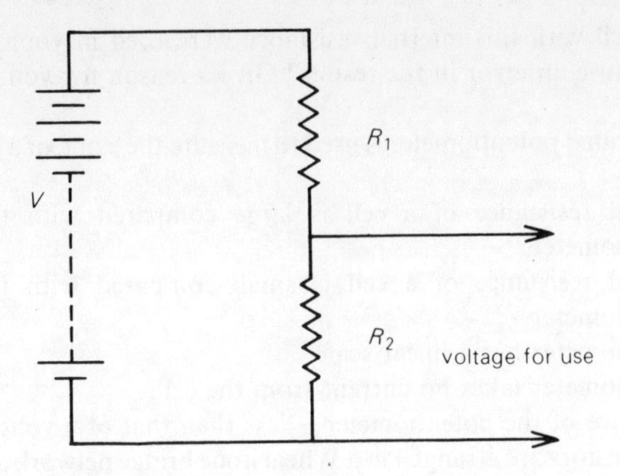

Fig. 24 The potential divider

current gives a balance-point at 36 mm in a potentiometer circuit. If a standard cell of e.m.f. 1.0186 V balances at 611 mm in the same circuit, find the value of the current passed through the resistor.

5 Two resistors are connected in series across a 12.0 V supply. Calculate the value required for the second resistor if the voltage drop across it is to be 8.0 V, taking the value of the first resistor to be $2 \times 10^6\ \Omega$.

Examination questions 5

1 (a) Explain, with the aid of circuit diagrams, how a potentiometer can be used to measure
(i) a current known to be of the order of 10 A,
(ii) an e.m.f. known to be of the order of 5 mV.
You may assume that any apparatus you require is available.
(b) In an experiment to measure the e.m.f. of a thermocouple, a potentiometer is used in which the slide wire is 2.000 m long and the resistance of the wire is 6.000 Ω. The current through the wire is 2.000 mA and the balance point is 1.055 m from one end. Calculate the e.m.f. of the thermocouple. [AEB Specimen paper, 1977]

2 (a) Describe how you would use a potentiometer incorporating a wire of resistance about 2 Ω and other necessary apparatus to calibrate an ammeter in the range 0.2 to 1.0 A. Give a labelled circuit diagram; specify, with the reason, the magnitude of any resistance components used; and show how the results are calculated from the observations.
(b) If the current in the potentiometer wire slowly decreased during the above experiment, how would you detect this change and what effect would the change have on your results?
(c) A standard cell has an e.m.f. of 1.0186 V. When a 0.5 M Ω resistor is connected across it, the potential difference between the terminals is 1.0180 V. Calculate a value for the internal resistance of the cell. If a

standard cell with this internal resistance were used in your experiment would it cause an error in the results? Give a reason for your answer.

[JMB]

3 A calibrated potentiometer is used to measure the e.m.f. of a cell because the

A internal resistance of a cell is large compared with that of the potentiometer.

B internal resistance of a cell is small compared with that of the potentiometer.

C potentiometer has a linear scale.

D potentiometer takes no current from the cell.

E resistance of the potentiometer is less than that of a voltmeter. [L]

4 Four resistors are arranged in a Wheatstone bridge network as in Fig. 19 (p. 41). $R_1 = 1\ \Omega$, $R_2 = 2\ \Omega$, $R_4 = 4\ \Omega$, and R_3 is a variable resistor. The current I entering the bridge is 1.5 A.

When R_3 is adjusted so that the bridge is balanced, the current in the 1 Ω resistor is

A 0.33 A B 0.5 A C 0.66 A D 1.0 A E 1.5 A [O]

6 D.C. measuring instruments

The moving-coil milliammeter

This is the instrument most generally used in circuit work. It consists of a rectangular coil of a number of turns of fine copper wire which is pivoted on jewelled bearings between the poles of a permanent horseshoe magnet (Fig. 25). The electromagnetic field produced around the coil when a current is passed through it is increased by the use of a soft-iron cylindrical core. The coil attempts to turn so that its magnetic axis will line up with the field of the horseshoe magnet; a spiral spring opposes this movement, and the coil will set in some intermediate position. A pointer is attached to the coil and moves over a scale marked in divisions.

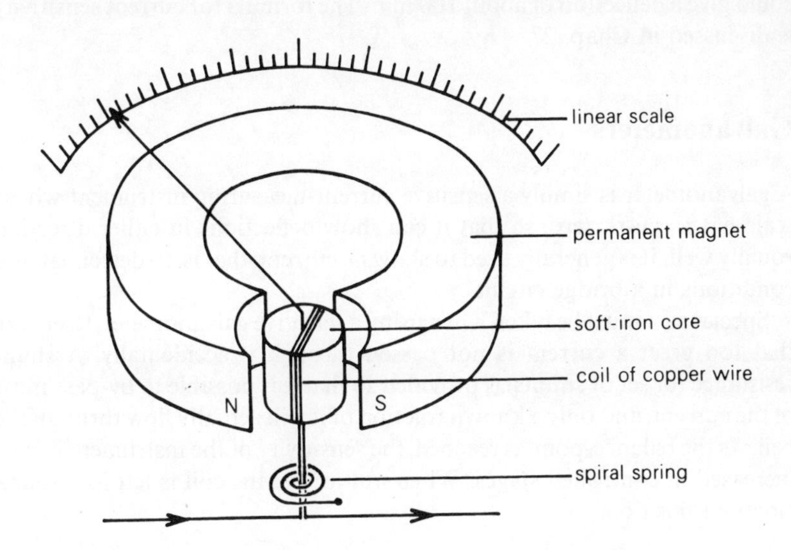

Fig. 25 The moving-coil milliammeter

By shaping the poles of the permanent magnet suitably around the soft-iron core, it is possible to arrange that the magnetic field is everywhere *radial.* The turning force is therefore constant, and this means that the scale will be linear, i.e. its divisions will all be the same size. This is a great convenience for a general-purpose measuring instrument.

An ammeter is always connected into a circuit *in series*, and so should have low resistance in order not to alter the current that is to be measured. The resistance of the coil can only be kept low by using few turns, and this makes the instrument less sensitive. A practical compromise for general purposes would be a meter of about 20 Ω resistance, giving a full-scale deflection for 5 mA.

The suspended-coil instrument

A more sensitive system can be produced if the coil is not pivoted, but suspended by a fine phosphor-bronze wire. A small mirror is attached to this wire, and light from a lamp within the instrument is reflected from the mirror on to a transparent scale. Such an arrangement, called a light-pointer, means that very small deflections can be measured.

A suspended system of this kind is generally clamped when not in use, in case of accidents, and it must be unclamped and the instrument levelled carefully each time it is required.

The *sensitivity* of a current-measuring instrument is defined by **the deflection obtained per unit current** (milliamp or microamp as appropriate). Using a scale at a distance of 1 m from the mirror, a current of 1 μA could give a deflection of about 100 mm. The formula for current sensitivity is discussed in Chap. 27.

Galvanometers

A galvanometer is simply a sensitive current-measuring instrument whose scale has a centre zero, so that it can show deflections in either direction equally well. It is generally used to show *no* current; that is, to detect balance conditions in a bridge circuit.

Special care must be taken when using a sensitive galvanometer, to ensure that too great a current is not passed through it accidentally. A shunt resistance (or set of shunts) is provided so that it is possible to by-pass most of the current, and only a known fraction of it will actually flow through the coil. As the balance-point is reached, the sensitivity of the instrument can be increased in controlled stages. When not in use, the coil is left in a short-circuited position.

Ammeters

If it is required to extend the range of a moving-coil milliammeter so that it can be used to measure larger currents, this is done by the use of a *shunt*, i.e. a low resistance connected in parallel with the coil (Fig. 26). This is generally only a short length of resistance wire, soldered in place inside the case of the instrument. Multi-meters are supplied with a set of shunts which fix externally across the terminals of the meter.

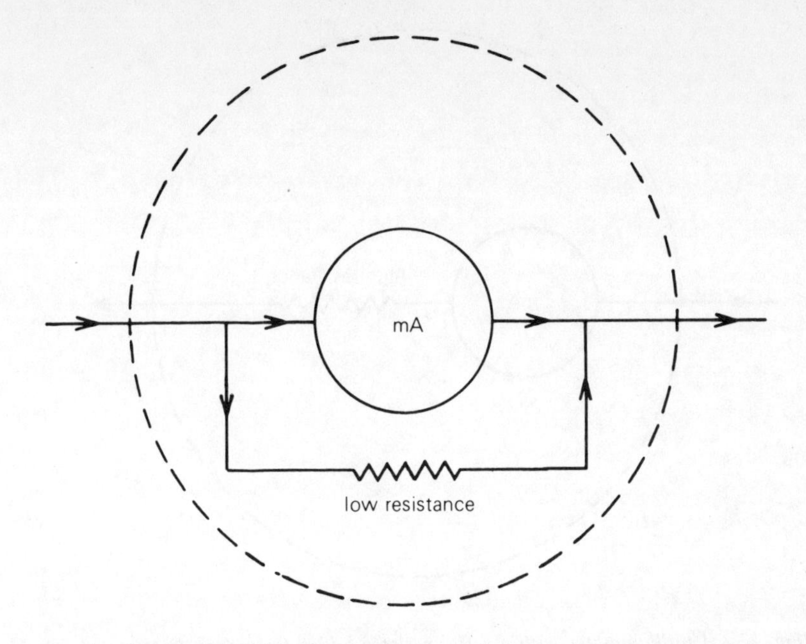

Fig. 26 A shunt used with a milliammeter

To calculate the value of the shunt needed for a particular conversion, it is best first to consider how much current may pass through the meter, and so how much must be by-passed through the shunt. Then, since the shunt and the meter are to be in parallel,

$$\frac{\text{resistance of shunt}}{\text{resistance of meter}} = \frac{\text{current through meter}}{\text{current through shunt}}.$$

Voltmeters

Moving-coil voltmeters are merely adaptations of moving-coil milliammeters, the scale of the instrument being marked appropriately in volts or millivolts by the use of Ohm's law.

For instance, suppose the scale of a particular milliammeter is marked up to 15 mA. If the resistance of the meter is 20 Ω, then a current of 15 mA through the meter will correspond to a potential difference of 300 mV across its terminals. The meter can thus be used directly to measure voltages up to this value.

The sensitivity (deflection per millivolt) of a meter cannot be *increased*, but its range can be extended by adding a high resistance in series with the coil (Fig. 27). This is generally a small carbon resistor soldered into the circuit inside the case of the instrument. Multi-meters are supplied with a bobbin-type resistor which plugs into one terminal of the meter.

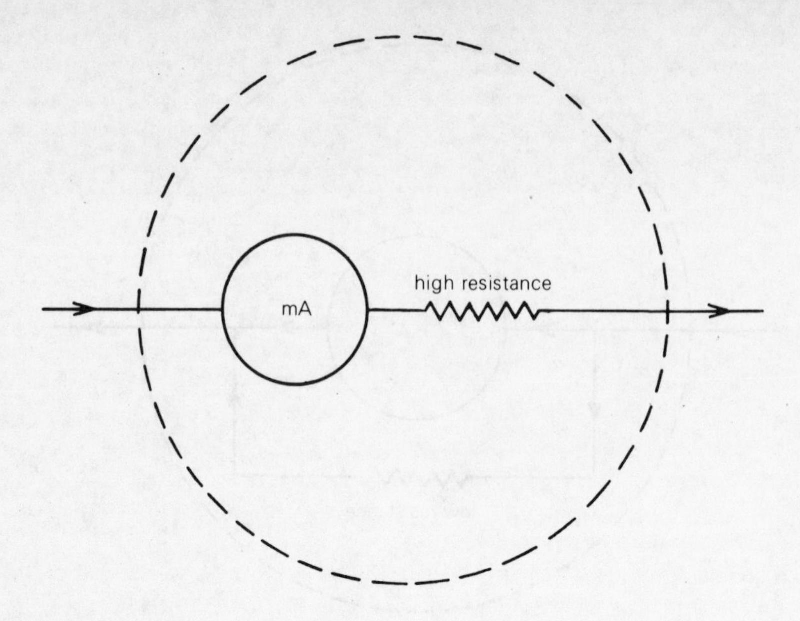

Fig 27   A milliammeter with a series resistor added for use as a voltmeter

To calculate the value of the series resistor which should be used for a particular conversion of a milliammeter into a voltmeter, consider the maximum voltage = (resistance of milliammeter) × (current for maximum deflection) which may be applied across the terminals of the meter. This will show how much of the p.d. to be measured must be 'stepped down' across the series resistor.

Since the two resistances are in series,

$$\frac{\text{value of series resistor}}{\text{resistance of meter}} = \frac{\text{voltage to be 'stepped down'}}{\text{voltage allowed across meter}}.$$

Having employed Ohm's law in this way to calibrate a moving-coil voltmeter, such an instrument must not be used in experiments designed to verify Ohm's law. Voltage measurements for this specific experiment must be made with instruments based on other principles, e.g. a potentiometer or an electrostatic voltmeter.

EXERCISES

1 A moving-coil milliammeter has resistance 15 Ω and gives a full-scale deflection for a current of 5 mA. Calculate the value of the shunt that must be used with this meter so that it can measure currents up to 500 mA.

2 When a current of 1 mA is passed through a suspended-coil galvanometer, a deflection of 200 mm is observed on a scale some distance away. A 390 Ω resistor is then placed in series with the galvanometer, which itself has a resistance of 10 Ω. Find the sensitivity of the galvanometer under these new conditions.

3 A moving-coil milliammeter has resistance 20 Ω and current sensitivity of 5 divisions per mA. How may it be converted to read in volts with sensitivity of (i) 5 divisions per volt, (ii) 10 divisions per volt?

The ballistic galvanometer

If the coil of a galvanometer is free to turn apart from the restoring force of the spring or suspension, i.e. if there are no damping forces such as air resistance or eddy currents, then it can be shown that **the angle which the coil turns through during its first swing is proportional to the quantity of electricity which passed through the coil as it began to move.**

The argument supporting this result requires that the passage of the electricity through the coil must take only a small fraction of the time of its swing, and the chief use of the instrument is in measurements connected with the rapid discharge of capacitors.

The construction of a ballistic galvanometer differs from an ordinary galvanometer in two main ways:
(a) the coil is heavier, so that its time of swing is longer, and
(b) the core of the coil is made of a non-conducting material, to avoid the production of eddy currents.

However, many instruments are designed to be suitable for use in either capacity.

Examination questions 6

1 (a) What quantity is measured by a ballistic galvanometer?
(b) State *two* essential requirements of a galvanometer intended for ballistic use, giving the reason for each.
(c) When using a ballistic galvanometer with a search coil, why is a series resistor often included in the circuit? [AEB Nov. 1979]

2 A milliammeter, of negligible resistance, gives a full scale deflection for 1 mA. It is to be used as a voltmeter giving a full scale deflection for 10 V. The correct circuit for this modification is
A a 10 Ω resistor in series with the meter.
B a 10^4 Ω resistor in series with the matter.
C a 100 Ω resistor in parallel with the meter.
D a 10 Ω resistor in parallel with the meter and a 10^3 Ω resistor in series with them.
E a 10 Ω resistor in series with the meter and a 10^3 Ω resistor in parallel with them. [C]

3 A certain thermocouple thermometer is calibrated by placing its hot and cold junctions in steam and melting ice respectively and measuring an e.m.f. of 5.6 mV with a potentiometer. Subsequently, the thermocouple, of resistance 10 Ω, is used in series with a millivoltmeter of resistance 100 Ω. If the millivoltmeter reads 2.8 mV when the cold junction is in melting ice and the hot junction is in a liquid bath, what is the temperature of the bath on the Celsius scale of this thermometer? [C]

4 A galvanometer of resistance 5 Ω gives full-scale deflection for a current of 2 mA. With a series resistance of 995 Ω it can be used as

A a microammeter reading up to 2 μA.

B an ammeter reading up to 0.2 A.

C a voltmeter reading up to 0.2 V.

D an ammeter reading up to 2 A.

E a voltmeter reading up to 2 V. [O]

7 Ferromagnetism

Theory of atomic magnetism

The naturally magnetic elements, iron, cobalt and nickel, are part of a group known as 'transition elements' in the periodic table. This table is an arrangement of all the elements in order of increasing atomic mass, beginning with hydrogen. Each element is assigned an *atomic number* according to its place in the periodic table.

The simple theory of atomic structure, based on chemical evidence at the start of the twentieth century, is that the atom consists of a massive positively charged nucleus surrounded by orbiting 'shells' of electrons. The total number of electrons differs for each element, and is equal to the element's atomic number. The innermost shell, called the K shell, can contain no more than two electrons, the second shell L, eight electrons, the third shell M, eighteen electrons, and so on.

The first group of transition elements, that containing the naturally magnetic elements, occurs in the middle of the fourth row of the periodic table. Iron, as an example, has atomic number 26, and its twenty-six electrons have the configuration 2.8.13.3. This indicates that the K and L electron shells are full, but the M shell has five gaps while the fourth shell already contains three electrons (Fig. 28). These gaps in the inner shell are the cause of the phenomenon called magnetism.

Since the electrons are negatively charged, as they move in orbits around the nucleus an electromagnetic effect is produced. Symmetrical arrangements of electrons cancel out as far as electromagnetism is concerned. The odd electrons in the outmost orbit are only loosely held and are lost if the atom is ionised or combined chemically with another atom. The gaps in the inner shell, however, produce a permanent unsymmetricality in the total electromagnetic field, and the atom as a whole is a magnetic *dipole*.

Other elements, close to iron in the periodic table, also have gaps in the inner shell of their atomic structure and show magnetic effects, though not as strongly as iron.

These magnetic dipoles correspond to the tiny magnets of the old molecular theory of magnetism. Generally in a specimen of magnetic material, the dipoles are already linked together to some degree into a *domain*.

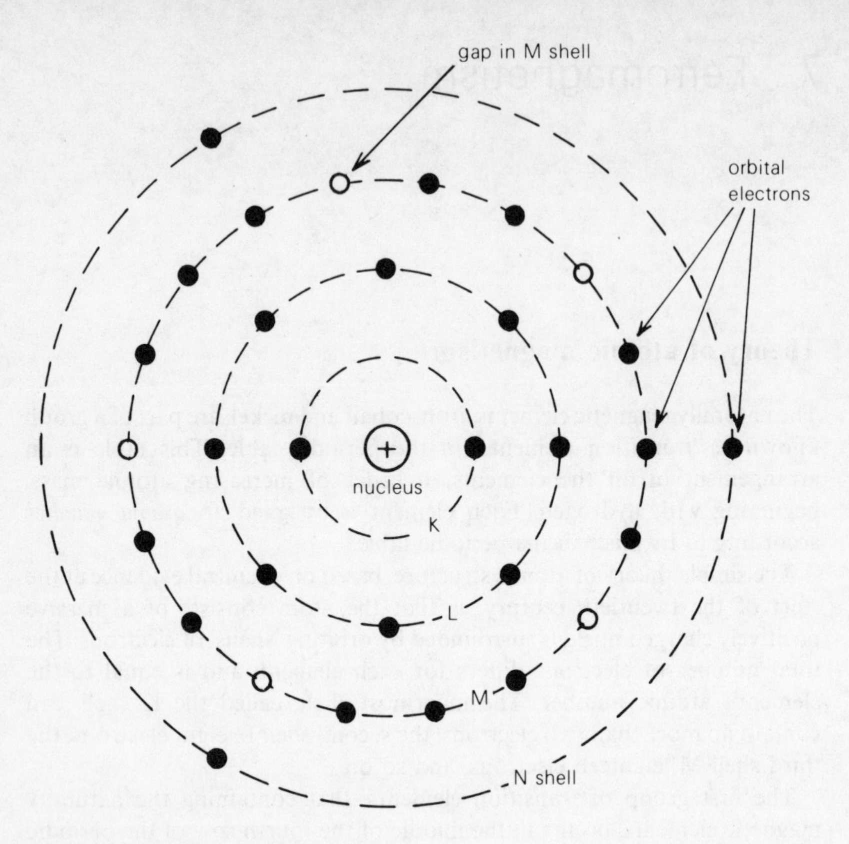

Fig. 28 The structure of an iron atom

Magnetisation

If a specimen of magnetic material is placed in an external magnetic field, the domains will tend to turn into line with the field, and the specimen will show induced magnetism.

Different magnetic materials vary in their response to external fields. Soft iron responds easily, but steel is less responsive since the processes of its manufacture (particularly the hardening process) have made it more difficult for the domains to turn within the lattice structure. To investigate the behaviour of different materials, it is necessary to compare the strength of the induced magnet with the strength of the external field. Various experimental techniques are available to do this, using either a deflection magnetometer, a vibration magnetometer, or a search coil with a ballistic galvanometer.

The results are generally as shown in Fig. 29. Starting with an unmagnetised specimen, its condition corresponds to point O on the graph. If the external field is gradually increased, some of the domains swing round into alignment and induced magnetism is observed. The results follow the

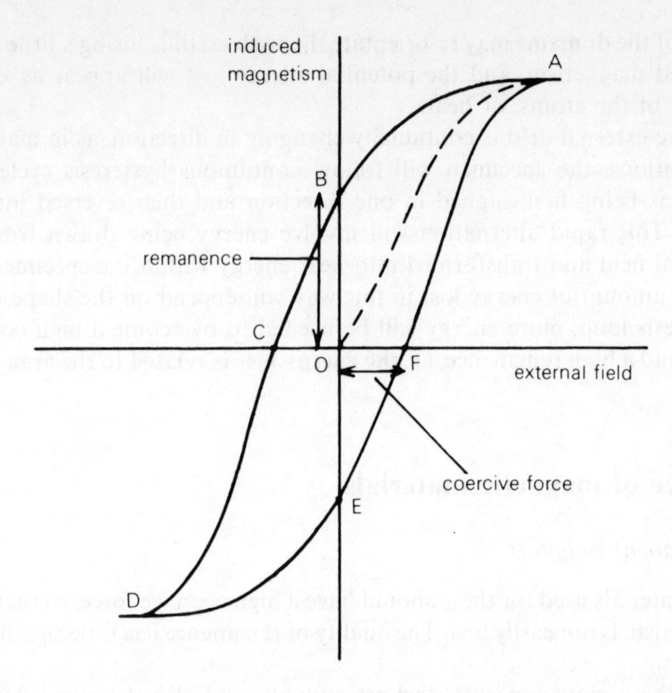

Fig. 29 Graph of external field and induced magnetism

dotted line up to point A, where the line flattens and becomes parallel to the field axis. This corresponds to magnetic saturation of the specimen.

If the field is now gradually decreased and then reversed, the observed results do not lie on the original line OA, but instead follow the line ABCD. This is because the attraction between adjacent domains opposes their immediate response to the external change.

Thus when the applied field has decreased to zero, there is still some magnetism remaining in the specimen, as shown by the distance OB on the graph. This effect is called *remanence*. The induced magnetism will not fall to zero until the external field reaches a certain value in the opposite direction, corresponding to the point C. The strength of this field is called the *coercive force* of the material.

Point D on the graph represents saturation in the opposite direction. If the field is again decreased to zero and then applied in the original direction, the results will follow the path DEF back to A, forming a symmetrical shape called a *hysteresis loop*.

Energy of magnetisation

Since work must be done to magnetise a specimen, it then possesses a form of potential energy. If the external field is removed after the magnetisation, some of the domains may re-orientate themselves, thus losing a little of the induced magnetism, and the potential energy lost will appear as kinetic energy of the atoms, or heat.

If the external field is continually changing in direction, as in many a.c. applications, the specimen will follow continuous hysteresis cycles, the domains being first aligned in one direction and then reversed into the other. This rapid alternation will involve energy being drawn from the external field and transformed into heat energy within the specimen.

The amount of energy lost in this way will depend on the shape of the hysteresis loop; more energy will be needed to overcome a high coercive force and a high remanence, i.e. **the energy loss is related to the area of the loop**.

Choice of magnetic materials

Permanent magnets

The materials used for these should have a high coercive force, so that their magnetism is not easily lost. The quality of remanence has little significance here.

All the various types of steel are suitable, and also the alloys Ticonal, Alcomax and Alnico, which are composed of varying proportions of iron, nickel, cobalt and aluminium.

Permanent magnets are required for moving-coil meters, loudspeakers and earphones, also in microphones.

Electromagnets

A low coercive force is required here, so that the electromagnet can be easily turned off. Low remanence is also an advantage if possible, particularly for magnetic switches, but this is not so important if the application is in a.c. work.

The natural elements, soft iron, nickel and cobalt, are all suitable, but recently developed alloys such as Mumetal and Stalloy have been especially designed for this purpose.

Uses of electromagnets for a.c. work include transformer cores, the armatures of generators, and telephone diaphragms. Until recently, the central processing units of computers were built with ferrite beads to serve as magnetic memories, but these are already obsolescent.

Demagnetisation

To remove induced magnetism from a specimen, the only sure method is to place it inside a coil through which a.c. is passing, and then slowly remove it. While inside the coil, the magnetisation of the specimen will follow

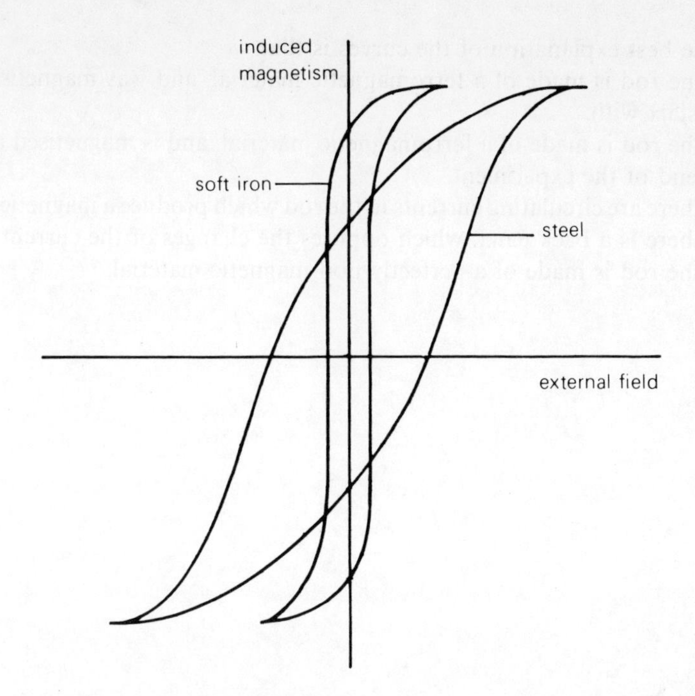

Fig. 30 Hysteresis curves of soft iron and steel

continuous hysteresis cycles, but these will get smaller and smaller as the specimen is withdrawn, and eventually the domains will lie completely at random.

This principle is used to remove unwanted magnetism from watches, and also from magnetic tapes and recording mechanisms, for which purpose a suitable form of the apparatus is marketed under the name of a 'de-gausser'.

Examination questions 7

1 Explain the following terms: *magnetic moment*, *intensity of magnetisation*, and *magnetic flux density*.

In terms of each of the properties (i) hysteresis loss, (ii) saturation magnetisation, (iii) remanent magnetisation, (iv) coercivity, (v) electrical resistivity, discuss the choice of materials for
(a) permanent magnets (b) electromagnet cores (c) transformer cores. [AEB June 1975]

2 A rod is placed inside a solenoid through which a direct current I can be passed, and the magnetic flux density B at a point on the axis of the solenoid outside the rod is measured. As the current is increased from zero up to I_1, the curve in Fig. 29 (p. 57) is followed from O to A, but as it is then reduced from I_1 to zero, the curve from A to B is followed.

The best explanation of the curves is that

A the rod is made of a ferromagnetic material, and was magnetised to start with.

B the rod is made of a ferromagnetic material, and is magnetised at the end of the experiment.

C there are circulating currents in the rod which produce a magnetic field.

D there is a back e.m.f. which opposes the changes of the current.

E the rod is made of a perfectly non-magnetic material. [O]

8 Electromagnetic induction

One of the most important discoveries in physics was made about one hundred and fifty years ago, when Faraday succeeded in producing electricity from magnetism. This is the phenomenon called electromagnetic induction, i.e. the production of an electromotive force in a conductor as it cuts across the lines of force of a magnetic field.

Faraday's experiments led to the following conclusions:
The e.m.f. induced increases with:
(a) the strength of the magnetic field,
(b) the area marked out in the field by the conductor,
(c) the speed of the relative motion.

Magnetic flux

The first two of these factors are combined in the idea of *magnetic flux*. The magnetic field B exists throughout the whole region, and if the shape of the conductor marks out an area A in the plane normal to the field, then the **magnetic flux through this area is equal to (field strength) × (area) = BA.**

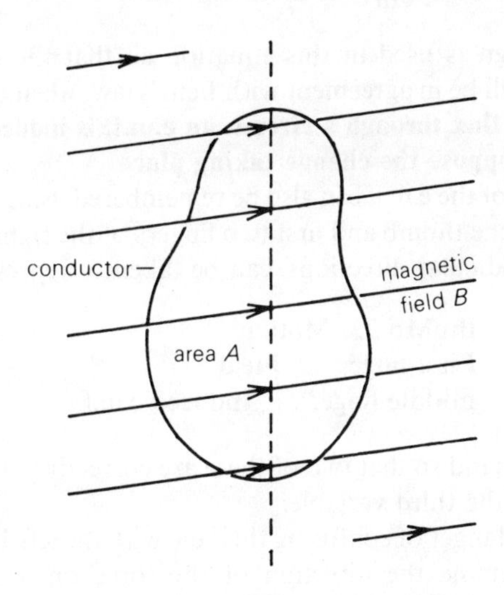

Fig. 31 Magnetic flux

If the strength of the magnetic field is measured in tesla (T) (defined later in Chap. 27) and the area is in square metres, then the magnetic flux Φ will be in *webers* (Wb).

It is sometimes preferable to think of the *field strength* in units of Wb m^{-2}, and it is then referred to by the phrase *magnetic flux density*.

If the normal to the plane of the area A makes an angle θ with the direction of the magnetic field, then the magnetic flux through A will be

$$\Phi = BA\cos\theta.$$

The induced e.m.f.

Faraday's three conclusions can be combined in the statement: **the induced e.m.f. is proportional to the rate of change of magnetic flux through the circuit.**

The unit of magnetic field strength, the tesla, was chosen so that this relation becomes an equality:

e.m.f. = rate of change of magnetic flux.

When the magnetic flux is in webers, and time is in seconds, the e.m.f. will then be in volts.

$$E = -\frac{\mathrm{d}\Phi}{\mathrm{d}t} \quad \text{using calculus notation.} \qquad (1)$$

The minus sign is used in this equation so that the direction of the induced e.m.f. will be in agreement with Lenz's law: **when there is a change in the magnetic flux through a circuit, an e.m.f. is induced in the circuit which tends to oppose the change taking place.**

The direction of the e.m.f. can also be remembered using Fleming's right-hand rule. Here the thumb and first two fingers of the right hand, if held in mutually perpendicular directions, can be taken to represent:

thuMb . . . Motion
First finger . . . Field
mIddle finger . . . Induced e.m.f.

Orientating the hand so that two of these are correctly placed will indicate the direction of the third variable.

Since there is danger of confusing this rule with the left-hand rule (which is used to determine the direction of the force on a current-carrying conductor in a magnetic field), it is probably wiser to rely on a straightforward application of Lenz's law in any particular situation.

Faraday's disc

Various simple forms of apparatus are available to demonstrate the application of Faraday's law.

In Fig. 32 a metal disc is rotated by a motor (not shown) within a long solenoid. The current through this solenoid can be altered by means of a variable resistor. Brush contacts to the rim of the disc at A and to the axle at B lead to a voltmeter.

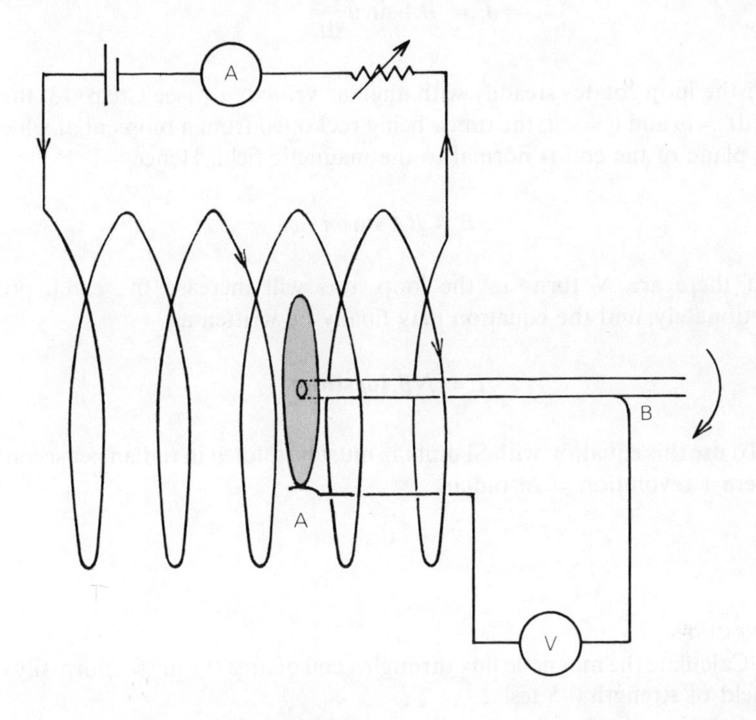

Fig. 32 Demonstration of Faraday's laws

The magnetic field inside the solenoid is longitudinal, and as the disc rotates, it cuts across the lines of force. The induced e.m.f. is indicated on the voltmeter, and it will be found that it is proportional to (a) the current passing through the solenoid, and (b) the rate of rotation of the disc. As this turns, each radius sweeps out the circular area of the disc, and this area must be multiplied by the number of revolutions of the disc per second if the rate of change of magnetic flux is to be calculated.

If the disc is rotating clockwise and the direction of the solenoid current is as shown, the induced e.m.f. will make the centre of the disc positive with respect to the rim.

The simple dynamo

Referring back to Fig. 31, if the conducting loop is rotated about the dotted line shown so that its plane cuts across the lines of magnetic flux, a varying e.m.f. will be induced around the loop, given by

$$E = -\frac{d}{dt}(BA\cos\theta)$$

$$\Rightarrow E = BA\sin\theta\frac{d\theta}{dt}.$$

If the loop rotates steadily with angular velocity ω, (see Chap. 13) then $d\theta/dt = \omega$ and $\theta = \omega t$, the time t being reckoned from a moment at which the plane of the coil is normal to the magnetic field. Hence

$$E = BA\sin\omega t\,.\,\omega.$$

If there are N turns in the loop, this will increase the e.m.f. proportionately, and the equation may finally be written as

$$\boldsymbol{E = NBA\omega \sin\omega t}.$$

To use this equation with SI units, ω must be quoted in radian per second, where 1 revolution = 2π radian.

EXERCISES

1 Calculate the magnetic flux through a coil of area 0.1 m^2 set normally to a field of strength 0.5 tesla.

2 Calculate the flux density of a magnetic field if it produces a magnetic flux of 4 weber through a square coil of side 0.2 m set with its plane at 60° to the field.

3 A wire of length 0.4 m is moved at right angles across a magnetic field of 0.1 tesla. If it moves with a speed of 1 m s^{-1}, find the e.m.f. induced between its ends.

4 A Faraday disc of radius 20 mm rotates at 10 rev/s normally to a magnetic field of 1.2×10^{-3} T. Calculate the e.m.f. induced between the axle of the disc and its rim.

5 A simple dynamo consists of a narrow coil of 50 turns, the area of each loop being 0.01 m^2. It is rotated at 5 rev/s about a diameter of the coil which is at right angles to a magnetic field of 0.1 T. Find the maximum value of the e.m.f. generated.

The alternator

The equipment used by large-scale suppliers such as the Central Electricity Generating Board differs in design from the simple dynamo, although still based on the same principle. Figure 33 illustrates one type of generator, called an *alternator*.

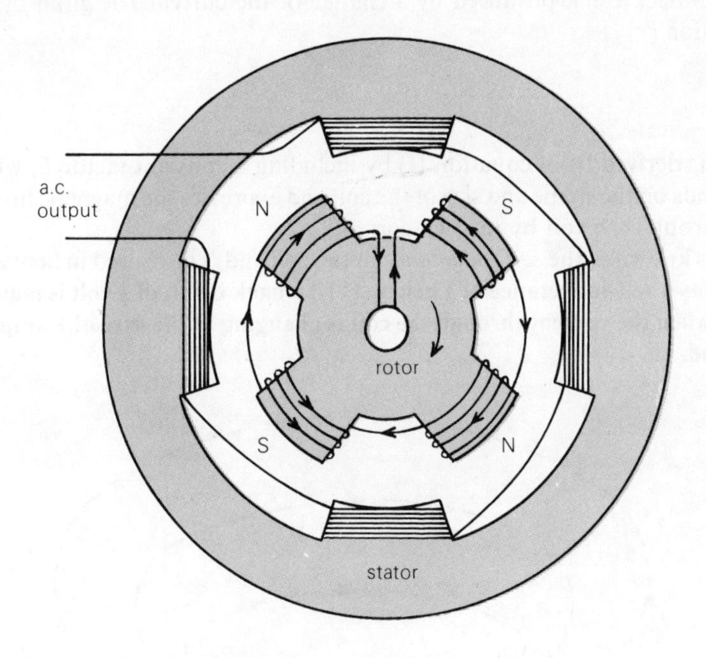

Fig. 33 Construction of an alternator

In this design, the magnetic field moves while the coil remains stationary. The magnetic field is created by an electromagnet, which may have the normal two poles, or four poles as shown in the diagram. The windings of the coils of the magnet must be in alternate directions, so that north and south poles are produced alternately. This assemblage is called the *rotor*. It is turned by a steam turbine driven by coal or nuclear power.

The coils in which current is to be induced are wound upon a vast soft-iron yoke and form the *stator*. This is much more massive than the rotor since a very large number of turns of wire must be used to produce sufficient current for the national supply, whereas only a small current is needed to energise the electromagnet.

Four or six stator coils are generally used. This is to smooth the motion, as the rotor would tend to turn unevenly between equilibrium positions if only two poles were involved.

Self-inductance

A current flowing in a cylindrical coil sets up a magnetic field around it. The coil is itself then a conductor in a magnetic field, and if the field changes, an e.m.f. will be induced in the coil tending to oppose the change. This is referred to as a *back e.m.f.*

The back e.m.f. produced by a change of the current I is given by the equation

$$E = -L\frac{\mathrm{d}I}{\mathrm{d}t} \qquad (2)$$

This is derived from equation (1) by including a constant factor L, which depends on the shape and size of the coil and expresses the magnetic flux set up through the coil by its own current.

L is known as the *self-inductance* of the coil and is measured in henrys. **A coil has a self-inductance of 1 henry (H) if a back e.m.f. of 1 volt is induced in it when the current through the coil is changing at the rate of 1 amp per second.**

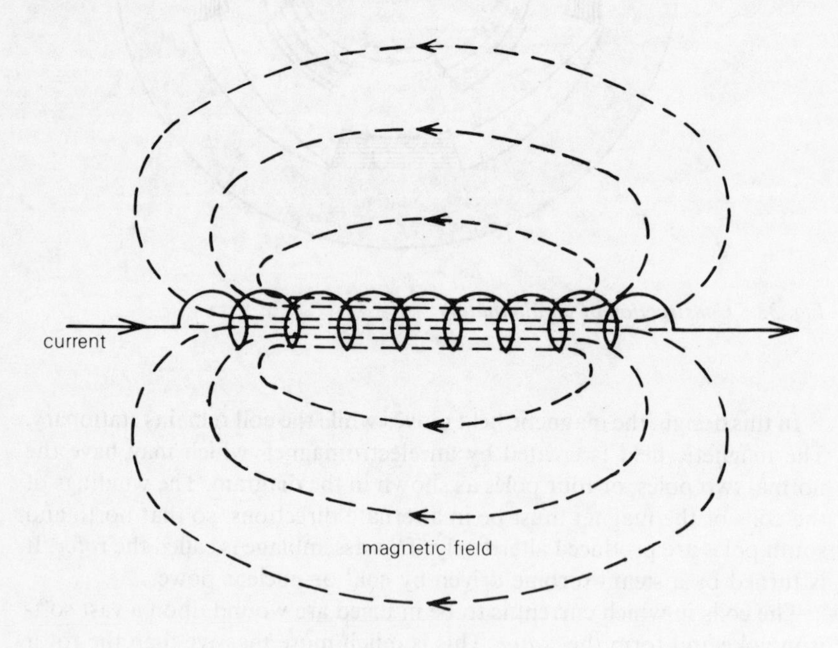

Fig. 34 The magnetic field around a solenoid

Coils made up of a large number of closely-wound turns produce strong magnetic fields and so have high inductance. This affects the rate at which a current becomes established in a circuit after switching on, which, being opposed by the back e.m.f. it creates, will follow an exponential curve as shown in Fig. 35.

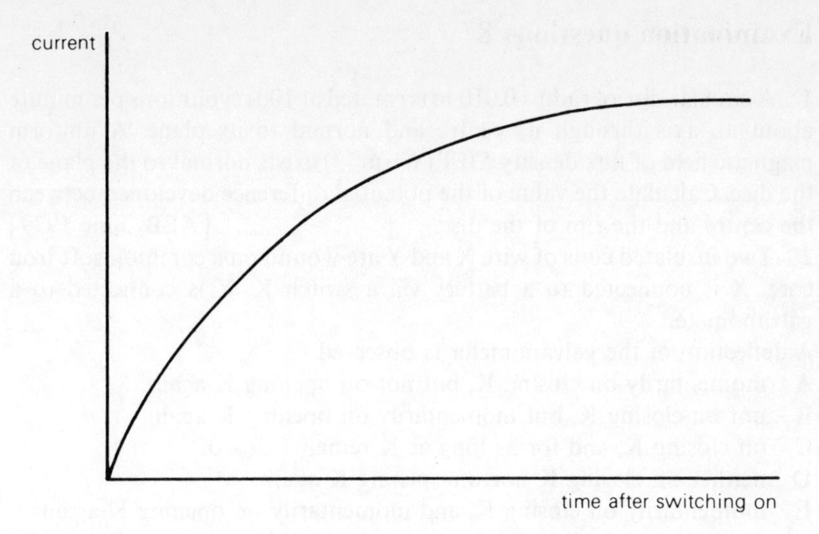

Fig. 35 Graph showing the growth of a current through a coil

To avoid this time-lag, the coils used as standard resistors are wound non-inductively; at the centre of the coil, the direction of the winding is reversed. Thus the magnetic effects of the two halves of the coil cancel out (Fig. 36).

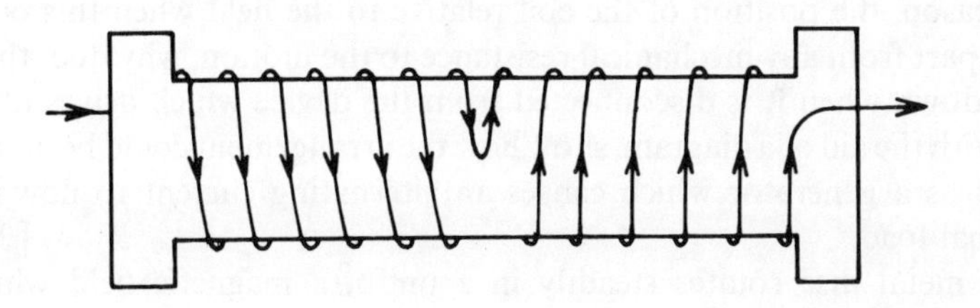

Fig. 36 Non-inductive winding of a resistance coil

A corresponding effect is seen on breaking a circuit containing a high inductance. The induced back e.m.f. tries to maintain the current, and may in fact be sufficient to cause a spark at the contacts of the switch. This demonstrates the amount of energy which is stored in the magnetic field of a coil.

EXERCISE

6 Find the back e.m.f. produced in a coil whose self-inductance is 8 henry, by a current growing at the steady rate of 0.1 amp s^{-1}.

Examination questions 8

1 A circular disc of radius 0.010 m is rotated at 100 revolutions per minute about an axis through its centre and normal to its plane. A uniform magnetic field of flux density 5.0 T ($Wb\,m^{-2}$) exists normal to the plane of the disc. Calculate the value of the potential difference developed between the centre and the rim of the disc. [AEB, June 1979]

2 Two insulated coils of wire X and Y are wound on a common soft-iron core. X is connected to a battery via a switch K. Y is connected to a galvanometer.

A deflection of the galvanometer is observed

A momentarily on closing K, but not on opening K again.

B not on closing K, but momentarily on opening K again.

C on closing K, and for as long as K remains closed.

D neither on closing K nor on opening K again.

E momentarily on closing K, and momentarily on opening K again.

[C]

3 A rectangular coil of N turns, each of dimensions a metres by b metres, has its ends short-circuited and is rotated at constant angular speed ω rad s^{-1} in a uniform magnetic flux density of B tesla ($Wb\,m^{-2}$). The axis of rotation passes through the midpoints of the sides, of length a, of the coil and is at right-angles to the direction of the magnetic field.

(a) Explain why there is an e.m.f. in the coil and derive an expression which shows how its magnitude varies with time.

(b) What is the frequency of the e.m.f. in Hz?

(c) Derive an expression for the maximum value of the e.m.f. and state, with the reason, the position of the coil relative to the field when this occurs.

(d) Apart from any mechanical resistance to the motion, why does the coil slow down when it is disconnected from the device which drives it?

(e) With the aid of a diagram, show how the arrangement could be modified to act as a generator which causes an alternating current to flow in an external load. [JMB]

4 A metal disc rotates steadily in a uniform magnetic field which is directed normally to the plane of the disc and covers the whole of it. A and B are brushes; A makes contact with the rim of the disc and B with its centre.

By considering an electron between the brushes, show that a potential difference exists between A and B.

Explain what changes, if any, will be needed in the torque required to keep the disc rotating at the same speed if A and B are connected by a resistor. [L]

Light I

9 Refraction at a plane surface

The two laws of refraction, which are experimental conclusions, are:
(1) The incident ray, the refracted ray, and the normal to the surface at the point of incidence all lie in the same plane.
(2) The ratio of the sine of the angle of incidence to the sine of the angle of refraction is constant for any two transparent materials.

The second law, known as Snell's law, was not discovered until 1620, although the phenomena of refraction had been studied for many centuries before that date. The establishment of Snell's law initiated much scientific discussion about the nature of light, and two rival theories developed, Newton's corpuscular theory and Huygens' wave theory.

Conflict between the two theories hinged on the question of whether light travelled faster or slower in a more dense medium. This question was not resolved until the speed of light was actually measured in the laboratory, in about 1862, when Foucault's experimental results proved that light travels more slowly in water than in air, thus supporting the wave theory.

There are still, however, certain phenomena which can only be explained by assuming the existence of 'corpuscles' of light (photons, Chap. 35) and the modern approach is to regard light sometimes as a wave and sometimes as a particle, using mathematics of a form which will accommodate both points of view.

Relation between refractive index and speed of light

Consider a beam of light approaching obliquely to an interface between two transparent media. PAB is the position of the wavefront when it first meets the interface, at P.

If the beam is refracted towards the normal, as shown in Fig. 37, then by the time the far edge of the beam reaches the interface at Q, the ray incident at P will have travelled on to C, where QC is at right angles to PC. The wavefront at that time will occupy the position CDQ.

The time taken for the wave to travel from B to Q in medium 1 must be the same as that taken for it to travel from P to C in medium 2. Since the ray has bent towards the normal, PC will be less than BQ, and the wavelength of the light in medium 2 must be shorter than its wavelength in medium 1.

Let the speeds of light in the two media be c_1 and c_2.

$$\text{time taken} = \frac{\text{distance}}{\text{speed}} = \frac{BQ}{c_1} = \frac{PC}{c_2}. \qquad (1)$$

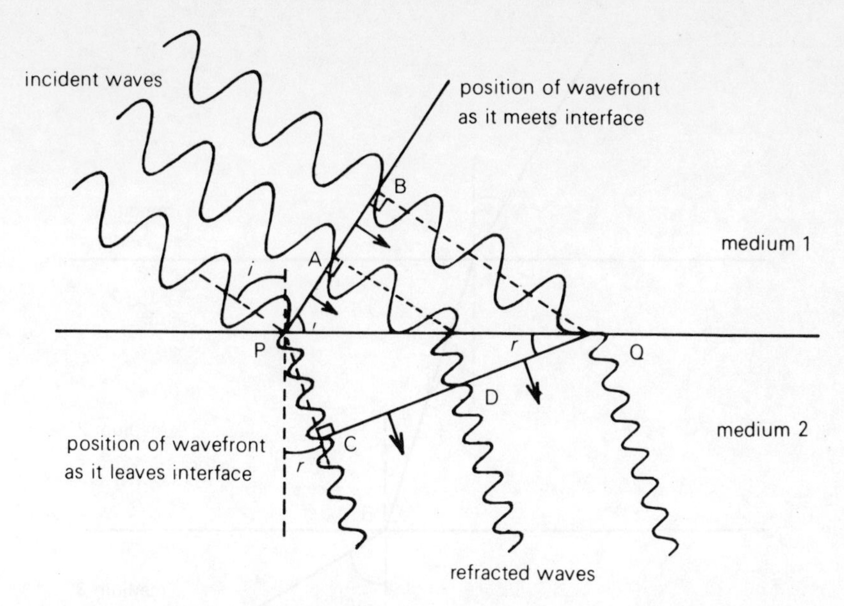

Fig. 37 Refraction of waves at a plane interface

Now the angle of incidence of the ray at P, marked i in Fig. 37, is equal to $\angle$ BPQ, and the angle of refraction at P, marked r, is equal to $\angle$ PQC.

So
$$\frac{\sin i}{\sin r} = \frac{BQ/PQ}{PC/PQ} = \frac{BQ}{PC}.$$

Combining this with equation (1)

$$\frac{\sin i}{\sin r} = \frac{c_1}{c_2}.$$

This is Snell's law, and the constant involved is the ratio of the speeds of light in the two media. If medium 1 is a vacuum (or for practical purposes, air) then the value of this constant is called the *refractive index, n,* of medium 2.

Refractive index of a medium = $\dfrac{\textbf{speed of light in vacuo}}{\textbf{speed of light in the medium}}$

Refraction in parallel layers

The argument may be extended to involve three separate media, as shown in Fig. 38, where medium 2 is in the form of a parallel-sided slab (such as the glass wall of a tank).

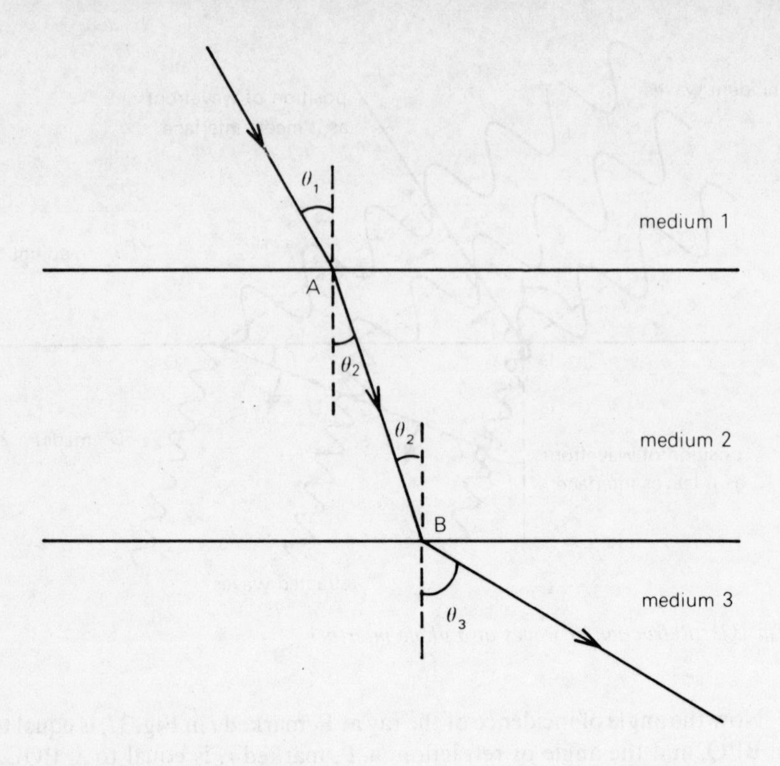

Fig. 38 Refraction in parallel layers

If the angles of incidence and refraction are denoted by θ_1, θ_2 and θ_3 as in the figure,

at A,
$$\frac{\sin \theta_1}{\sin \theta_2} = \frac{c_1}{c_2}$$

and at B
$$\frac{\sin \theta_2}{\sin \theta_3} = \frac{c_2}{c_3},$$

the two angles marked θ_2 being alternate angles.

These equations can be combined in the form

$$\frac{\sin \theta_1}{c_1} = \frac{\sin \theta_2}{c_2} = \frac{\sin \theta_3}{c_3}$$

or
$$n_1 \sin \theta_1 = n_2 \sin \theta_2 = n_3 \sin \theta_3,$$

since the refractive index n is inversely proportional to the speed of light in the medium, i.e. **$n \sin \theta$ is constant for any particular ray**.

In this expression, θ is the angle made by the ray with the normal within the medium whose refractive index is n.

Huygens' wavelets

Huygens' explanation of the phenomena of reflection and refraction was that every point on a wavefront should be considered as the centre of a set of secondary wavelets. These spread out in all directions, but interfere destructively with each other everywhere except in the direction in which the wave is travelling. The envelope of the wavelets after a given small interval of time shows the new position of the wavefront at that moment.

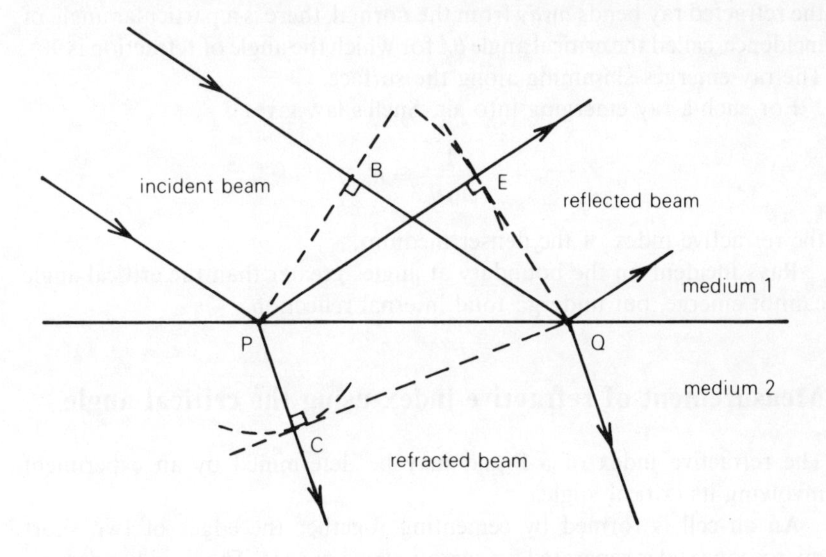

Fig. 39 Huygens' construction for refraction and reflection

Figure 39 shows the geometrical construction used by Huygens as part of his theory. The diagram should be compared with Fig. 37 (p. 71). As before, an incident beam of light approaches obliquely to the interface between two transparent media. P is the point at which the beam first strikes the interface, and Q is the point at which the far edge of the beam will reach it. PB is drawn perpendicular to the ray travelling to Q.

The distance BQ is measured, and a length $PC = BQ(n_1/n_2)$ is calculated, where n_1 and n_2 are the refractive indices respectively of the two media.

The arc of a circle, centre P and radius PC, is drawn and the tangent to this from Q is constructed to meet the arc at C. PC is then the refracted ray from P, and the refracted ray from Q can be drawn parallel to PC.

It can be seen that this construction is justified by the argument presented with Fig. 37.

The position of the beam caused by *partial reflection* at the interface can also be found by Huygens' construction. With centre P, draw an arc of a circle with radius PE = BQ and construct the tangent to this from Q to meet the arc at E. PE is then the reflected ray from P, and the reflected ray from Q can be drawn parallel to PE.

Huygens' theory is particularly useful in understanding the more complicated phenomenon of diffraction (Chap. 37).

The critical angle

If light is travelling from a more dense to a less dense medium, in which case the refracted ray bends *away* from the normal, there is a particular angle of incidence, called the critical angle θ_c, for which the angle of refraction is 90°. The ray emerges skimming along the surface.

For such a ray emerging into air, Snell's law gives

$$\frac{1}{\sin \theta_c} = n,$$

the refractive index of the denser medium.

Rays incident on the boundary at angles greater than the critical angle cannot emerge, but undergo total internal reflection.

Measurement of refractive index using the critical angle

The refractive index of a liquid may be determined by an experiment involving its critical angle.

An air-cell is formed by cementing together the edges of two short microscope slides separated by narrow strips of card. The liquid under test should be placed in a transparent straight-sided container; the air-cell is suspended in the liquid so that it may be turned about a vertical axis. Figure 40 shows the apparatus seen from the top (the thickness of the air space being exaggerated for clarity).

An observer looking through the side MN of the container will be able to see through the air-cell when the slide AB is parallel to MN. If the air-cell is turned slowly so that the angle of incidence of the light on AB increases, the ray emerging into the air space from the glass slide will be at a greater angle of refraction, since air is less dense than the liquid. Eventually this ray will be so much bent that it does not reach the far side of the air space, but strikes the cement seal at A (Fig. 40(b)). The observer will find his field of view suddenly cut off, and will instead see light entering the liquid through the face KN and being *reflected* into his eye.

A pointer attached to the support of the air-cell and moving over a circular scale will give a reading corresponding to this position. Suppose the air-cell has turned through an angle α from the straight-through position. If the cell is rotated backwards to the cut-off on the other side, then the angle between the two positions of cut-off will be 2α. It is easier and more accurate to determine α in this way than from a reading on one side only.

At the cut-off position, the ray of light leaves the liquid making the angle

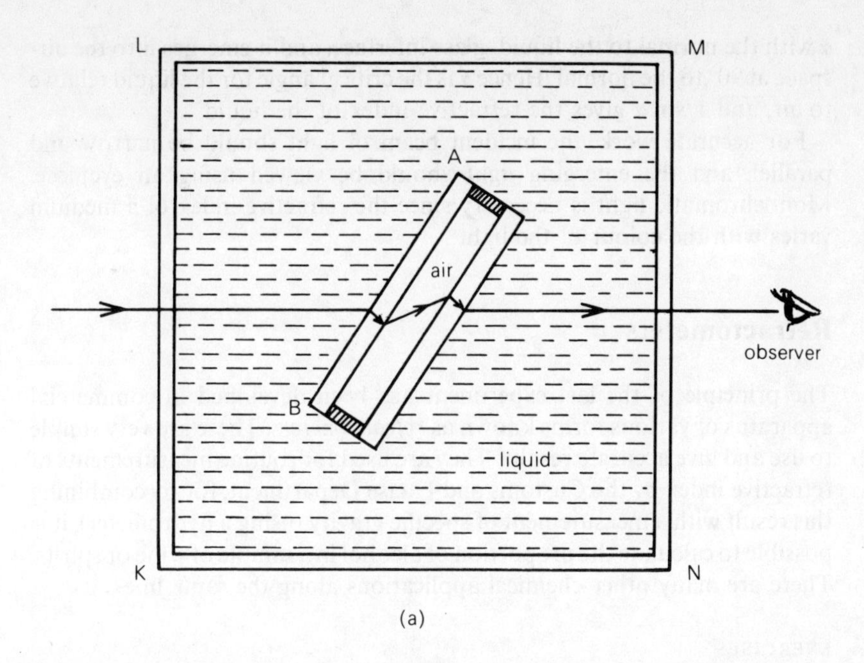

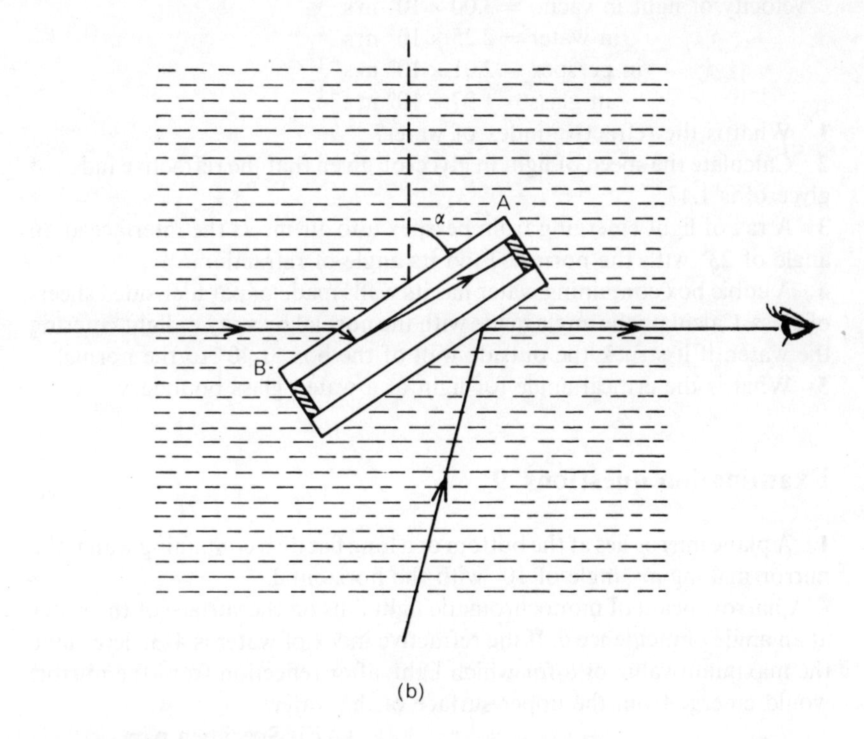

Fig. 40 Air-cell method for determination of the refractive index of a liquid (a) general view (b) position at cut-off

α with the normal to the liquid–glass interface, and it emerges into the air-space at 90° to the normal. Hence α is the critical angle for the liquid relative to air, and $1/\sin\alpha$ gives the refractive index of the liquid.

For accurate work, the incident beam of light should be narrow and parallel, and the emerging light should be viewed using an eyepiece. Monochromatic light is necessary since the refractive index of a medium varies with the colour of the light.

Refractometers

The principle of the last experiment has been developed in commercial apparatus of various forms known as *refractometers*. These are very simple to use and give accurate results. They are used for routine measurements of refractive index by the Customs and Excise Department, for by combining this result with a measurement of specific gravity (using a hydrometer), it is possible to calculate the proportion of alcohol in a sample of wine or spirits. There are many other chemical applications along the same lines.

EXERCISES

Use the following data where applicable:

velocity of light in vacuo = 3.00×10^8 m s^{-1}
in water = 2.25×10^8 m s^{-1}
in perspex = 2.01×10^8 m s^{-1}
in glass = 1.97×10^8 m s^{-1}.

1 What is the refractive index of water?

2 Calculate the speed of light in glycerol, given that the refractive index of glycerol is 1.47.

3 A ray of light emerging from perspex into air meets the interface at an angle of 23° with the normal. Find its angle of refraction.

4 A cubic box containing water has its walls made of parallel-sided sheets of glass. Calculate the angle made with the normal by a ray of light entering the water, if it struck the outside wall of the box at 40° to the normal.

5 What is the critical angle for light at a water–glass boundary?

Examination questions 9

1 A plane mirror lies at the bottom of a long flat dish containing water, the mirror making an angle of 10° with the horizontal.

A narrow beam of monochromatic light falls on the surface of the water at an angle of incidence θ. If the refractive index of water is 4/3, determine the maximum value of θ for which light, after reflection from the mirror, would emerge from the upper surface of the water.

[AEB Specimen paper, 1977]

2 A wave of wavelength 3 m travelling with speed 12 m s^{-1} in a certain medium enters another medium of refractive index 1.5 times that of the first medium. The correct speed and wavelength in the second medium are given by

	wavelength/m	speed/m s^{-1}
A	2	8
B	2	18
C	3	18
D	4.5	8
E	4.5	12

[C]

3 'Observed differences in the speed of light in different media support the wave theory of light.' Discuss this statement qualitatively by considering the deviation of a beam of light on passing from one medium to another. [L]

4 Light is refracted on passing across the boundary between two media in which its velocities are v_1 and v_2. Show that the observed facts are consistent with the statement that 'all points on a wavefront take the same time to travel to a new position on that wavefront.' [L]

5 Light travelling through a vacuum with velocity c has wavelength λ and frequency f. It then enters a medium of refractive index n. Which one of the following describes the characteristics of the wave in that medium?

A velocity c/n, wavelength λ/n, frequency f.
B velocity cn, wavelength λn, frequency f.
C velocity cn, wavelength λ, frequency f/n.
D velocity cn, wavelength λ, frequency fn.
E velocity c/n, wavelength λ/n, frequency f/n. [O]

10 Refraction through a prism

When a beam of white light passes through a prism, it is both *deviated* from its original direction, and *dispersed* into a coloured spectrum, as shown in Fig. 41.

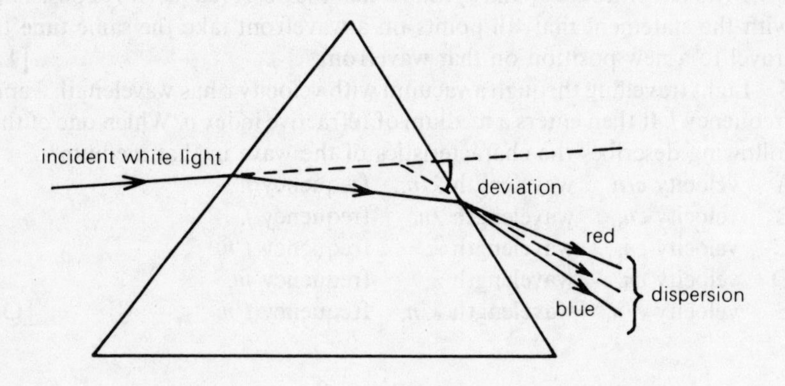

Fig. 41 Deviation and dispersion produced by a prism

The deviation is found to vary with the angle of incidence of the light on the first face of the prism, in the way shown in Fig. 42. This graph may be obtained experimentally, either using a ray-box which gives a narrow beam, or by using four optical pins, two to define the incident ray and another two to define the emergent ray, located by viewing the first two pins through the prism.

The graph is not symmetrical. It has a shallow minimum which is not easy to determine by these experiments, but may be found accurately using a spectrometer, as described below.

Minimum deviation

From considerations of symmetry, the minimum deviation must occur when the light passes symmetrically through the prism. In that case (Fig. 43) the angles made by the incident ray and the emergent ray with their respective normals will be equal; call these angles i. The two angles made

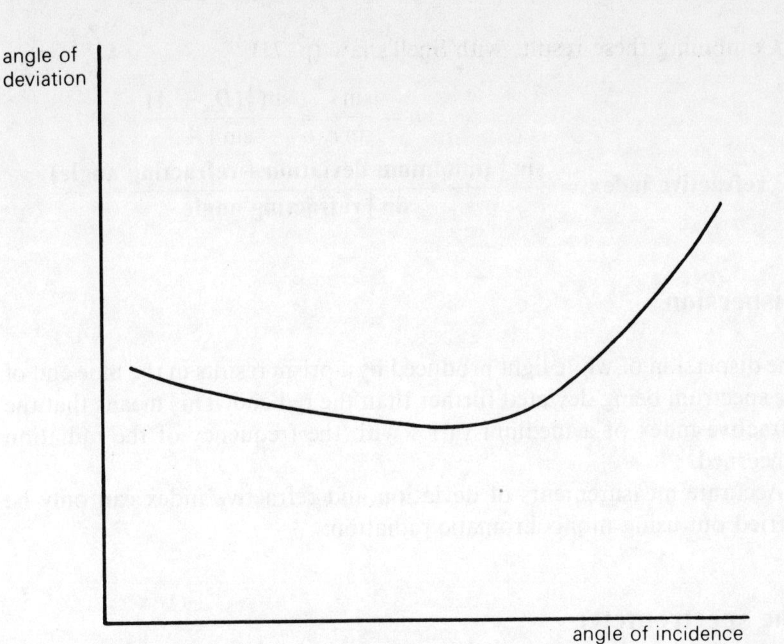

Fig. 42 Graph of angle of incidence and angle of deviation

with the normals by the ray of light inside the prism will be equal; write these as r.

Then, by geometry,

angle of minimum deviation $D_m = 2(i - r)$
and the refracting angle of the prism $A = 2r$.
Hence $r = \frac{1}{2}A$
and $i = \frac{1}{2}(D_m + A)$.

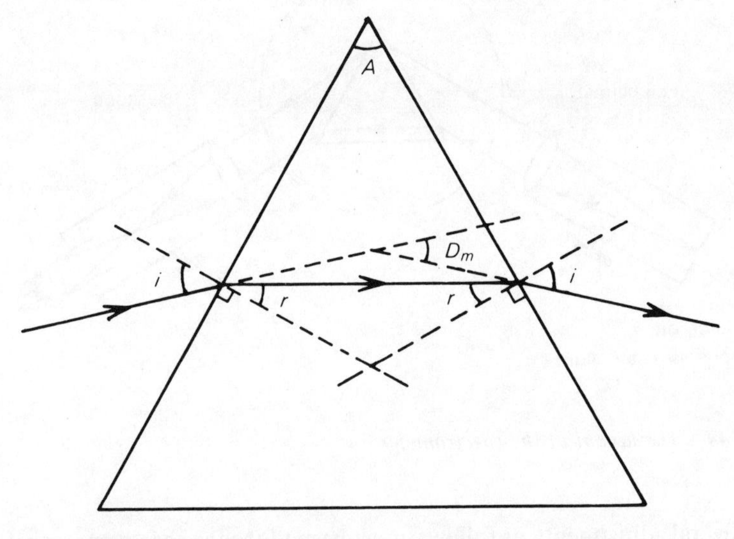

Fig. 43 Minimum deviation position

Combining these results with Snell's law (p. 71),

$$n = \frac{\sin i}{\sin r} = \frac{\sin \frac{1}{2}(D_m + A)}{\sin \frac{1}{2}A},$$

or **refractive index** $= \dfrac{\textbf{sin}\,\frac{1}{2}\,\textbf{(minimum deviation + refracting angle)}}{\textbf{sin}\,\frac{1}{2}\,\textbf{refracting angle}}$

Dispersion

The dispersion of white light produced by a prism results in the blue end of the spectrum being deviated further than the red end. This means that the refractive index of a medium varies with the frequency of the radiation concerned.

Accurate measurements of deviation and refractive index can only be carried out using monochromatic radiation.

The spectrometer

This is a precision instrument comprising an optical system (the collimator) which sends out a parallel beam of light, a telescope for receiving this light, and a turntable set in between the two. The position of the telescope can be located on a circular scale marked in degrees which is fitted with verniers to give readings accurate to one-tenth of a degree. The layout of the optical system when the spectrometer is used with a prism to measure deviation is shown in Fig. 44.

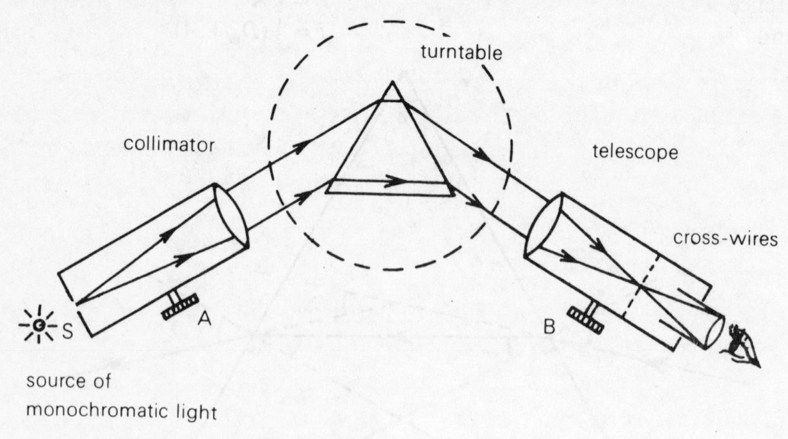

Fig. 44 The layout of the spectrometer

Several adjustments, as follows, must be made to the spectrometer before the prism is placed on the table:

(a) *The telescope.* Cross-wires set inside the tube of this should be rotated into the desired position, and the eyepiece then focused so that the cross-wires appear sharp.

The telescope is next turned towards a distant object, preferably well outside the laboratory, and knob B is adjusted until the object is seen clearly in the same plane as the cross-wires. This means that the telescope is now set to receive parallel light.

(b) *The collimator.* A source of monochromatic light is set close to the slit S of the collimator, shielded so that stray light from the source does not reach the observer. The telescope is turned into the straight-through position and the slit is viewed through it. Knob A, on the collimator, is adjusted until the slit appears sharp. In this position the collimator must be giving out parallel light, since the telescope has already been set to this condition.

(c) *The turntable.* The spectrometer normally stands so that the beam of light in use is horizontal, and the turntable top should also be horizontal. This can be adjusted by levelling screws, either using a spirit level or by an optical method. One way of doing this is to place the prism on the table and view light from the collimator reflected from each side of the prism in turn, rotating the table so that the reflected beams pass into the telescope which is kept in a fixed position. If one image appears higher in the field of view than another, the level of the turntable should be adjusted.

Use of the spectrometer to measure minimum deviation

The prism which is to be studied is placed on the centre of the turntable. The slit S is nearly closed so that only a fine beam of light is in use; the source must be positioned carefully to give the brightest possible image.

The direction of the deviated beam should first be found using the naked eye. If the arrangement is as shown in Fig. 44, then rotating the turntable clockwise will increase the angle of incidence of the light on the first face of the prism, and an anticlockwise rotation will decrease this angle.

The table should be turned slowly and steadily with one hand and the telescope moved with the other hand to keep the image of the slit in the field of view. A position will be found in which the movement of the image across the field of view changes direction. This corresponds to minimum deviation.

The position should be determined accurately so that small movements of the table in alternate directions cause the image to come exactly up to the cross-wires and then recede. The turntable should be fixed in this position using the clamp attached to it, and the scale reading of the telescope noted.

The prism is then removed, and the telescope turned back to the straight-through position. The reading of the scale is again noted, and the difference between the two readings gives the angle of minimum deviation.

If the source of light gives several discrete frequencies (as does a neon or mercury vapour lamp), these will produce separate coloured images. Each may be studied on its own, and will give results varying by a fraction of a degree.

Measurement of the deviating angle of the prism

The spectrometer can be used to give an accurate measure of the angle of a prism as follows.

The slit S should be opened fairly widely and the particular angle to be measured should be pointed towards the collimator as shown in Fig. 45. Light then falls on both faces of the prism, and reflected images can be seen on both sides. The turntable is kept fixed, and the telescope moved to receive first one image and then the other, centring them accurately using the crosswires. The two positions of the telescope are noted from the scale and the difference between the readings equals twice the angle of the prism.

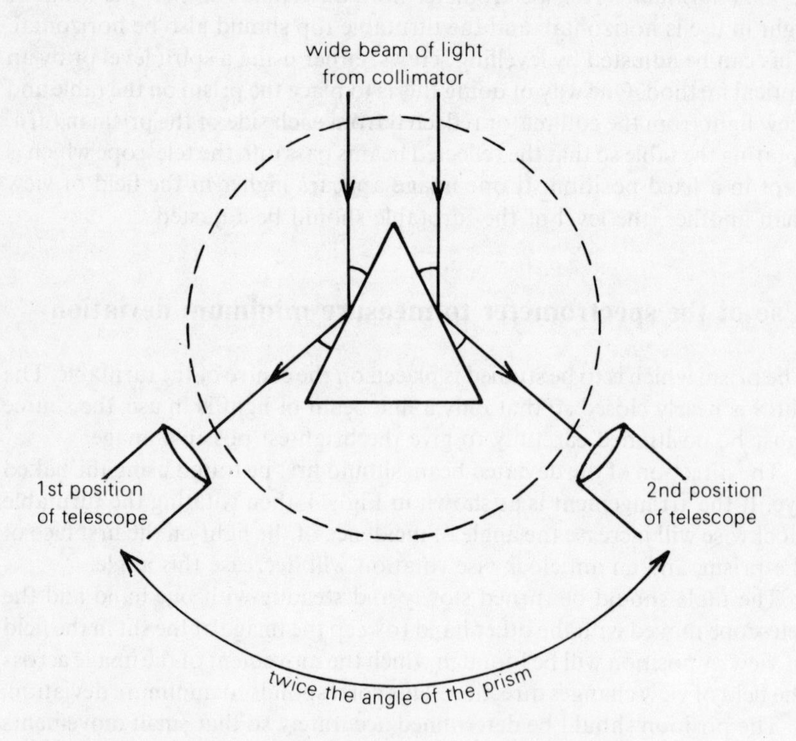

Fig. 45 Optical measurement of the angle of a prism

This result can then be combined with the value obtained for the angle of minimum deviation to give the refractive index of the material of the prism for the particular colour of light used. Such results will normally give four-figure accuracy.

It is also possible to use the spectrometer to measure the refractive index of a liquid, by placing the liquid inside a specially made hollow prism whose sides are thin parallel-sided pieces of glass.

EXERCISES

1 What angle of incidence corresponds to an angle of minimum deviation of 24° for a prism of refracting angle 40°?

2 Find the angle of minimum deviation for a prism of refracting angle 60° and refractive index 1.53.

3 Calculate the refractive index of the material of a prism of refracting angle 59.8° if the corresponding angle of minimum deviation is 39.8°.

Spectra

The main use of the spectrometer is in the study of the spectra produced by different sources of radiation. These fall into two distinct classes:

(1) *The continuous spectrum.* The spectrum of white light is the best known example of this, familiar since Newton's days. In general, such a spectrum is produced from an incandescent solid or liquid.

The visible part of this spectrum is made up of a continuously varying band of colour from deep red to dark blue. Investigation shows that radiation of longer wavelength is also present, called the *infrared*, and there may also be some extension of the spectrum into shorter wavelengths, the *ultraviolet* region.

The laws governing the variation of intensity with wavelength in the spectra from different sources will be considered in Chap. 11.

(2) *The line spectrum.* When viewed through a spectrometer, a line spectrum can be seen to consist of discrete lines or bands of different wavelengths. These give separate images, each of a different colour.

This type of spectrum is produced from a hot gas, and the lines are characteristic of the elements present in that gas. Figure 46 shows the lines in the visible part of the spectrum of (a) sodium and (b) mercury. Other lines also exist in the infrared and ultraviolet. The theory of the line spectrum is considered in Chap. 35.

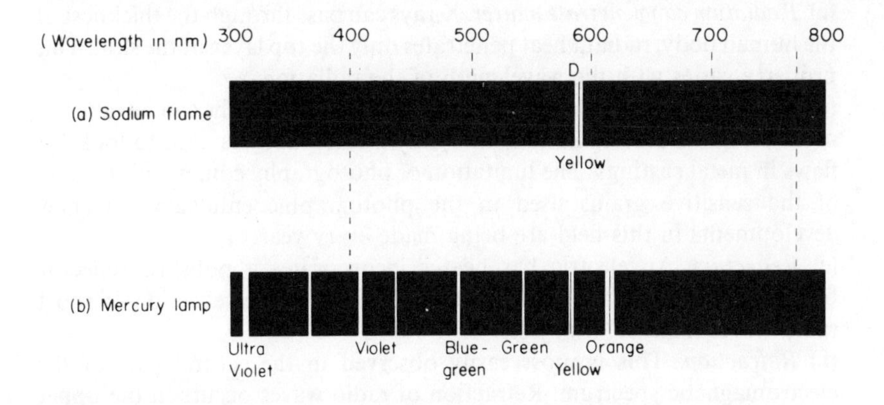

Fig. 46 Line spectra

The electromagnetic spectrum

The full range of the spectrum is shown in Fig. 47. This can be considered either in terms of *wavelength* (longer waves occur on the left-hand side of the diagram) or in terms of *frequency* (low frequencies on the left). The various sections named are not rigorously limited but merge into each other, and in some cases overlap.

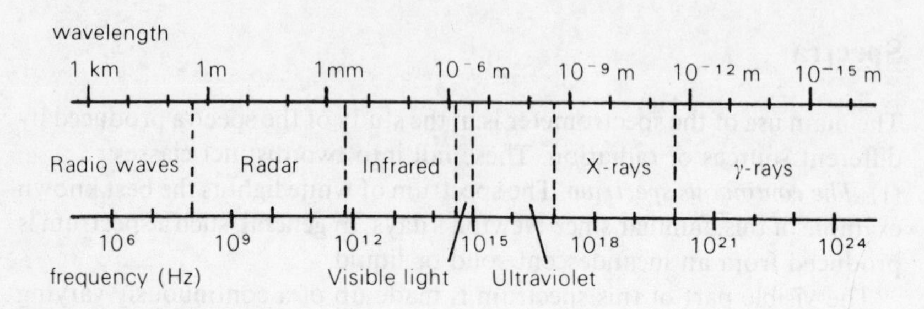

Fig. 47 The electromagnetic spectrum

All types of radiation included in this spectrum have the same properties, and illustrations of these properties can be found in every field.

(a) *All radiation travels with the same velocity through a vacuum.* Standard measurements of this velocity have been carried out using both visible light and microwaves, and the result is at present quoted as $c = 2.9979250 \times 10^8$ m s^{-1}. The fact that radiant heat reaches us through space from the sun and that radio communication was possible with the astronauts on the moon demonstrate that this property applies to other types of radiation.

(b) *Radiation travels in straight lines.* Shadows are cast not only by visible light, but also by X-rays and by the heat from the sun, as can be seen as the frost melts on a lawn on a bright winter's day.

(c) *Radiation can penetrate matter.* X-rays can pass through the thickness of the human body; radiant heat penetrates only the top layer of the skin. This property varies with the wavelength of the radiation.

(d) *Radiation will affect a photographic plate.* Photographs can be taken at night using infrared radiation, and γ-ray photography is used to look for flaws in metal castings. The limitation of photography comes with the size of the sensitive grains used in the photographic emulsion, but new developments in this field are being made every year.

(e) *Reflection.* An electric bar heater incorporates a polished reflector behind the element; the basic principle of radar is the reflection of short radio waves from solid objects.

(f) *Refraction.* This is most easily observed in the central part of the electromagnetic spectrum. Refraction of radio waves occurs in the upper layers of the atmosphere and results in the improved reception of radio signals noticeable after sunset.

(g) *Interference.* See Chap. 36.
(h) *Diffraction.* See Chap. 37.
(i) *Polarisation.* See Chap. 39.

Examinations questions 10

1 The range of wavelengths of infrared radiation is approximately

A 10^{-9} m to 10^{-7} m D 10^{-4} m to 10^{-1} m
B 10^{-7} m to 10^{-6} m E 10^{-1} m to 10^{+2} m
C 10^{-6} m to 10^{-3} m [C]

2 An equilateral plastic prism is found to have an angle of minimum deviation of 30°. What is the refractive index of the material?

A 1.33 B 1.41 C 1.51 D 1.62 E 1.71 [O]

11 The continuous spectrum

As mentioned in the previous chapter, this is the name given to the spectrum of radiation obtained from hot solids and liquids. It is studied by the use of a source of thermal radiation known as a *black body*.

A coloured surface appears coloured because it reflects certain wavelengths of the spectrum selectively while absorbing the remainder. This makes it unsuitable as a source of radiation for use in general experimental work. A black body is specially designed so that it completely absorbs all wavelengths of radiation falling upon it, and similarly radiates all wavelengths unselectively.

A typical form of black body consists of a thick stainless steel hollow cylinder, blackened inside, and having a small hole in one end which forms the radiant source (Fig. 48). Any radiation entering the hole will be absorbed by the black interior walls with very little chance of being reflected out of the hole again.

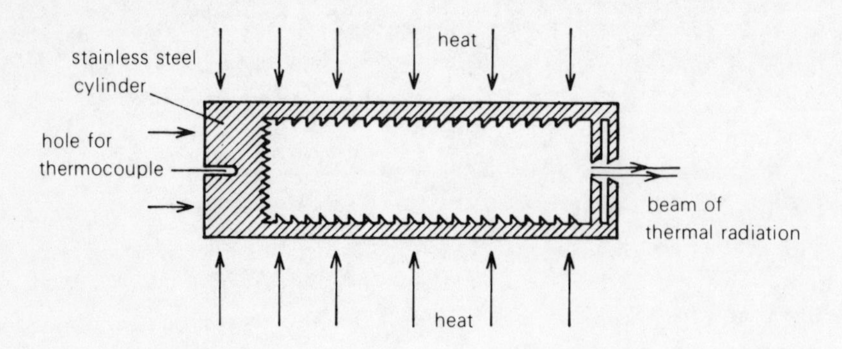

Fig. 48 A black body

When the cylinder is strongly heated in a furnace, it can become red-hot or white-hot, and the corresponding spectrum of radiation will be emitted. Experimental work is, of course, confined to the narrow beam of black body radiation coming from the hole, the rest of the hot apparatus being screened off.

To study the radiation in detail, it is first passed through one of a set of filters which each transmit only a small band of wavelengths. The selected radiation then falls on to a photoelectric surface (Chap. 35) and the photoelectrons emitted from this are multiplied electronically to give a reading on a counter. This method supersedes the old-fashioned experiments using thermopiles or bolometers.

Figure 49 shows the form of the results for black bodies at various temperatures.

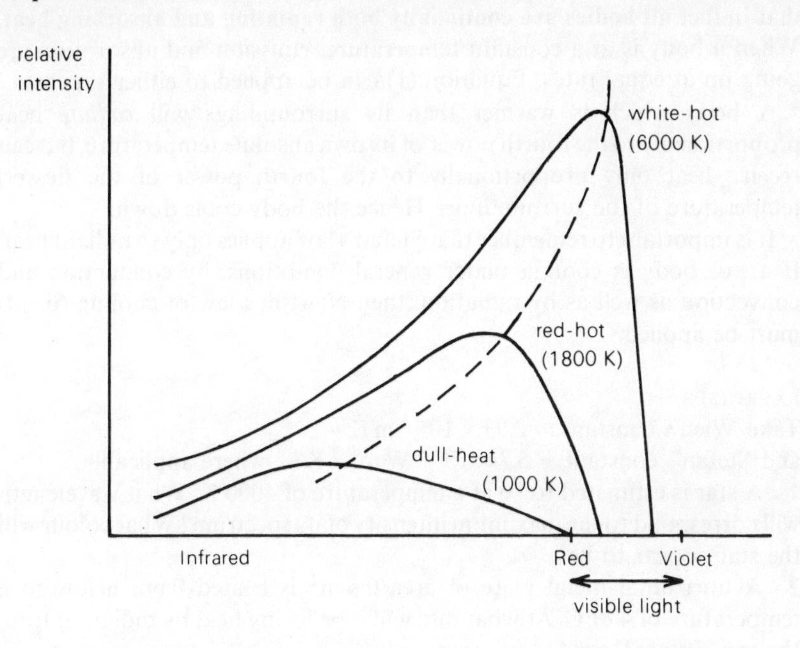

Fig. 49 Graphs showing the spectral distribution of thermal radiation from black bodies

The curves are all similar, rising to a rounded peak and then falling away quite rapidly to a cut-off point on the side of the shorter wavelengths. Measurements on the longer wavelength side show no cut-off but tail away indeterminately as the limit of sensitivity of the detector is reached.

The main conclusions of these experimental observations are stated in two laws:

(1) **(wavelength corresponding to peak of curve) × (absolute temperature associated with curve) = constant,**

i.e. $$\lambda_m T = \text{constant}.$$

This is *Wien's displacement law*; the value of the constant is 2.93 $\times 10^{-3}$ m K.

(2) **Total energy radiated per second per square metre of the source $\propto$ T^4**

or $$E = \sigma T^4. \quad (1)$$

This is *Stefan's law*; the value of Stefan's constant is $\sigma = 5.7 \times 10^{-8}\ \mathrm{W\,m^{-2}\,K^{-4}}$. E is represented by the area under the particular curve for T.

Prevost's theory of exchanges

Although we notice heat radiation only from bodies warmer than ourselves, it would be ridiculous to suppose that the process takes place only at temperatures above that of the human body. Prevost, in 1792, suggested that in fact **all bodies are continually both radiating and absorbing heat**. When a body is at a constant temperature, emission and absorption are going on at equal rates. Equation (1) can be applied to either process.

A body which is warmer than its surroundings will *radiate* heat proportionally to the fourth power of its own absolute temperature, but can *receive* heat only proportionally to the fourth power of the (lower) temperature of the surroundings. Hence the body cools down.

It is important to remember that Stefan's law applies only to radiant heat. If a hot body is cooling under general conditions, by conduction and convection as well as by radiation, then Newton's law of cooling (p. 14) must be applied.

EXERCISES

Take Wien's constant = 2.93×10^{-3} m K
and Stefan's constant = 5.7×10^{-8} W m^{-2} K^{-4} where applicable.

1 A star is estimated to be at a temperature of 4000 K. What wavelength will correspond to the maximum intensity of its spectrum? What colour will the star appear to be?

2 A horizontal metal plate of area 0.5 m^2 is heated from below to a temperature of 400°C. At what rate will it be losing heat by radiation from the top surface?

3 An astronaut working in space is losing heat at a rate of 20 W. If his total surface area is 2 m^2, calculate the temperature of his outer layer of clothing.

4 A hot block of metal of total surface area 1 m^2 is suspended in a vacuum at the centre of a sphere whose walls are kept at a temperature of 300 K.

Find (a) the rate at which the block is radiating heat when it is at a temperature of 500 K, and (b) the rate at which it is receiving radiant heat from its surroundings.

The inverse square law

The intensity of any radiation which travels in straight lines from a point source decreases as the square of the distance from the source.

This is illustrated in Fig. 50. The quantity of radiation (flux) which falls on the area A would spread out to cover the larger area B at a greater distance. Since intensity is defined in terms of energy received per second *per square metre*, if B is at twice the distance of A from the source of radiation, its area is four times the area of A, and the energy received per square metre will be reduced by a factor of four. In general,

$$\frac{\text{intensity of radiation over area A}}{\text{intensity of radiation over area B}} = \left(\frac{\text{distance of B from source}}{\text{distance of A from source}}\right)^2.$$

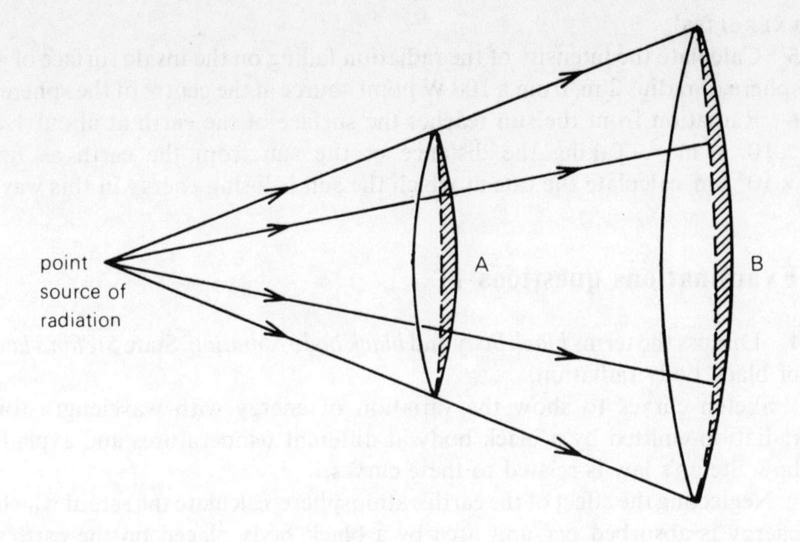

Fig. 50 The spreading-out of radiation from a point source

This is a simple geometrical result which applies, not only to radiation, but also to any force-field which can be described in terms of straight lines of force radiating from a point source. Examples of these are (a) all forms of electromagnetic radiation, (b) sound waves, (c) magnetic field strength, (d) electric field strength, (e) gravitation.

When applying the inverse square law to any of these situations, two points should be watched:

(a) The source must be a point one; alternatively, observations must be made only at considerable distances from the source, so that it appears relatively very small.

(b) Any measuring device must always admit the whole of the cone along which the radiation is travelling. In Fig. 51, the beam of radiation fills the mouth of the detector shown in position A, but some will bypass the detector if it is moved back to position B. Hence the readings obtained at A and B will not show a correct inverse square law relation.

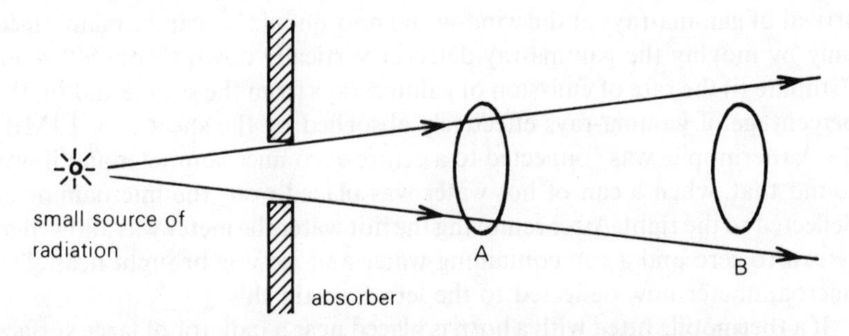

Fig. 51 An experimental set-up in which the inverse square law will not apply

5 Calculate the intensity of the radiation falling on the inside surface of a sphere of radius 2 m, from a 100 W point source at the centre of the sphere.
6 Radiation from the sun reaches the surface of the earth at about 1.4 $\times 10^3$ W m^{-2}. Taking the distance of the sun from the earth as 1.5 $\times 10^8$ km, calculate the rate at which the sun is losing energy in this way.

Examinations questions 11

1 Discuss the terms *black body* and *black body radiation.* State *Stefan's law* of black body radiation.

Sketch curves to show the variation of energy with wavelength for radiation emitted by a black body at different temperatures and explain how Stefan's law is related to these curves.

Neglecting the effect of the earth's atmosphere, calculate the rate at which energy is absorbed per unit area by a black body placed on the earth's surface so that the sun's rays fall normally on it. (Assume that the black body temperature of the sun = 6000 K, the radius of the sun = 7.0×10^8 m, the mean distance of the earth from the sun = 1.5×10^{11} m, and that the Stefan constant = 5.7×10^{-8} W m^{-2} K^{-4}). [AEB, June 1976]
2 If the thermodynamic temperature of a body doubles, the frequency at which the emitted radiation is most intense is multiplied by

A $\frac{1}{16}$ B $\frac{1}{2}$ C 2 D 4 E 16 [C]

3 Only a very small fraction of the energy supplied to a domestic light bulb is emitted as light principally because
A the filament is not a black body
B the filament surface area is too small
C the filament temperature is too low
D most of the energy is given out at shorter wavelengths
E energy is absorbed by the gas in the bulb. [C]
4 The window of a gamma-ray detector has an area of 4.0×10^{-4} m^2 and is placed horizontally so that it lies 2.0 m vertically above and on the axis of an effective point source of gamma-rays. When a sheet of gamma-ray absorber is introduced between source and detector, the initial rate of arrival of gamma-rays at the window, 60 photon min^{-1}, can be maintained only by moving the gamma-ray detector vertically down through 0.20 m. Estimate (i) the rate of emission of gamma-rays from the source and (ii) the percentage of gamma-rays effectively absorbed by the sheet. [JMB]
5 A thermopile was connected to a centre-zero microammeter and it was found that, when a can of hot water was placed near, the microammeter deflected to the right. After removing the hot water the meter was allowed to return to zero and a can containing water and ice was brought near. The microammeter now deflected to the left. Explain this.

If a thermopile fitted with a horn is placed near a radiator of large surface area and then gradually moved away, the microammeter readings change very little until a certain distance is reached and, beyond this, the readings begin to fall. Explain this.

The solar radiation falling normally on the surface of the earth has an intensity $1.40\ \mathrm{kW\,m^{-2}}$. If this radiation fell normally on one side of a thin, freely suspended blackened metal plate and the temperature of the surroundings was 300 K, calculate the equilibrium temperature of the plate. Assume that all heat interchange is by radiation.

(The Stefan constant $= 5.67 \times 10^{-8}\ \mathrm{W\,m^{-2}\,K^{-4}}$). [L]

very little until a certain distance is reached and, beyond this, the readings begin to fall. Explain this.

The solar radiation falling normally on the surface of the earth has an intensity 1.40 kW m^{-2}. If this radiation fell normally on one side of a thin, freely suspended, blackened, metal plate and the temperature of the surroundings was 300 K, calculate the equilibrium temperature of the plate. Assume that all heat interchange is by radiation.

(The Stefan constant = 5.67×10^{-8} W m^{-2} K^{-4}.) [L]

Mechanics and Properties of Matter

12 Motion in a straight line

The equations governing the motion of a body in a straight line are best derived by calculus. If any of the three relations distance–time, velocity–time and acceleration–time are known in mathematical form, the others may be obtained by differentiation or integration as appropriate, since

$$\frac{\mathrm{d}}{\mathrm{d}t}\text{(distance)} = \text{velocity}$$

and

$$\frac{\mathrm{d}}{\mathrm{d}t}\text{(velocity)} = \text{acceleration.}$$

Alternatively, graphical methods may be used, especially if a precise mathematical relation cannot be found. The set of graphs shown in Fig. 52 refer to the motion of a horse during a race, and it may be seen how the significant features of each graph relate to the others.

Air resistance

The four simple forms of the equations of motion under uniform acceleration [$v = u + at$, $s = \frac{1}{2}(u+v)t$, $s = ut + \frac{1}{2}at^2$ and $v^2 = u^2 + 2as$] are of little use in practical situations as they ignore the resistance to motion offered by the air. Figure 53 shows how the true path taken by a projectile differs from the theoretical parabola.

A body moving through the air, as through any denser fluid, finds a resisting force opposing its motion. Much experimental work has been done to find how this force varies with the shape and speed of the moving body, and the streamlined design of racing cars, motor boats, aeroplanes and rockets stems from this research.

The results concerning the variation of the force of air resistance with speed are as follows:

For low speeds, up to a few $\mathrm{m\,s^{-1}}$, force $\propto$ speed.

This relation holds if the air flows smoothly past the body, and the resisting force is produced by the viscosity of the air.

For higher speeds, up to about $250\ \mathrm{m\,s^{-1}}$, force $\propto$ (speed)^2. At these speeds the air flow becomes turbulent. An observer would feel the wind caused by the body's passing.

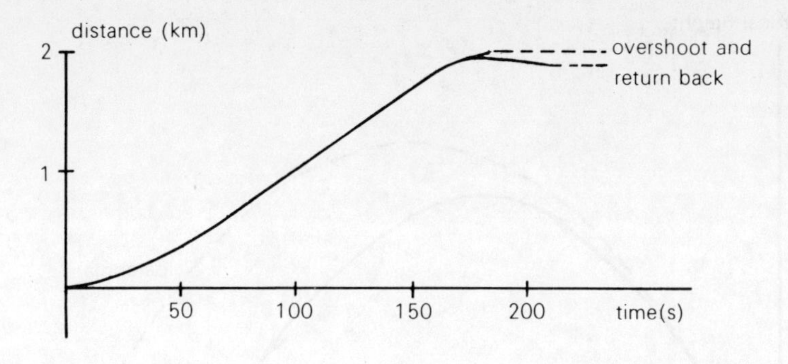

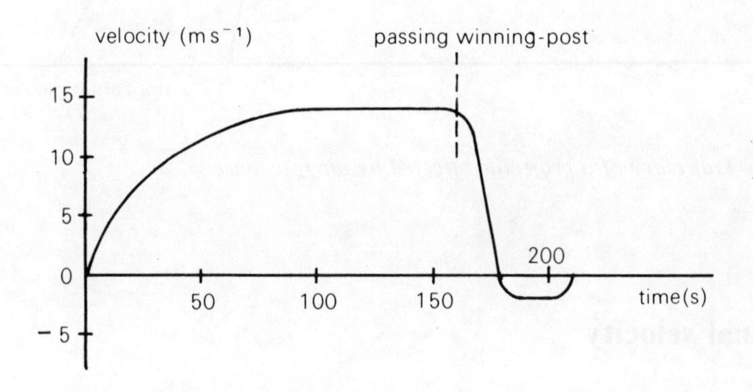

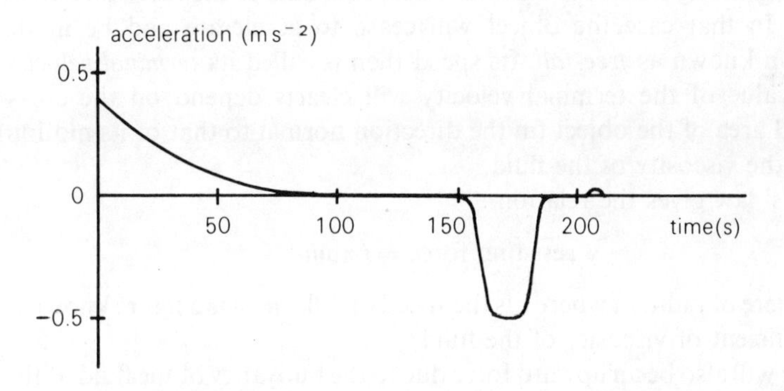

Fig. 52 A set of graphs describing the motion of a horse during a race

As the velocity of sound (approximately 330 m s^{-1}) is approached, the resistance increases very rapidly indeed. For a long time it was thought impossible for a body to overcome this resistance and travel at supersonic speed, hence the phrase 'to break the sound barrier'.

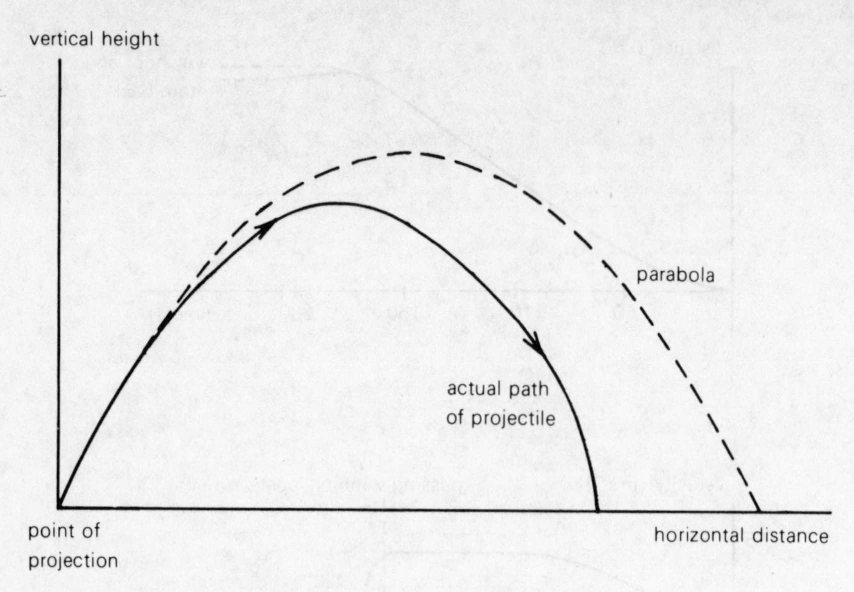

Fig. 53 Trajectory of a projectile affected by air resistance

Terminal velocity

As an object falls under gravity through air or any other fluid, the resisting force of the fluid gradually increases with the speed of the object, and if the fall is long enough, the resistance may become equal to the force exerted by gravity. In that case the object will cease to accelerate and be in the condition known as *free-fall.* Its speed then is called its *terminal velocity.*

The value of the terminal velocity will clearly depend on the cross-sectional area of the object (in the direction normal to that of its motion) and on the viscosity of the fluid.

Stokes' law gives the relation

$$\text{resisting force} = 6\pi a\eta v$$

for a sphere of radius a where v is the speed of fall, and η is a figure known as the coefficient of viscosity of the fluid.

There will also be an upward force due to the buoyancy of the fluid. If this can be neglected, then the force exerted by gravity on the sphere will be simply its weight.

Hence **its terminal velocity is given by** $\dfrac{\textbf{weight}}{\boldsymbol{6\pi a\eta}}$.

For a man falling from an aircraft, the velocity of free-fall is about 50–60 m s^{-1}, thus there is time during a fall from 3000 m to carry out various acrobatic feats. If the man's virtual area is increased by the use of a parachute, his terminal velocity will be reduced to about 5 m s^{-1}.

Simple harmonic motion

Simple harmonic motion (s.h.m.) is defined as **the motion of a body whose acceleration is always directed towards a fixed point and is proportional to the distance of the body from that point.**

If the distance is denoted by x, then the equation of simple harmonic motion is

$$\frac{d^2x}{dt^2} = -\omega^2 x \qquad (1)$$

where ω is a numerical constant, and the minus sign shows that the acceleration is opposite in direction to the measurement of x.

The solution of (1) is

$$x = a \sin \omega t$$

if the time t is measured from a moment when the body is passing the fixed point. This represents **an oscillation with the fixed point as the centre of motion, amplitude a and period** $\frac{2\pi}{\omega}$.

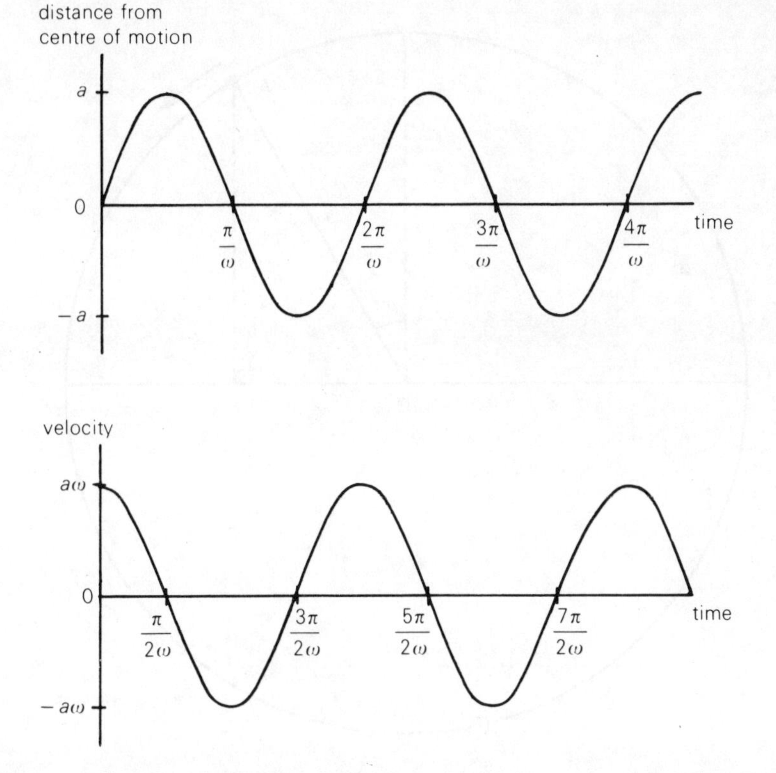

Fig. 54 Graphs illustrating simple harmonic motion

The velocity of the body at time t is given by

$$\frac{dx}{dt} = a\omega \cos \omega t$$

which can also be written as

$$\frac{dx}{dt} = \omega\sqrt{(a^2 - x^2)}$$

since $\sin \omega t = x/a$. This shows that the velocity is zero when $x = a$ or $-a$, the extreme limits of the oscillation, which occurs when $\omega t = \pi/2, 3\pi/2$, etc.

The velocity is a maximum for $\omega t = 0, \pi$ etc., for which times $x = 0$, and the body is passing through the centre of motion.

Simple harmonic motion as a projection of circular motion

Although the numerical value of ω need have no physical significance in a simple harmonic motion, it is possible to think of it as the angular velocity of a point moving in a circle of radius a (Fig. 55).

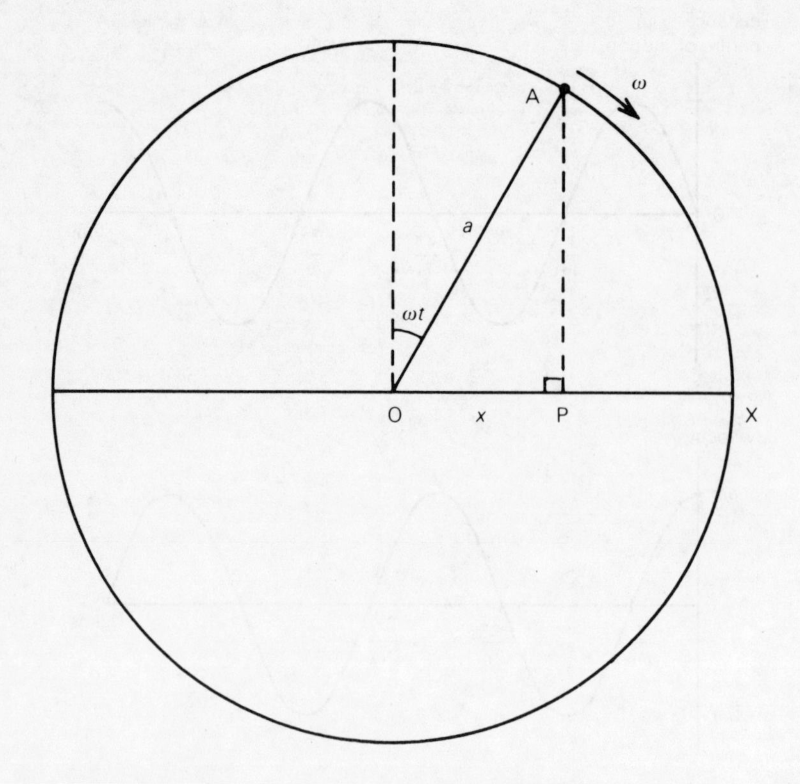

Fig. 55 The significance of ω in simple harmonic motion

In that case, ωt is the angle moved through in time t, and $x = a \sin \omega t$ is the projection of the radius to the point A on a particular diameter (OX) of the circle.

As A moves around the circle with constant speed, the point P moves with s.h.m. along the diameter.

Examples of simple harmonic motion

The motion of objects oscillating under actual physical conditions is in general only simple harmonic as a first approximation.

The simple pendulum

A theoretical simple pendulum consists of a particle of negligible volume at the end of an inelastic weightless string, as shown in Fig. 56. Let the mass of the particle be m and the length of the string be l.

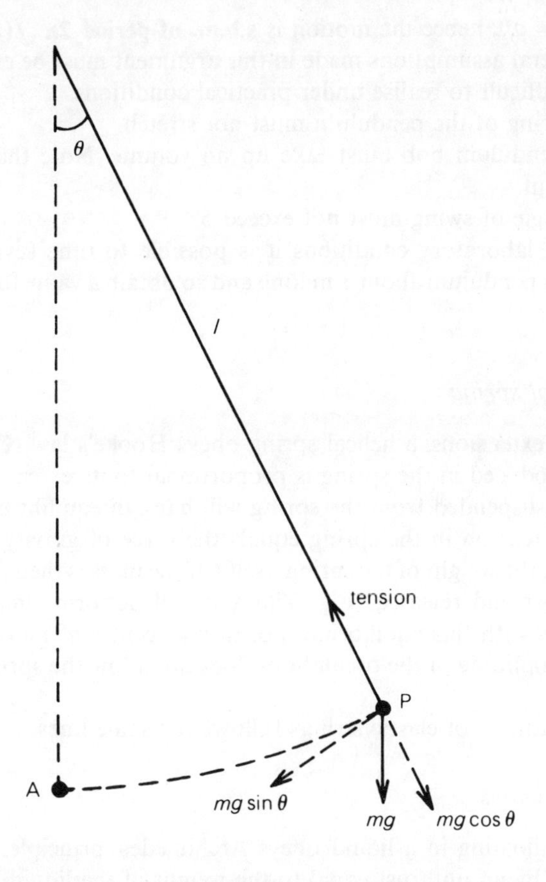

Fig. 56 The simple pendulum

The particle is drawn aside from its rest position at A to a position P in which the string makes an angle θ with the vertical. The forces acting on the particle are then (i) its weight mg and (ii) the tension in the string.

Resolving these in the direction of the string, their components must balance since the string is inelastic. Resolving at right angles to the string, the component $mg \sin \theta$ of the weight is not balanced, and the particle will, if free, move back towards A with acceleration $g \sin \theta$.

Now the length of the arc AP = $l\theta$ if θ is measured in radians. *For small angles* (less than 5°), $\sin \theta = \theta$, and the small arc AP may be considered as a straight line of length x.

Then the acceleration of the particle is

$$g \sin \theta = g\theta = (g/l)x$$

towards A.

This equation is of the form

$$\frac{d^2 x}{dt^2} = -\omega^2 x$$

where $\omega^2 = g/l$, hence the motion is **s.h.m. of period $2\pi \sqrt{(l/g)}$.**

The several assumptions made in this argument must be emphasised as they are difficult to realise under practical conditions.

(a) The string of the pendulum must not stretch.

(b) The pendulum bob must take up no volume. Note that its mass is unimportant.

(c) The angle of swing must not exceed 5°.

In good laboratory conditions it is possible to time several hundred swings of a pendulum about 1 m long and so obtain a value for g correct to 0.1%.

The helical spring

For small extensions, a helical spring obeys Hooke's law (Chap. 16): the tension produced in the spring is proportional to its extension.

A mass suspended from the spring will hang in equilibrium at a point where the tension in the spring equals the force of gravity on the mass (neglecting the weight of the spring itself). If the mass is then pulled down a little further and released (Fig. 57(a)), it will perform simple harmonic oscillations with this equilibrium position as centre of motion, provided that the amplitude of the oscillations does not allow the spring ever to go slack.

The treatment of elastic strings follows the same lines.

Floating bodies

An object floating in a liquid obeys Archimedes' principle: its weight is supported by an upthrust equal to the weight of the liquid it displaces.

If the object is pushed a little deeper into the liquid, a greater upthrust will be produced, and when released, the object will oscillate with s.h.m. vertically about its equilibrium position (Fig. 57(b)).

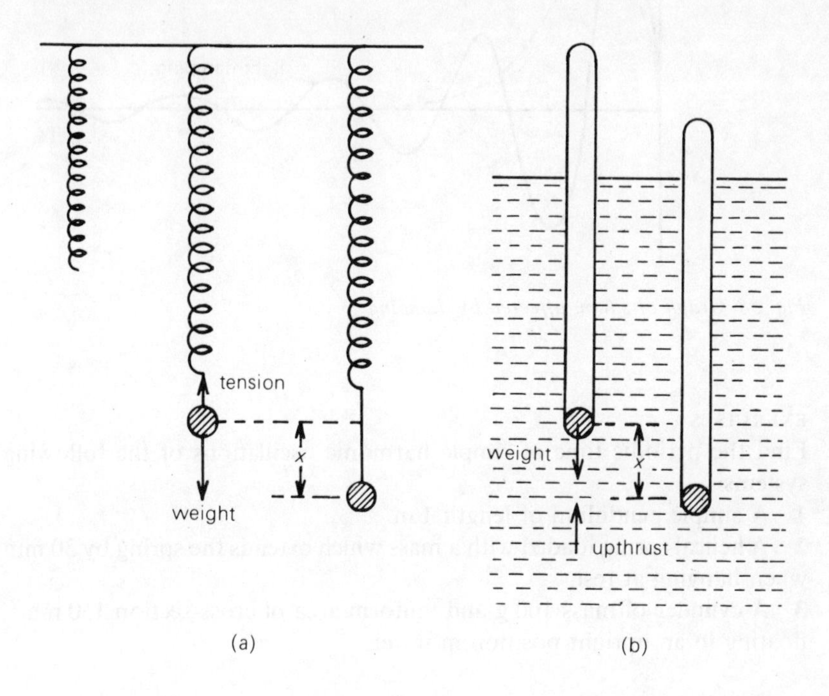

Fig. 57 Examples of s.h.m. (a) a helical spring, (b) a floating body

Analysis of these cases under particular conditions will lead to equations of the same form as equation (1), from which the period of the oscillations can be deduced straight away.

Damping

In these examples, the resistance to motion offered by the air or liquid has been ignored, and the oscillations could go on for ever. In all practical cases, the amplitude of the motion gradually dies away as energy is transferred to the surrounding medium (Fig. 58).

If the damping is deliberately increased, 'dead-beat' conditions can be obtained, in which a disturbance causes only a single swing of the system before it settles steadily into its equilibrium position. This motion is shown by the dotted line in Fig. 58. Such *critical damping* is produced by the use of an oil dash-pot on a balance, or by eddy currents in the core of a moving-coil galvanometer.

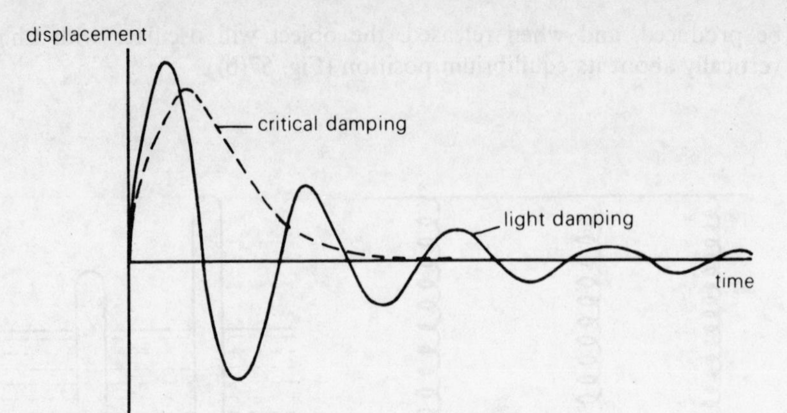

Fig. 58 Graph of s.h.m. affected by damping

EXERCISES

Find the periodic time of simple harmonic oscillations of the following systems:

1 A simple pendulum of length 1 m.

2 A helical spring loaded with a mass which extends the spring by 30 mm when hanging at rest.

3 A cylinder of mass 100 g and uniform area of cross-section 150 mm^2, floating in an upright position in water.

Lissajous' figures

If two simple harmonic forces are applied to an object simultaneously in directions at right angles to one another, the object will follow a path known as a Lissajous' figure.

This is often noticed when watching the swing of the bob of a long simple pendulum; it moves in one of the two alternative modes shown in Fig. 59. In (a) since the trace passes through the centre of motion, it follows that the two component s.h.m.'s are in step (in phase). Contrarily, in (b) the two motions are half a cycle out of phase.

Lissajous' figures can be demonstrated by applying a.c. signals to both the X- and Y-plates of an oscilloscope. If the frequencies of the two signals are in a simple ratio, a simple curve can be produced. Figure 59(c) shows the case in which the ratio is 2 : 1; this can be deduced by comparing the number of turning-points of the figure horizontally and vertically, since these occur during the same time interval.

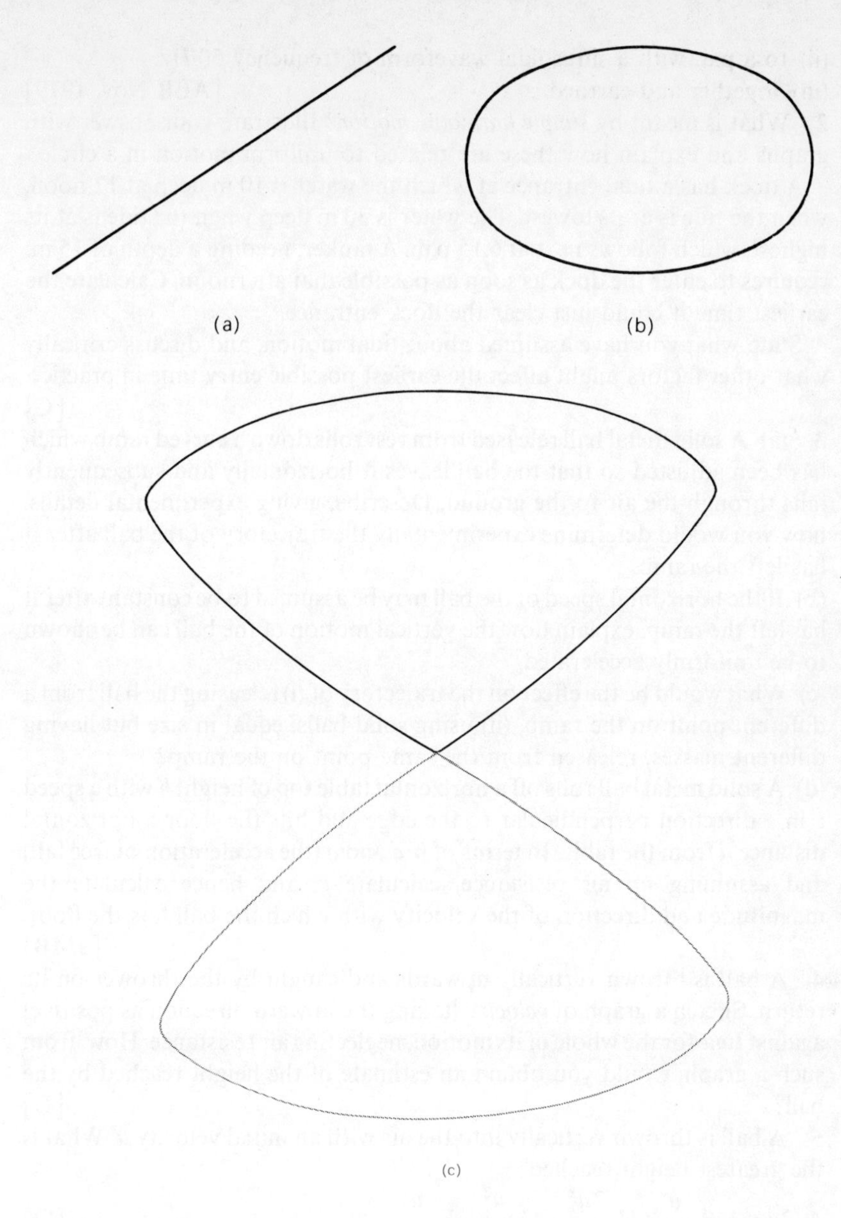

Fig. 59 Lissajous' figures (a) components in phase, (b) components half a cycle out of phase, (c) frequencies of components 2:1

Examination questions 12

1 Sketch the waveform you would expect to see if the Y plates of an oscilloscope were connected to a sinusoidally varying p.d. of frequency 100 Hz and the X plates were connected
(i) to a p.d. with a saw-tooth waveform of frequency 50 Hz,

(ii) to a p.d. with a sinusoidal waveform of frequency 50 Hz,
(iii) together and earthed. [AEB Nov. 1979]

2 What is meant by *simple harmonic motion*? Illustrate your answer with graphs and explain how these are related to uniform motion in a circle.

A dock has a tidal entrance at which the water is 10 m deep at 12 noon, when the tide is at its lowest. The water is 30 m deep when the tide is at its highest, which follows next at 6.15 p.m. A tanker, needing a depth of 15 m, requires to enter the dock as soon as possible that afternoon. Calculate the earliest time it could just clear the dock entrance.

State what you have assumed about tidal motion, and discuss critically what other factors might affect the earliest possible entry time in practice. [C]

3 (a) A solid metal ball released from rest rolls down a curved ramp which has been adjusted so that the ball leaves it horizontally and subsequently falls through the air to the ground. Describe, giving experimental details, how you would determine experimentally the trajectory of the ball after it has left the ramp.
(b) If the horizontal speed of the ball may be assumed to be constant after it has left the ramp, explain how the vertical motion of the ball can be shown to be uniformly accelerated.
(c) What would be the effect on the trajectory of (i) releasing the ball from a different point on the ramp, (ii) using solid balls, equal in size but having different masses, released from the same point on the ramp?
(d) A solid metal ball rolls off a horizontal table top of height h with a speed v in a direction perpendicular to the edge and hits the floor a horizontal distance d from the table. In terms of h, d and g (the acceleration of free fall) and assuming no air resistance, calculate v, and hence calculate the magnitude and direction of the velocity with which the ball hits the floor. [JMB]

4 A ball is thrown vertically upwards and caught by the thrower on its return. Sketch a graph of *velocity* (taking the upward direction as positive) against *time* for the whole of its motion, neglecting air resistance. How, from such a graph, would you obtain an estimate of the height reached by the ball? [L]

5 A ball is thrown vertically into the air with an initial velocity u. What is the greatest height reached?

A $2gu^2$ B $\frac{u}{2g}$ C $\frac{3u^2}{2g}$ D $\frac{u^2}{g}$ E $\frac{u^2}{2g}$ [O]

6 State Newton's first and second Laws of Motion and describe briefly how you would try to verify the second law experimentally.

A loaded freight train, including the locomotive, may be assumed to constitute a mass of 10^6 kg uniformly distributed over a length of 500 m. It is subject to a constant frictional force of 5×10^4 N and the locomotive exerts a constant tractive force of 10^5 N. The train proceeds down a long uniform gradient of 1 in 100, followed by a level section of track of length 700 m. It then ascends a long uniform gradient of 1 in 100. Draw a graph showing the *acceleration* of the train against the position of the front of the

locomotive. The graph should start 200 m before the locomotive reaches the level section and end when the locomotive is 700 m beyond the start of the ascent. Numerical values of acceleration should be shown on the graph at appropriate points. (Take the acceleration due to gravity, g, as $10\ \mathrm{m\,s^{-2}}$.)
[OC]

13 Motion in a circle

The expressions and equations relating to motion in a circle are very similar to those already familiar for rectilinear motion. The following table shows a comparison between the two cases:

Table 4

Rectilinear motion		Circular motion	
Distance travelled	s	Angle turned through	θ
Velocity	v	Angular velocity	ω
Acceleration	a	Angular acceleration	$\frac{d\omega}{dt}$
Force	F	Torque	T
Mass	m	Moment of inertia	I
Momentum	mv	Angular momentum	$I\omega$
Kinetic energy	$\frac{1}{2}mv^2$	Kinetic energy	$\frac{1}{2}I\omega^2$

Angles

Angles are usually measured in *radians* as this unit has a more logical basis than the degree (which derives from the ancient belief that the earth travels around the sun in 360 days).

One radian (written 1^c) is the angle subtended at the centre of a circle by an arc equal in length to the radius.

Angular velocity

An object moving in a circle of radius r with a constant speed v will have a constant angular velocity ω measured in radians per second (rad s^{-1}).

Since the object moves a distance v around the circle in one second,

$$\omega = \frac{v}{r}$$

which can alternatively be written $v = r\omega$.

Angular acceleration

If the object is accelerating around the circle, its angular acceleration is $\frac{d\omega}{dt}$ rad s^{-2}.

Torque

Angular acceleration must be produced by a force acting on the object. The *moment* of this force (or the total moment of all the forces acting) about the centre of rotation is called its *torque*. The units of torque are N m.

Moment of inertia

This is an expression used in connection with the rotation of a rigid body about a fixed axis.

A force applied to a particular particle of mass *m* will give it a linear acceleration according to Newton's second law of motion:

$$\text{force} = \text{mass} \times \text{acceleration}.$$

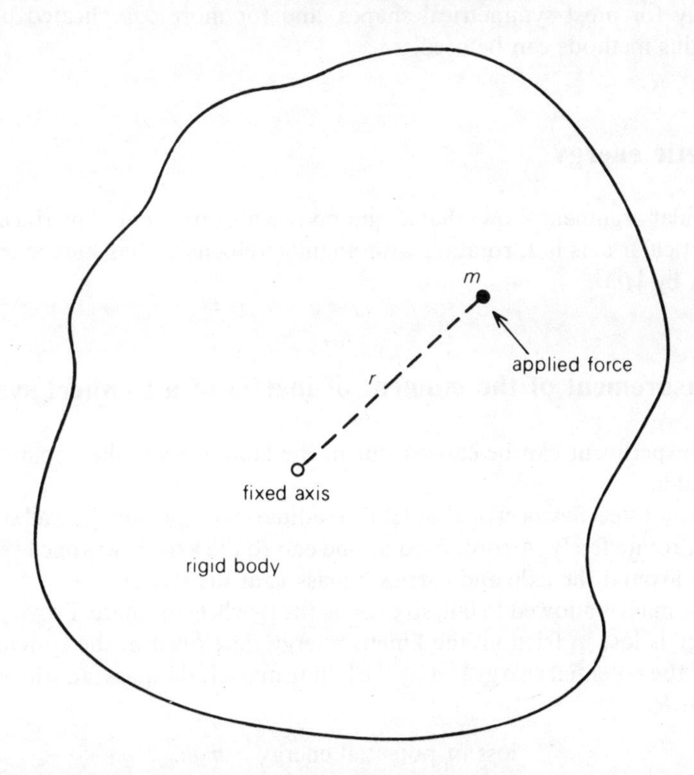

Fig. 60 Calculation of the moment of inertia of a rigid body

If the particle is fixed at a distance r from an axis of rotation,

$$\text{force} \times r = \text{mass} \times \text{acceleration} \times r.$$

But (force $\times$ r) represents the torque exerted by the force, and the linear acceleration is equal to ($r \times$ angular acceleration). Substituting in the previous equation,

$$\text{torque} = mr^2 \times (\text{angular acceleration}).$$

The rigid body is made up of a large number of particles constrained by internal forces to move all together with the same angular velocity ω. Extending the equation to include all the particles of the body,

$$\text{total torque} = \Sigma mr^2 \times (\text{angular acceleration}). \qquad (1)$$

Comparing this with the second law of motion, it can be seen that the term 'mass' has been replaced by Σmr^2. This expression is called the *moment of inertia* of the body and is given the symbol I.

Equation (1) can then be written in symbols as

$$T = I\,\frac{\mathrm{d}\omega}{\mathrm{d}t}.$$

The values of moments of inertia about particular axes can be calculated simply for most symmetrical shapes, and for more complicated bodies calculus methods can be used.

Kinetic energy

A similar argument shows that a rigid body whose moment of inertia about a particular axis is I, rotating with angular velocity ω, has kinetic energy given by $\frac{1}{2}I\omega^2$.

Measurement of the moment of inertia of a flywheel system

This experiment can be carried out in the laboratory if the apparatus is available.

A stout steel flywheel of about 0.1 m radius is mounted on the wall so that it can rotate freely. A cord, fixed at one end to the axle, is wrapped several times around the axle and carries a mass m at the free end.

The mass is allowed to fall, so causing the flywheel to rotate. Provided no energy is lost in friction, the kinetic energy developed in the system will equal the potential energy lost by the falling mass. If the mass falls through a height h,

$$\text{loss of potential energy} = mgh.$$

The end of the cord is attached to the axle by a loop over a hook, so that it

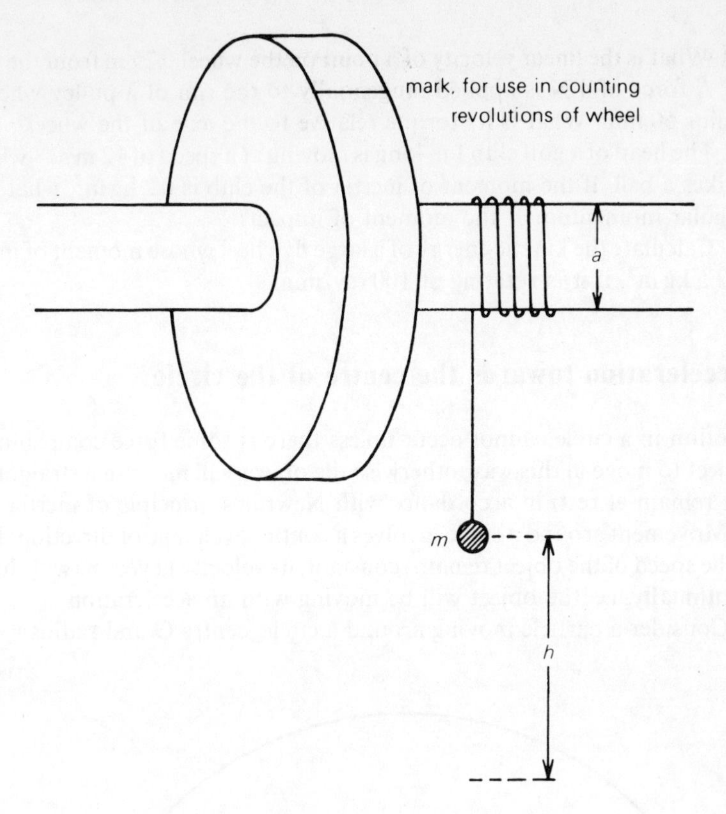

Fig. 61 Determination of the moment of inertia of a flywheel

will fall free as soon as the cord has completely unwound. The flywheel will continue to turn, and its speed of rotation can be measured by counting and timing a number of revolutions. This will give ω, the maximum angular velocity of the flywheel. The corresponding linear velocity v of the falling mass can be calculated, since $v = a\omega$ where a is the radius of the axle around which the cord is wound. The radius a should be measured using vernier callipers.

$$\text{Total kinetic energy of the system} = \tfrac{1}{2}I\omega^2 + \tfrac{1}{2}mv^2.$$

Equating the expressions for potential energy lost and kinetic energy developed will give a value for I, the moment of inertia of the flywheel.

The effect of friction in the bearings can usually be neglected, but this can be checked by allowing the flywheel to continue rotating and observing if it appears to lose speed quickly. It is possible to make a correction for friction if this seems necessary.

A flywheel of laboratory dimensions would have a moment of inertia of about 0.1–0.3 kg m^2.

EXERCISES

1 A wheel is revolving at a steady rate of 120 rev/min.

(i) What is its angular velocity, in rad s^{-1}?

(ii) What is the linear velocity of a point on the wheel 0.25 m from the axle?

2 A force of 4 N is applied tangentially to the rim of a pulley wheel of radius 50 mm. What is its torque relative to the axle of the wheel?

3 The head of a golf club 1 m long is moving at a speed of 12 m s^{-1} when it strikes a ball. If the moment of inertia of the club is 0.2 kg m^2, what is its angular momentum at the moment of impact?

4 Calculate the kinetic energy of a large flywheel whose moment of inertia is 2.5 kg m^2, if it is rotating at 100 rev/min.

Acceleration towards the centre of the circle

Motion in a circle cannot occur unless there is some force compelling an object to move in this way; otherwise the object will move in a straight line (or remain at rest) in accordance with Newton's principle of inertia.

Movement around a circle involves a continual change of direction. Even if the speed of the object remains constant, its velocity (a vector) will change continually, i.e. the object will be moving with an acceleration.

Consider a particle moving around a circle, centre O and radius r, with

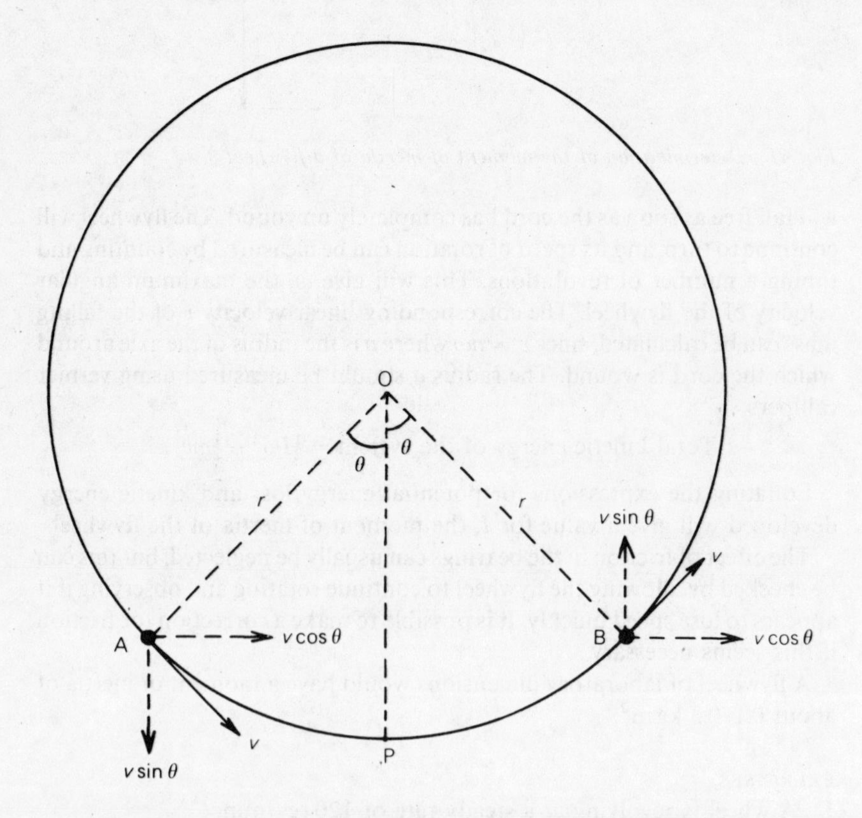

Fig. 62 Calculation of the acceleration of a particle moving in a circle

uniform speed v. To calculate its acceleration at the point P, consider two points A and B at equal angles θ on opposite sides of the radius OP.

The velocity v of the particle at A may be resolved into components $v \sin\theta$ parallel to OP and $v \cos\theta$ perpendicular to OP. Similarly its velocity at B may be resolved into $v \sin\theta$ parallel to PO and $v \cos\theta$ perpendicular to PO. So the change of velocity between A and B is $2v \sin\theta$ in the direction PO. The distance AB $= 2r\theta$

$\Rightarrow$ time taken to travel from A to B $= 2r\theta/v$

$\Rightarrow$ average acceleration of particle between A and B

$$= \frac{\text{change of velocity}}{\text{time taken}}$$

$$= \frac{2v \sin\theta}{2r\theta/v}$$

$$= \frac{v^2}{r}\frac{\sin\theta}{\theta} \text{ parallel to PO.}$$

Now if θ is very small, then $\sin\theta = \theta$ and the points A and B are very close to P. Then

average acceleration between A and B

$$= \text{actual acceleration at P} = \frac{v^2}{r} \text{ towards O} \qquad (2)$$

$$= r\omega^2 \text{ substituting } \omega = v/r.$$

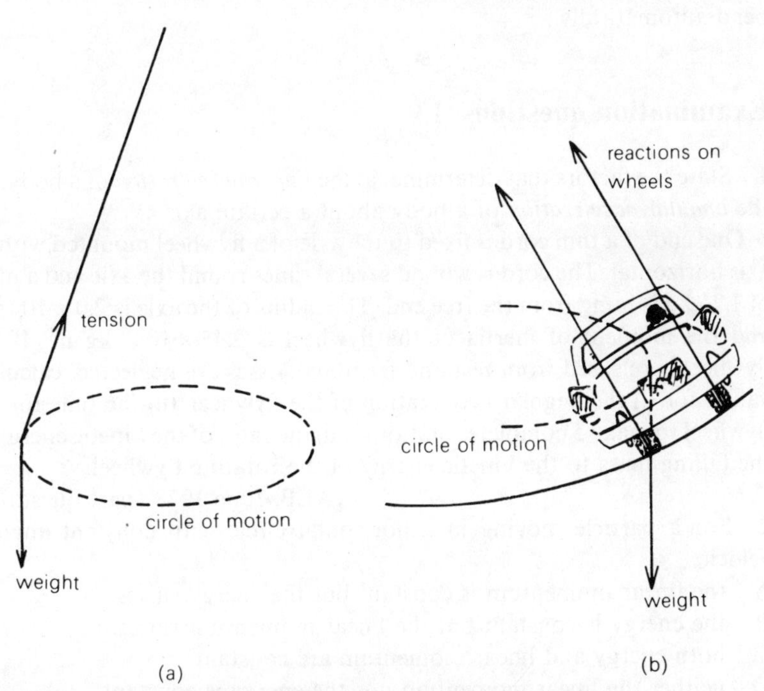

Fig. 63 (a) The conical pendulum (b) A car rounding a bend

Figure 63 illustrates two everyday situations in which an object is induced to move in a circle. (a) represents a small mass whirling at the end of a string; such a system is known as a *conical pendulum.* If the path of the mass is a horizontal circle, then its weight is completely balanced by the vertical component of the tension in the string, while the horizontal component of this tension provides the necessary force towards the centre of the circle.

Figure 63 (b) shows a car (rear view) moving around a banked track. The components of the reactions of the track on the wheels provide a vertical force to balance the weight of the car and a horizontal 'force-towards-the-centre'. If the car is travelling at the appropriate speed for that particular slope of banking, it will move around the track without needing to be steered. This principle is applied in the construction of racing circuits, motorway bends (which are designed for a speed of 70 m.p.h.), and many fairground attractions.

EXERCISES

5 A small object is moving at a rate of 8 rev/s in a horizontal circle of radius 0.2 m. What acceleration does it have towards the centre of the circle?

6 A small object is moving as a conical pendulum at the end of a string 0.3 m long. If its path is a horizontal circle of radius 0.15 m, at what speed must the object be moving?

7 A motorway is to include a circular arc of radius 200 m. If it is to be designed for vehicles moving at a speed of 70 m.p.h. (31.3 m s^{-1}), calculate the angle required for the banking of the surface so that a car will take the bend automatically.

Examination questions 13

1 State the factors that determine (a) the *linear acceleration* of a body, (b) the *angular acceleration* of a body about a certain axis.

One end of a thin cord is fixed to the axle of a flywheel mounted with its axis horizontal. The cord is wound several times round the axle and a mass of 1.25 kg is hung from the free end. The radius of the axle is 2.0×10^{-2} m and the moment of inertia of the flywheel is $2.45 \times 10^{-2} \text{ kg m}^2$. If the flywheel is released from rest and frictional losses are neglected, calculate values for (i) the angular acceleration of the flywheel, (ii) the time for the flywheel to make 5 complete rotations, (iii) the ratio of the kinetic energy of the falling mass to the kinetic energy of the rotating flywheel.

[AEB, June 1978 (part question)]

2 For a particle moving in a horizontal circle with constant angular velocity

A the linear momentum is constant but the energy varies.
B the energy is constant but the linear momentum varies.
C both energy and linear momentum are constant.
D neither the linear momentum nor the energy is constant.
E the speed and the linear velocity are both constant. [C]

3 Define the terms *angular acceleration, torque, moment of inertia.* Upon what factors does the moment of inertia of a body depend?

A toy car (described by the manufacturers as 'friction-powered') is of mass 0.15 kg and contains a flywheel of moment of inertia 2.4×10^{-5} kg m^2 that may be set into rotation ('charged') by pushing the car rapidly along the floor. When the car is released after charging, it is found to travel a distance of 1.6 m up a plane inclined at 10° to the horizontal before it comes to rest.
(a) Trace the energy transformations that take place during the motion of the car.
(b) At what speed was the flywheel rotating immediately before the car was released?
(c) If the charging process took 2.0 s, find the torque (assumed constant) exerted on the flywheel during this period.
(d) To what extent is it correct to describe the toy as 'friction-powered'?

$(g = 10\ \text{m s}^{-2})$ [C]

4 A skater is spinning with her arms outstretched. When she lowers her arms to her sides her rate of rotation increases. This is due to
A a decrease in her moment of inertia.
B a decrease in her potential energy.
C an increase in her angular momentum. [L]

5 A particle is connected to a fixed point by a light inextensible string of length l. The particle moves in a horizontal circle of radius r with speed v. The angle the string makes with the vertical is given by

A $\tan^{-1}\dfrac{v^2}{gl}$ B $\tan^{-1}\dfrac{v^2}{gr}$ C $\tan^{-1}\dfrac{gr}{v^2}$

D $\tan^{-1}\dfrac{gl}{v^2}$ E $\tan^{-1}\dfrac{rv^2}{g}$ [O]

14 Gravitation

All the developments of modern astronomy and space travel are founded on Newton's universal law of gravitation. This was proposed in 1666 as a unifying theory to bring together the great quantity of observational data that had been accumulated since the invention of the telescope.

It was agreed that some force of attraction must exist to hold the planets in their orbits about the sun. According to Newton's theory, this force F would depend on

(a) the two masses, m of the planet and M of the sun, and
(b) the distance r between them.

He suggested that the second factor might obey an inverse square law, so that

$$F \propto \frac{1}{r^2}$$

and combining the two factors,

$$F \propto \frac{mM}{r^2}.$$

If F is to be in newtons when m and M are in kilograms and r is in metres, a numerical constant G can be introduced so that the relation becomes an equality:

$$F = G\frac{mM}{r^2}. \tag{1}$$

This is *Newton's law of gravitation*, and G is called the *gravitational constant*.

The beauty of this theory is that it applies not only to astronomical bodies, but to all masses even in terrestrial situations.

Measurement of G

The gravitational constant G was first determined by Cavendish in 1798. Figure 64 shows a modification of this experiment, carried out in 1895 by Boys.

The gravitational attraction between two masses was measured by balancing it against the torsional twist set up in a quartz fibre suspension 0.5 m long. The moveable mass was a small gold ball 5 mm in diameter,

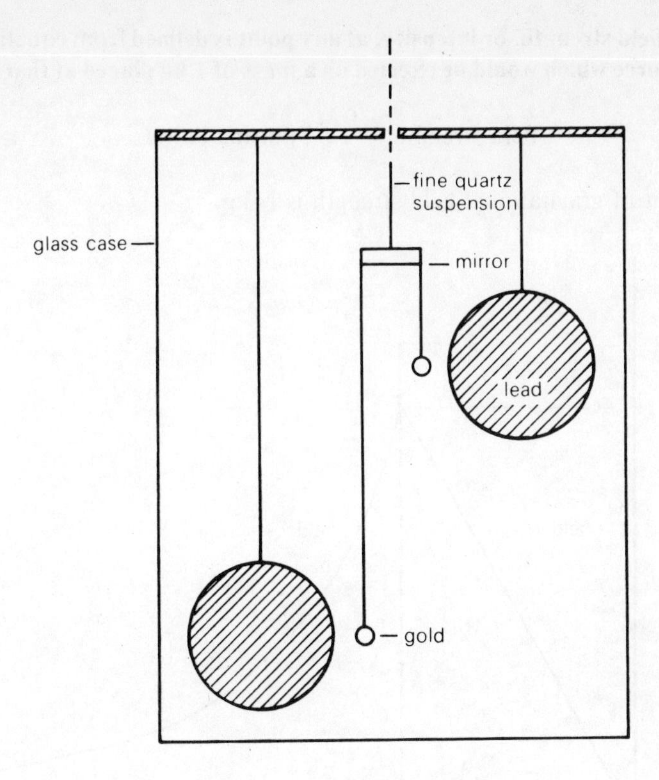

Fig. 64 Boys' apparatus for the determination of G

weighing 3 g, which was attracted by a 7 kg sphere of lead placed close to it. The deflection was doubled by using two similar masses suspended at a lower level, sufficiently distant to permit their gravitational effect on the upper masses to be ignored.

The angle of deflection θ was measured by means of a light-pointer, a beam of light reflected from the polished bar from which the two gold balls hung, moving over a scale some distance away.

The torsional couple $k\theta$ set up in the quartz fibre by this deflection was determined by a separate experiment, which involved timing small-angle oscillations of the suspended system, having first removed the two lead spheres. This motion could be considered to be simple harmonic (p. 97) with a period of $2\pi\sqrt{(\text{moment of inertia}/k)}$.

The whole apparatus was screened from air currents during use, by enclosing it in a glass case.

The value of G is taken at present to be $6.67 \times 10^{-11}\ \text{N m}^2\,\text{kg}^{-2}$.

Gravitational fields

The *gravitational field* of a planet or star is the region in which it would exert an appreciable gravitational force on any other mass.

The field strength, or intensity, at any point is defined from equation (1) **as the force which would be exerted on a mass of 1 kg placed at that point.**

i.e. $$\text{field strength} = \frac{GM}{r^2}, \text{ putting } m = 1.$$

The unit of gravitational field strength is $\mathrm{N\,kg^{-1}}$.

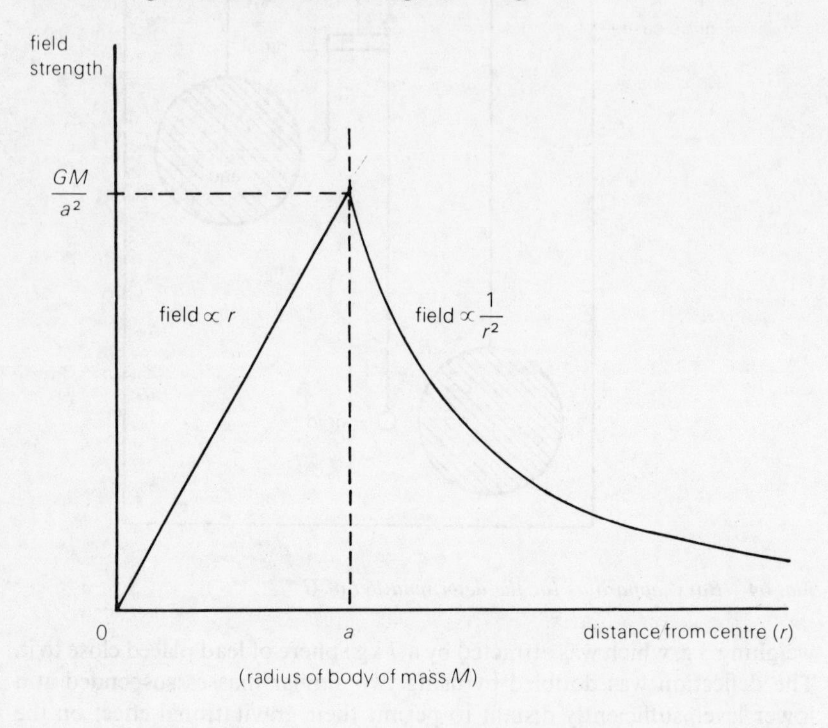

Fig. 65 Graph of distance and gravitational field for a sphere

Figure 65 shows how the field strength due to a body would theoretically vary with distance. At distances greater than the radius a of the body, the field falls off according to the inverse square law. Within the substance of the body itself, the field is lower than at the surface, as some of the mass of the body is then above the point considered, and it can be shown mathematically that this part of the graph is a straight line. At the centre of the body, considerations of symmetry dictate that the field must be zero.

The acceleration due to gravity

Near to the surface of the earth, an object allowed to fall is observed to accelerate at a constant rate of about $9.8\ \mathrm{m\,s^{-2}}$. This is the result of the force of gravity acting between the earth and the object, causing it to have *weight*.

The weight of an object of mass m at a distance r from the centre of the earth, mass M, is given by equation (1) as $G(mM/r^2)$. So its acceleration if free to fall is

$$g = G\frac{M}{r^2} \tag{2}$$

This is independent of the mass of the object itself, as was demonstrated first by Galileo at the Leaning Tower of Pisa, and recently by Neil Armstrong on the moon.

It should be noted that this expression GM/r^2 is the same as that for gravitational field strength (p. 116) although different units are involved in the two cases. The graph of Fig. 65 also applies to the variation of g with distance from the centre of the earth.

For an object out in space, r is large and g is very small. This is the condition observed inside space-rockets.

For an object on the surface of the moon, the symbol M in equation (2) must stand for the mass of the moon, not that of the earth, and r for the radius of the moon. This leads to a local value for g equal to about $\frac{1}{6}$ of its value on earth.

Measurement of *g*

Direct method: timing of free fall

This experiment can be carried out with a fair degree of accuracy if an electrically operated timer is available. An electric circuit starts the timer and releases an object at the same moment, and after the object has fallen a measured distance, it operates a switch in the circuit which then stops the timer.

In one version, the falling object is a steel ball which is held in position by an electromagnet. Switching off the current to the electromagnet also starts the timer, and the ball falls on a spring switch which stops the timer (Fig. 66(a)).

Alternatively, as shown in Fig. 66(b), the falling body is a metre rule which is suspended by a loop of cotton so that the lower edge of the rule hangs just clear of a narrow beam of light which is focused on a photoelectric device. This controls the timer so that it operates only when the light is cut off. The cotton thread is burned through and the rule allowed to fall; the timer will record how long it took to fall through 1 metre.

In both versions of this experiment, the time involved is about $\frac{1}{4} - \frac{1}{2}$s.

Indirect method: using a simple pendulum

The mathematics behind this experiment has been covered in the chapter on simple harmonic motion, p. 97.

A simple pendulum, consisting ideally of a point mass at the end of a weightless unstretchable string of length l, will have a period of oscillation

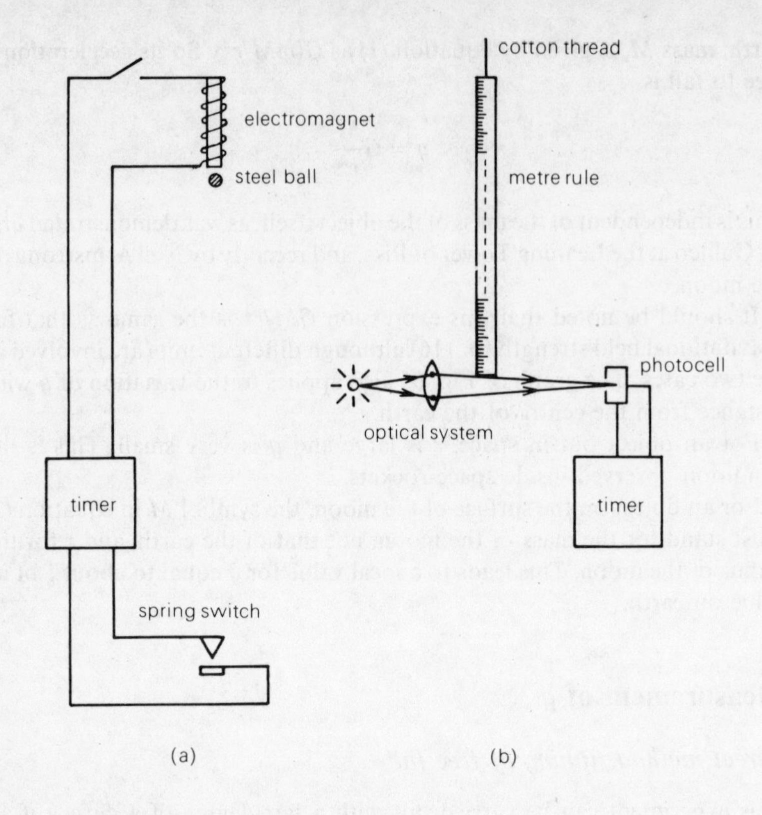

Fig. 66 Determination of g by free fall

equal to $2\pi\sqrt{(l/g)}$, provided that the angle of swing is not greater than 5°. A reasonable approximation can be made using a small heavy bob, if the length l is measured to the centre of the bob.

Swings should be counted as the string crosses the midline position. For accuracy, a 'large number' of swings, at least one hundred, must be timed. This will give a value of g correct to about 0.25 %.

Variation of g with latitude

The measured value of g is found to vary slightly from place to place over the surface of the earth, being greater at the poles than at the equator by about $0.052\ \mathrm{m\,s^{-2}}$.

Two separate factors contribute to this difference:

(a) The earth is not a perfect sphere, but is flattened over the polar regions (Fig. 67). An object on the ground near one of the poles is closer to the earth's centre of mass than if the object were at the equator.

(b) The earth rotates about an axis through the poles. Objects at the equator are moving in a circle with considerable speed, and hence a force-

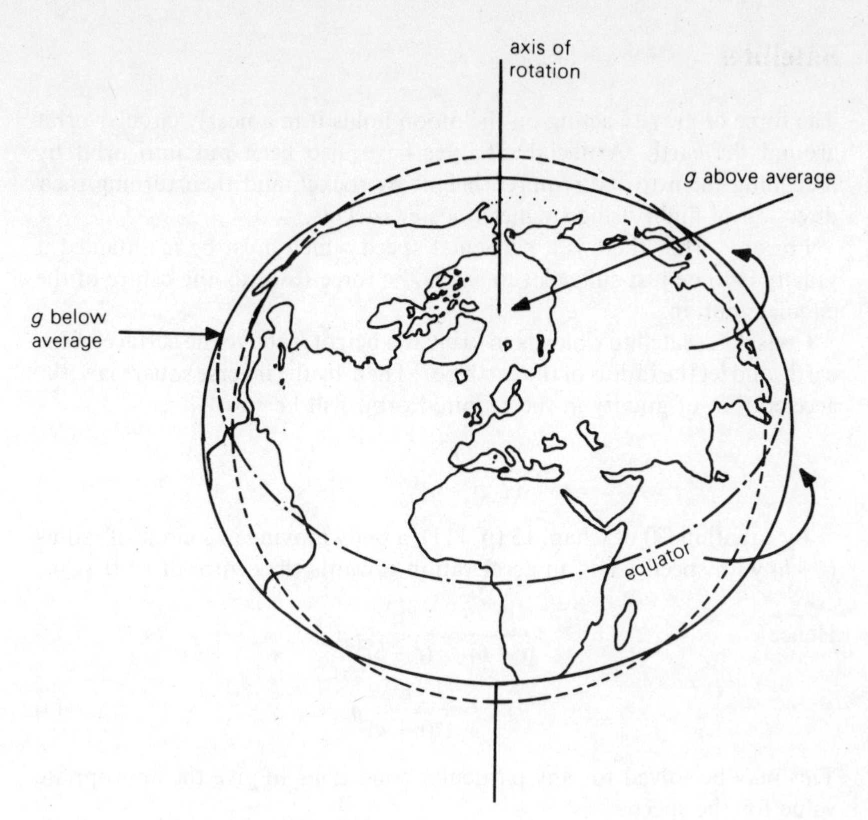

Fig. 67 Variation of g with latitude

towards-the-centre must cause this motion (p. 110). This force would not appear as weight. At the poles, the circular speed of an object is low.

The angular velocity of the earth is $2\pi^c$ in 24 hours. Taking its radius at the equator as 6.38×10^6 m gives the required acceleration-towards-the-centre as 0.0337 m s^{-2}.

The remainder of the observed discrepancy (0.052 − 0.0337) m s^{-2} must be attributed to factor (a).

EXERCISES

Take $G = 6.67 \times 10^{-11}$ N m^2 kg^{-2} where required.

1 Calculate the gravitational attraction between two masses of 1 kg and 2 kg, placed 1 m apart.

2 A small mass of 40 g is carried at the end of a light pivoted bar, 60 mm from its axis. A fixed mass of 100 kg is placed with its centre 0.2 m from the centre of the 40 g mass. Find the torque set up by the attraction between the two masses.

3 Assuming that the earth is a sphere of radius 6.35×10^6 m, use the observed value of $g = 9.81$ m s^{-2} to deduce a value for the mass of the earth.

4 If g at a point on the earth's surface is 9.8081 m s^{-2}, what would be the acceleration of a freely falling object at a height of 2000 m above that point? Take the radius of the earth as 6400 km.

Satellites

The force of gravity acting on the moon holds it in a nearly circular orbit around the earth. Artificial satellites have also been put into orbit by launching them to the required height by rocket, and then turning their direction of flight using smaller booster rockets.

For any orbit, there is a particular speed which must be maintained if gravity is to be just sufficient to act as the force-towards-the-centre of the circular motion.

Consider a satellite which is to orbit at a height h above the surface of the earth, and let the radius of the earth be r. Then, by the inverse square law, the acceleration of gravity in the required orbit will be

$$\frac{r^2}{(r+h)^2}g.$$

By equation (2) of Chap. 13 (p. 111), a body moving in a circle of radius $(r+h)$ with speed v has an acceleration towards the centre of $v^2/(r+h)$.

Hence
$$\frac{v^2}{(r+h)} = \frac{r^2}{(r+h)^2}g$$
$$\Rightarrow v^2 = \frac{r^2}{(r+h)}g. \tag{3}$$

This may be solved for any particular conditions to give the appropriate value for the speed.

When travelling in orbit, every object within the satellite is moving at this speed independently and will be apparently *weightless*. This is not true weightlessness, since the objects are still within the gravitational field of the earth.

The time taken by the satellite to orbit the earth once is given by

$$\frac{2\pi \times (\text{radius of orbit})}{\text{speed}}.$$

If the satellite is to circle the earth once in 24 hours, so remaining always over the same point of the earth's surface, then it must have angular velocity equal to $2\pi^c$ in 24 hours, i.e. $v/(r+h)$ is fixed at this value.

Combining this requirement with equation (3) gives an equation for h which will have only one solution. This means that only one orbit (known as a 'parking orbit') satisfies the given condition. All the relay satellites nowadays used in world communication must be placed in this particular orbit.

If the body is moving at a slower speed than that necessary for its height, it will not be able to maintain the same orbit, but will fall gradually back to earth. This is the case with most of the present artificial satellites, which are being slowed by friction, their orbit being comparatively close to the earth and within the bounds of its atmosphere. Small bodies will probably burn up as they fall, as most meteorites do because of the heat produced by the friction, but fragments of larger satellites do reach the earth's surface.

EXERCISES

5 Taking g at the surface of the earth as 9.8 m s^{-2} and the radius of the earth as 6.35×10^6 m, calculate (i) the speed and (ii) the height at which a relay satellite must circle the earth if it is to remain always over the same point of the surface.

6 The moon orbits the earth in 656 hours at an average distance of 3.83×10^5 km. Use this data to obtain a value for the mass of the earth. (Take $G = 6.67 \times 10^{-11}$ N m^2 kg^{-2}).

Examination questions 14

1 (a) State Newton's law of gravitation.
(b) Describe either (i) an accurate free fall method for the determination of the gravitational acceleration g, or (ii) the basic principles of a laboratory method for the determination of the gravitational constant G.
(c) If the mean density of the moon is 0.60 times the mean density of the earth and the mean diameter of the earth is $\frac{11}{3}$ times the mean diameter of the moon, calculate a value for the length of a simple pendulum of period one second on the surface of the moon.

($g = 10$ m s^{-2}) [AEB, June 1975]

2 A space capsule is travelling between the earth and the moon. Find the distance from the earth at which it is subject to zero gravitational force. (Consider only the gravitational fields of the earth and the moon.)
(Mass of the earth = 6.0×10^{24} kg,
mass of the moon = 7.4×10^{22} kg,
distance between the centres of the earth and moon = 3.8×10^8 m)

[C]

3 A satellite, of mass m, moving in a circular orbit of radius r around the earth with a uniform speed v has an acceleration. Explain why this is so and derive an expression for the magnitude of the acceleration. What is its direction?

Explain, in non-mathematical terms, why it is not possible to deduce the mass of the satellite from observations of its period and the radius of its path.

Discuss the energy changes which occur as the satellite approaches and enters the earth's atmosphere.

Explain why an astronaut in an orbiting spacecraft may be said to be in a state of weightlessness. It has been suggested that an astronaut may be given an artificial 'weight' by causing the craft to rotate. Explain why this is so.

[L]

4 Two earth-satellites S_1 and S_2 describe circular orbits of radii r and $2r$ respectively. If the angular velocity of S_1 is ω, that of S_2 is

A $\frac{\omega}{2\sqrt{2}}$ B $\frac{\omega\sqrt{2}}{3}$ C $\frac{\omega}{2}$ D $\frac{\omega}{\sqrt{2}}$ E $\omega\sqrt{2}$ [O]

5 Assuming that the earth is a uniform sphere of radius 6400 km, the calculated value of g (in $\mathrm{m\,s^{-2}}$) at a point 1600 km above the surface of the earth is

A 2.5 B 4 C 6.4 D 8 E 10 [O]

6 State Newton's law of gravitation and describe an experiment to measure the gravitational constant G.

Derive a relation between g, the acceleration due to gravity at the earth's surface, the mass M and radius r of the earth, and the gravitational constant G. You should assume the gravitational force acts as if the mass of the earth were concentrated at its centre.

Show that the period T of a satellite in circular orbit around the earth is independent of the mass of the satellite. Calculate the period of such an orbit of radius 8000 km, taking the earth to be a sphere of radius 6400 km.

Why is it impossible to orbit the earth, under gravitational forces only, in 10^3 s? [OC]

15 Momentum

The importance of the concept of *momentum* is in its application to the study of collisions and explosions. Here the principle of the conservation of momentum may be used where the principle of the conservation of energy is often impossible to apply, owing to the difficulty of measuring how much energy has been transformed into heat.

Momentum is a vector quantity, defined for a particular body as equal to the product mass × velocity. There is no special name for the unit of momentum, which should be quoted as kg m s^{-1}.

The principle of the conservation of momentum states that, for any self-contained set of moving bodies, the total momentum remains constant.

Since momentum is a vector, the principle may be applied to the sum of the components of momentum in any chosen direction, and so can be used to give three independent equations for a particular system of bodies.

EXERCISES

1 What momentum in the direction S is possessed by a car of mass 1.5 tonnes moving SW at 50 km hr^{-1}?

2 A ball of mass 1 kg moving at 0.1 m s^{-1} collides directly with a similar ball of mass 2 kg moving at 0.2 m s^{-1} in the opposite direction. In which direction must the two balls move after the impact?

3 A snooker player wishes to pot a ball in a direction at 120° to the direction of his shot with the cue ball. Draw a diagram to illustrate how the two balls must touch at the moment of the collision.

Experimental work on collisions

To study the changes in momentum during a collision, it is essential to reduce the effects of friction on the moving bodies, so that their velocities will remain constant long enough to be measured. There are several ways in which this may be done:

(a) using trolleys on a gently inclined board, carefully adjusted before the experiment so that a trolley moving down the board will do so with uniform velocity. This will be true if the component of the weight of the trolley down the board exactly cancels out the frictional resistance.

(b) using trolleys on a linear air-track. The track is supplied with a large number of small holes through which air is blown, so that the trolleys are actually supported just above the track. It must be levelled carefully before use.

(c) using carbon-dioxide pucks on a smooth horizontal board. These circular pucks are hollow; solid carbon dioxide is put into the space and as this warms up and vaporises, the vapour escapes through small holes in the base so that the pucks ride on individual air-cushions.

(d) using pucks which slide over the surface of a layer of tiny polystyrene balls.

The speeds of the trolleys in methods (a) and (b) can be found by timing their run over a measured distance. This is most accurately done using a photoelectrically controlled timer operated by two fine beams of light. Failing this, reasonable accuracy can be obtained using ticker-timers, one for each trolley. Figure 68 shows a typical result.

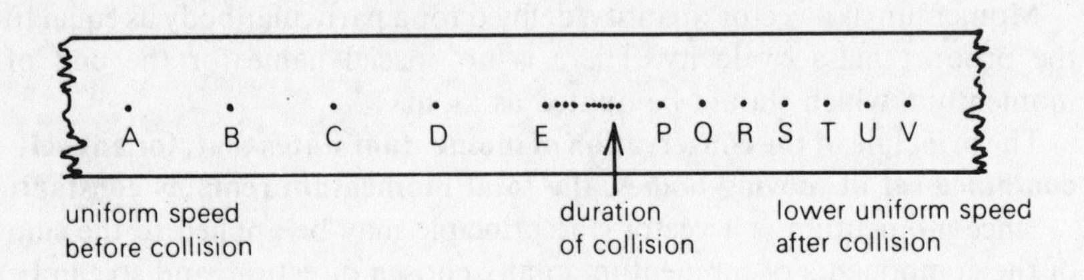

Fig. 68 A ticker-tape result showing a collision

The trolley was clearly travelling faster while marking out the section A → E, which must have been before the collision. As the time interval between successive dots is known (usually 1/50 s), the speed of the trolley can be found in mm s^{-1}.

Similarly, its lower velocity after the collision can be determined by measuring the distance between the dots in the section P → V.

A confused group of dots is often produced *during* the collision, as shown just to the left of P. This indicates that the collision was not a clean one, the two trolleys rebounding a little before finally separating. This does not affect the final results.

A second tape, attached to the other trolley, is of course needed to determine its speed in order to calculate the momenta of both the colliding bodies.

Oblique impacts, which can be studied using the experimental methods (c) and (d), must be observed photographically. If a cine-camera is not available, a series of pictures can be obtained using an ordinary camera set for a time exposure, and a stroboscopic lamp. Figure 69 shows a typical result. Commercial photographic kits are available which make it quite feasible to process the film in front of the class so that it may be studied within about a quarter of an hour.

The actual units of mass and velocity are not important in interpreting such pictures. Distances may be measured on any convenient scale, while angles, of course, will have remained constant. If the developed film is

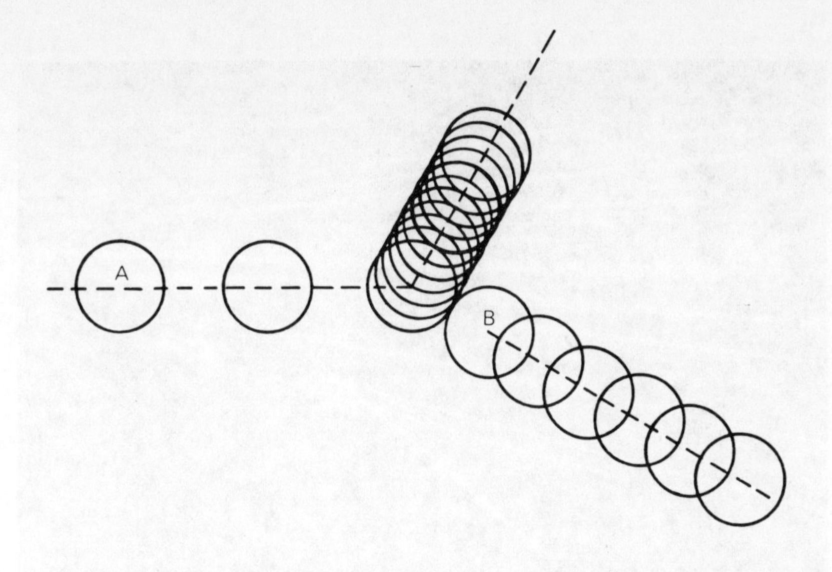

Fig. 69 An oblique collision recorded stroboscopically

projected on to a screen (or a blank wall), the relative speeds of the pucks may be determined by measurement of the apparent distances travelled in equal intervals of time.

Considering Fig. 69 in this way, it can be seen that the puck A approached the collision from the left at a speed of 8 units per flash, and moved off to the upper right at only 1 unit per flash. Puck B was stationary before the collision, and moved away along the line of centres at 3 units per flash. If the angles of the picture are measured with a protractor, the laws of momentum may be applied to deduce the ratio of the masses of the two pucks.

Collisions between sub-atomic particles

An immense amount of work has been carried out in investigating collisions between electrons, protons, neutrons, α-particles, etc. etc. Study of the results of such collisions yields information about the nature of these particles, and has also led to discoveries of many new types of sub-atomic particle.

Most of the experimental work results in photographs of impacts, of which Fig. 70 is an example. The moving particles have formed tracks of tiny bubbles in a superheated liquid, and these tracks may be analysed exactly as described above.

Photographs showing radioactive disintegrations are also considered in terms of momentum. The existence of the *neutrino* (a neutral particle otherwise identical to an electron) was first proposed in order to maintain the conservation of momentum in such a case.

Fig. 70 Collision of a moving α-particle with a stationary helium nucleus

Explosions

These may be simulated in the laboratory by placing two trolleys in contact with a compressed spring trapped between them. Real explosives, even percussion charges, are not suitable except in specialist hands.

Rocket propulsion

A rocket is caused to move by the continuous burning of its liquid fuel, the gases produced forcing themselves at high speed out of the exhaust, giving a corresponding amount of momentum to the rocket. Naturally this process is most efficient when outside the atmosphere, for otherwise the pressure of the atmosphere opposes the escape of the gases.

Similarly, space rockets are steered by short bursts of fire from small steering jets. A rocket will have several of these, mounted at different angles at different points around the body.

Impulse

The concept of momentum is used to find the force exerted by one body on another during collision.

Newton's second law of motion can be stated in the form:
The rate of change of momentum of a moving body is proportional to the

force acting upon it. This law is used to define the unit of force, the newton, so that

$$\text{force} = \frac{\text{change of momentum}}{\text{time}} \qquad (1)$$

when SI units are used.

The product (force × time) is defined as the *impulse* of the force, so that

impulse = force × time = change of momentum.

The units of impulse are N s.

Applications to problems on flow

When considering a stream of identical moving objects such as water, fuel, sand or atomic particles, it is easier to deal with a large convenient mass rather than a single individual object.

The *change of momentum per second* is calculated by finding the total mass passing a point in one second and multiplying by the velocity of the stream. Equation (1) is then applied.

This method is used to determine the forces exerted by a jet of water striking a wall, a load of sand falling on to the ground, the burning of fuel in a rocket, etc.

EXERCISES

4 A golf-club striking a ball of mass 80 g gives it a speed of 70 m s^{-1}. Calculate the impulse of the blow.

5 A pile driver of mass 500 kg falls freely through a distance of 2 m and is in contact with the pile for 2 s before coming to rest. Find the average force exerted on the pile, taking $g = 10$ m s^{-2}.

6 A space rocket after launch is burning fuel at the rate of 5 kg s^{-1}, and the exhaust gases are expelled at a speed of 2×10^3 m s^{-1} relative to the rocket. What forward thrust is given to the rocket?

7 A jet of water of diameter 40 mm is delivered by a hose-pipe at a speed of 10 m s^{-1}. If it strikes normally against a wall without rebounding, find the force exerted on the wall.

Angular momentum

For rotating bodies, *angular momentum* is defined as equal to the product (moment of inertia) × (angular velocity).

Accurate experimental work on angular momentum is generally beyond the scope of the school laboratory, but the principle of the conservation of angular momentum may be applied to theoretical problems where appropriate.

A concept of *impulsive torque* replaces the linear term 'impulse', and is defined as equal to

$$\textbf{torque} \times \textbf{time} = \textbf{change of angular momentum.}$$

Examination questions 15

1 (a) State (i) the law of conservation of linear momentum, and (ii) the law of conservation of angular momentum.
(b) Describe and explain an experiment which shows quantitatively the law of conservation of linear momentum. Discuss briefly any sources of error which, if ignored, might make it appear that the law does not hold.
(c) A stone of mass 2.0 kg is dropped from a height of 5.0 m and becomes embedded in a layer of mud on the surface of the earth. Calculate (i) the momentum of the stone immediately before it reaches the ground, (ii) the average retarding force exerted by the mud as the stone is brought to rest if the depth of penetration is 0.04 m, (iii) the time taken, after it first makes contact with the mud, for the stone to be brought to rest.
[AEB, June 1975]

2 A space-research rocket stands vertically on its launching pad. Prior to ignition, the mass of the rocket and its fuel is 1.9×10^3 kg. On ignition, gas is ejected from the rocket at a speed of 2.5×10^3 m s^{-1} relative to the rocket, and fuel is consumed at a constant rate of 7.4 kg s^{-1}. Find the thrust of the rocket and hence explain why there is an interval between ignition and lift-off.
(Acceleration of free fall, $g = 10$ m s^{-2}.) [C]

3 A motor car collides with a crash barrier when travelling at 100 km hr^{-1} and is brought to rest in 0.1 s. If the mass of the car and its occupants is 900 kg, calculate the average force on the car by a consideration of momentum.

Because of the seat belt, the movement of the driver, whose mass is 80 kg, is restricted to 0.20 m relative to the car. By a consideration of energy calculate the average force exerted by the belt on the driver. [JMB]

4 A block of wood of mass 1.00 kg is suspended freely by a thread. A bullet, of mass 10 g, is fired horizontally at the block and becomes embedded in it. The block swings to one side, rising a vertical distance of 0.50 m. With what speed did the bullet hit the block?
(Acceleration of free fall, $g = 10$ m s^{-2}.) [L]

5 For a particle of mass m which is initially moving vertically downwards with velocity u, obtain expressions for the change in momentum and the change in kinetic energy after:
(a) it has moved freely under gravity for time t;
(b) it has moved freely under gravity for a vertical distance s;
(c) it has collided with a fixed smooth plane surface which is perfectly elastic and is inclined at 60° to the horizontal. [O]

6 Define *linear momentum* and state the law of conservation of linear momentum. Describe an experiment to verify this law.

A rocket of mass M is moving at a constant speed in free space and initially has kinetic energy E. An explosive charge of negligible mass divides it into three parts of equal masses $\frac{1}{3}M$ in such a way that one part moves in the same direction as the parent rocket with kinetic energy $\frac{1}{3}E$, and the other two portions move off at an angle of 60° to this direction. Determine the total energy W imparted to the parts of the rocket in the explosion.

[OC]

16 Elasticity

The stretching of a helical spring

Suitable springs for this experimental investigation are obtainable about 100 mm long and 4 mm in diameter. The ends of these springs are looped so that one end may carry the load while the other is hooked over the rod of a clamp. A wire pointer secured near the lower end of the spring moves over a vertical millimetre scale. (Fig. 71.)

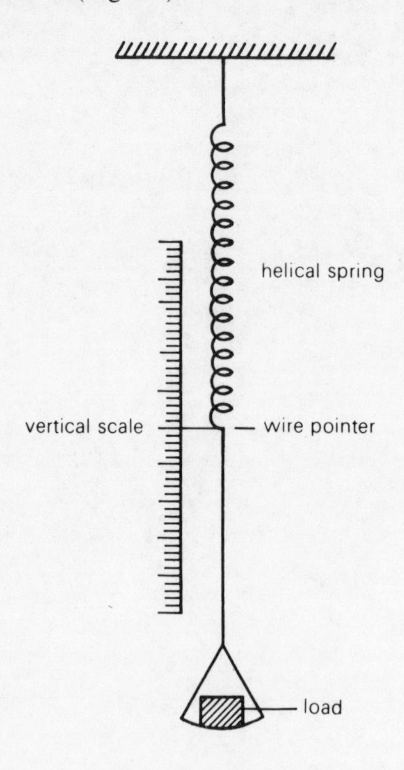

Fig. 71 Investigation of the stretching of a helical spring

Readings of the position of the pointer are to be taken for gradually increasing loads. There must be no preliminary stretching of the spring, and the load must never be decreased during alteration unless this is intentional as part of the study. If sufficient vertical space is available, a spring of the dimensions quoted above can stretch to about 4 m under a load of 10 kg or so. A cushion should be placed below such a load since the spring may break or come loose from its support.

The results should be shown on a graph of extension against load, as shown in Fig. 72. At the beginning, there may be a short section O → A where little extension occurs. This is observed only when a new spring is first used, and is similar to the behaviour of a new balloon on being inflated.

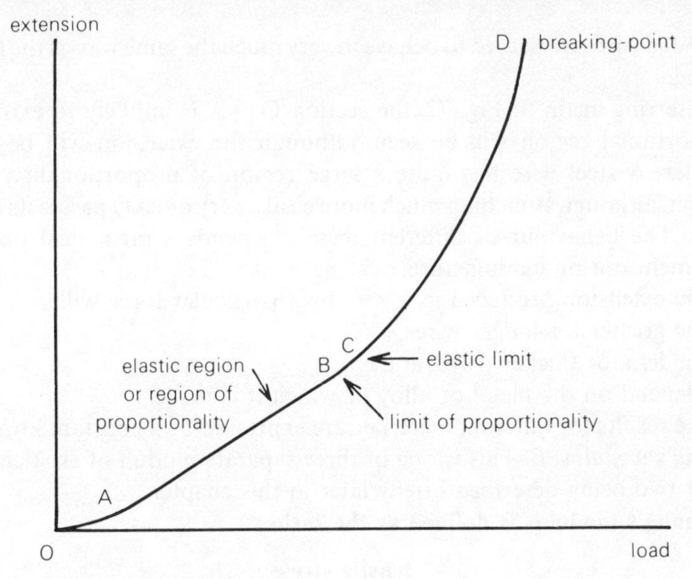

Fig. 72 Graph of load and extension

The next section of the graph, A → B, is clearly a straight line. In this region, if part of the load is removed and then replaced, the pointer will repeat its previous reading. This is called the *region of proportionality*.

As the load is increased, the readings will eventually cease to lie on this straight line, greater extensions being observed. The *elastic limit* occurs at some point C close to the *limit of proportionality*, B. At the point C, if the spring is studied carefuly, it will be possible to find a section where the turns have separated more than elsewhere, and this section is no longer *elastic*; that is, the turns will not close up together again if the load is removed.

Greater loads increase this effect. The part of the spring which has lost its elasticity will stretch increasingly, and other sections will be found to become inelastic as well. It is difficult to observe the breaking-point D as this generally arrives without warning, and there is little advantage in repeating the experiment since a different spring will probably display quite different regions of weakness.

Hooke's law

In the region of proportionality, Hooke's law may be applied:

extension is proportional to load.

Under such conditions, the spring is suitable for use as a spring balance or dynamometer, and it will perform simple harmonic motion if slightly disturbed when loaded.

The stretching of a wire

A loaded wire is observed to behave in very much the same way as the helical spring.

Referring again to Fig. 72, the section O → A is unlikely to exist. The proportional region will be seen, although the extension will be much smaller. A steel wire has quite a large region of proportionality, while copper, although stretching much more easily, very quickly passes its elastic limit. The behaviour of different metals depends a great deal on their treatment during manufacture.

The extension produced in a wire by a particular force will

(a) be greater for longer wires,
(b) be less for thicker wires, and
(c) depend on the metal or alloy of which it is made.

The results for different materials are expressed by a constant known as *Young's modulus, E*. This is one of three separate moduli of elasticity, the other two being described briefly later in this chapter.

Young's modulus is defined as the ratio

$$\frac{\textbf{tensile stress}}{\textbf{tensile strain}}$$

where
$$\text{stress} = \frac{\text{force}}{\text{area of cross-section of wire}} \tag{1}$$

and
$$\text{strain} = \frac{\text{extension}}{\text{original length}} \tag{2}$$

measured under elastic conditions.

The units of Young's modulus are $\mathrm{N\,m^{-2}}$. Some values for metals and alloys are given in Table 5.

Table 5

Material	Young's modulus/$\mathrm{N\,m^{-2}}$
Brass	1.01×10^{11}
Copper	1.30×10^{11}
Steel	2.1×10^{11}
Platinum	1.68×10^{11}
Aluminium	0.70×10^{11}
Lead	0.16×10^{11}
Plastic	$\sim 0.02 \times 10^{11}$

Young's modulus is used in connection with many problems of mechanical engineering. It also occurs in the theoretical treatment of sound waves in solids, since these are longitudinal vibrations.

Determination of Young's modulus

Figure 73 shows a simple form of the apparatus for this experiment. Two identical wires are required; one is to be loaded and the other, hanging alongside from the same support, carries the scale against which the extension will be measured. There are two reasons for this:

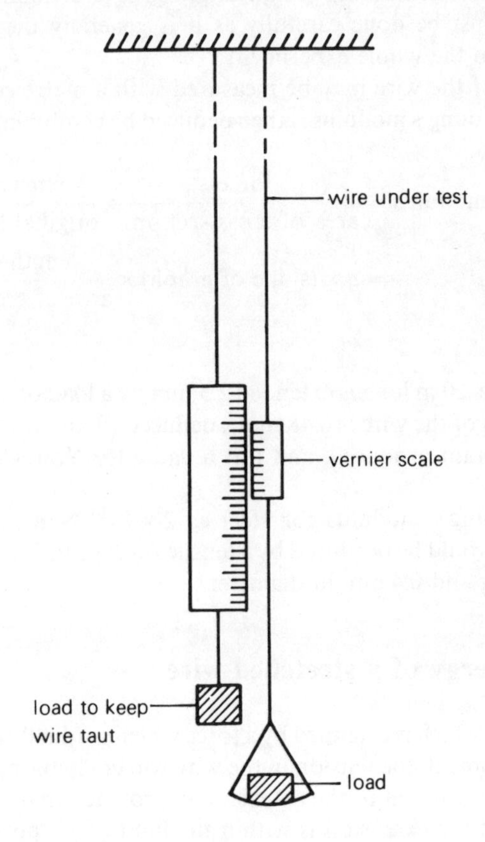

Fig. 73 Determination of Young's modulus for a wire

(a) to counteract any bending of the supporting beam when it takes the load,
(b) to counteract any expansion or contraction of the test wire due to a change of temperature during the experiment.

Before attempting to take any readings, the hanging wires should be loaded with about 1 kg and unloaded a few times to remove any kinks.

Experimental loads are added in steps of 0.5 kg (if the wires are of steel). Smaller steps would be time-wasting. To measure the extension of the wire accurately, the suspended scale is provided with a vernier which hangs from the test wire and slides freely in a vertical groove alongside the scale. This avoids any errors of parallax which might arise from the use of a separate pointer.

A graph of scale reading against load should be plotted as the readings are taken. Each load should be removed before adding a larger one, to check that the readings return to zero each time, i.e. that the wire has not passed its elastic limit. The slope of the graph is calculated in $\mathrm{m\,kg^{-1}}$.

The diameter of the wire should be measured using a micrometer screw-gauge, taking pairs of readings at right angles at several points along the test length. This must be done carefully as it is generally the least accurate measurement in the whole experiment.

The length of the wire may be measured with a metre rule.

A value for Young's modulus is then deduced by combining equations (1) and (2):

$$\text{Young's modulus} = \frac{\text{force}}{\text{area of cross-section}} \div \frac{\text{extension}}{\text{original length}}$$

$$= g \times (\text{slope of graph}) \times \frac{\text{length of wire}}{\text{area of cross-section}}.$$

EXERCISES

1 A brass wire 2.0 m long is extended 2.5 mm by a load of 6 kg. If the area of cross-section of the wire is $0.48\ \mathrm{mm^2}$, deduced (i) the stress set up in the wire, (ii) the strain produced, and (iii) a value for Young's modulus for brass.

2 Taking Young's modulus for steel as $2 \times 10^{11}\ \mathrm{N\,m^{-2}}$, calculate the extension that would be produced by a tensile force of 10 N acting on a steel wire 1.5 m long and 0.4 mm in diameter.

Potential energy of a stretched wire

Consider a wire which is extended by a force which gradually increases from zero. At any moment, the tension in the wire will equal the applied force F, and will be proportional to the extension e, provided that Hooke's law is obeyed, i.e. that the extension is within the limit of proportionality.

$$F = ke, \text{ where } k \text{ is a constant.}$$

If this force extends the wire a small amount δe, then the potential energy stored in the wire will be increased by an amount equal to the work done by the force, $F\delta e = ke\delta e$.

Using calculus, the **total increase in potential energy of the wire** is

$$\int F\,\mathrm{d}e = \int ke\,\mathrm{d}e$$

$= \frac{1}{2}ke^2$
$= \frac{1}{2}Fe \quad \text{since } ke = F$
$= \frac{1}{2} \times$ **(applied force)** $\times$ **(extension produced).**

Referring to Fig. 74, the increase in potential energy produced when the wire is extended from A to B is given by the area shown shaded, between the line of the graph and the axis of extension. This result may be applied even when the extension exceeds the elastic limit, provided that the shape of the graph is known.

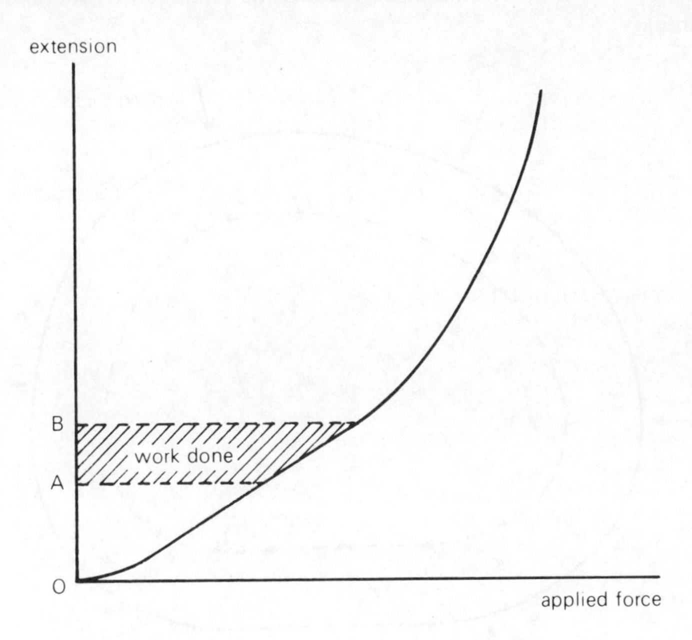

Fig. 74 Increase of potential energy due to stretching a wire

Note that it has been assumed throughout this argument that the extension has taken place slowly, so that no energy has been lost in the form of heat.

EXERCISES

3 How much potential energy has been given to a wire which has been slowly stretched a distance of 2 mm by a force of 50 N?

4 A steel wire 2 m long and 0.6 mm in diameter is slowly extended by a force of 120 N. If Young's modulus for steel is 2×10^{11} N m^{-2}, calculate the increase in the potential energy of the wire.

5 Using the following results from an experiment in stretching a wire, plot a graph of extension against load.

Load/kg	0	1	2	3	4	5	6
Extension/mm	0	0.75	1.5	2.25	3.0	3.85	5.0

From your graph, determine the potential energy which was given to the specimen as it was stretched from its original length (i) by 3 mm (ii) by 5 mm.

Bulk modulus

If a body changes in *volume* through the action of some external force, this produces a volume or *bulk strain* defined as (change in volume/original volume).

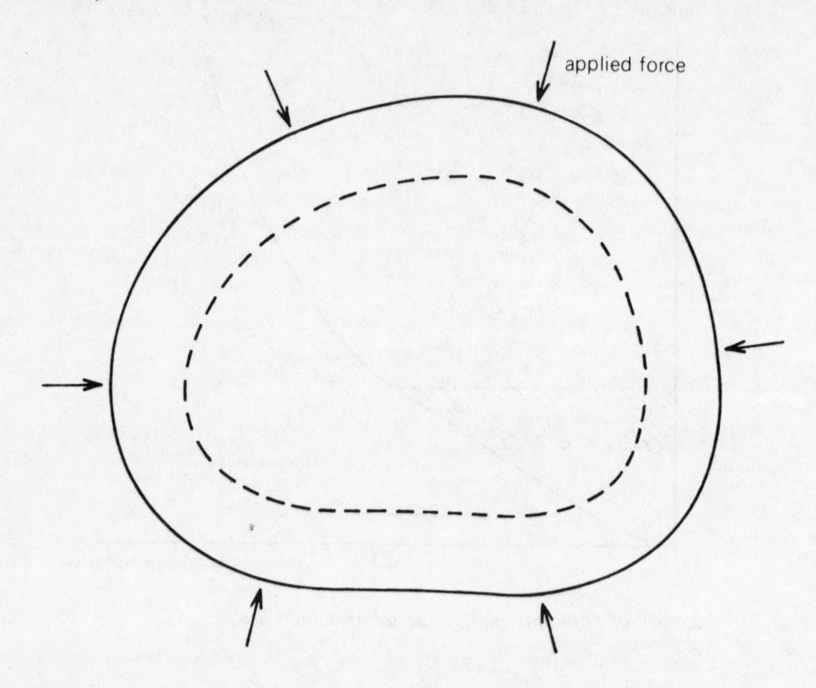

Fig. 75 Compression of a body by an applied force

The *bulk stress* producing this strain is measured by (applied force/area of surface on which it acts), which is generally equal to the *applied pressure*.

The *bulk modulus of elasticity* of a substance is defined as (bulk stress/bulk strain), provided that the deformation is within the elastic limit.

This modulus has many applications in hydrodynamics, and is also used in connection with sound waves in fluids since these are changes of pressure.

Modulus of rigidity

If a body is twisted or sheared, the *shear strain* is defined as the angle of shear, measured in radians.

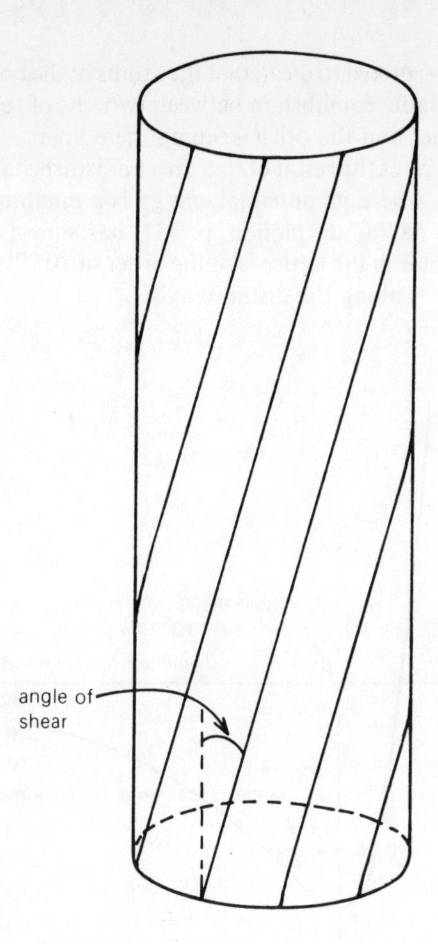

Fig. 76 Twisting of a body by an applied couple

The *shear stress* causing this strain is difficult to define simply, since its measurement depends much on the way in which it is applied. In all cases, the stress must, as before, be calculated in terms of force per unit area and measured in $N\,m^{-2}$.

Then the *modulus of rigidity* is defined as (shear stress/shear strain), provided that the deformation does not exceed the elastic limit.

This has applications in such fields as the suspension of the coil in a galvanometer, and the rotation of the pendulum in a torsional clock. The extension of a helical spring, considered at the beginning of the chapter, is in reality an example of a shear strain, not a tensile one.

The atomic theory of elasticity

The fact that any extended, compressed or twisted body will recover its original shape and size, provided that the deformation has not passed the elastic limit, is interpreted to mean that the atoms of the body normally exist in a position of stable equilibrium between two sets of forces, one holding the atoms together and the other keeping them apart.

Figure 77 illustrates the result of this. In a undisturbed state, an atom will settle at point A, where its potential energy is a minimum. Experimental work on crystals (X-ray diffraction, p. 332) has shown that the average spacing of the atoms in the lattice is of the order of 10^{-10} m. This value has been assigned to A along the distance axis.

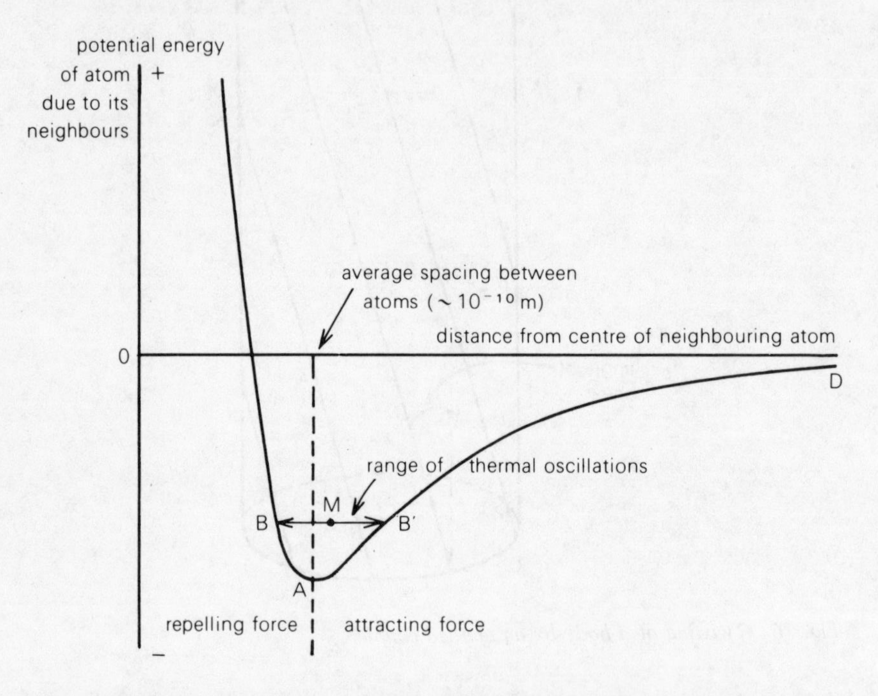

Fig. 77 Theoretical graph showing the variation in potential energy of an atom in a lattice due to the forces exerted by its neighbours

Forces of attraction are known to exist between neighbouring atoms or ions in a lattice structure. These are mainly chemical bonds of various types, but also there are the weak Van der Waals' forces, which are electrical in nature. These are considered to be due to temporary unsymmetricality in the positions of the electrons orbiting the nucleus. Calculations based on this concept lead to a force of attraction proportional to $1/(\text{separation})^6$.

Short-range forces of repulsion must also exist between atoms since it is so difficult to compress matter in the solid state, but at present little is known about the nature of these forces. In order to arrive at the shape of curve shown in Fig. 77, a relation of the form $1/(\text{separation})^{12}$ has been proposed for this repelling force.

The unsymmetrical shape of the curve on the two sides of the minimum is deliberate. If a small amount of heat energy is supplied to the atom, it will be able to exist in either of the positions B or B′, and may be supposed to oscillate between the two. This oscillation represents the thermal agitation of the atom in the solid state. Now the mid-point of BB′, M, is at a slightly greater distance from the neighbouring atom than is the point A, i.e. the substance has *expanded*.

Considering the depth of the minimum below the axis, if the atom were supplied with sufficient energy from outside to take it right along the curve to point D, it would then have broken away completely from its neighbouring atom. The distance of A below D on the vertical scale must correspond to the latent heat capacity of the substance *per atom*.

Examination questions 16

1 It is modern practice for railways to use long welded steel rails. If these are laid without any stress in the rails on a day when the temperature is 15°C, estimate the internal force when the temperature of the rails is 35°C. You may assume that the rails are so held that there is no longitudinal movement, and that the cross-sectional area of the rails is $7.0 \times 10^{-3}\ m^2$.

(The Young modulus for steel = $2.0 \times 10^{11}\ N\,m^{-2}$,
linear expansivity of steel = $1.0 \times 10^{-5}\ K^{-1}$.)

[AEB, Nov. 1979]

2 Draw labelled sketch graphs to show the way in which (a) the potential energy V, (b) the interatomic force F, of two atoms depend on their separation r.

Explain the shape of the curves and the relation between them, and mark on the r-axis the point corresponding to the equilibrium separation of the atoms.

Give a qualitative explanation of the origin of thermal expansion in a solid, making reference to your potential energy graph.

Why are the atoms of a crystalline solid arranged in a regular pattern, rather than being packed at random? [C]

3 (a) A heavy rigid bar is supported horizontally from a fixed support by two vertical wires, A and B, of the same initial length and which experience the same extension. If the ratio of the diameter of A to that of B is 2 and the ratio of Young's modulus of A to that of B is 2, calculate the ratio of the tension in A to that in B.

(b) If the distance between the wires is D, calculate the distance of wire A from the centre of gravity of the bar. [JMB]

4 Describe in detail an experiment which would enable you to determine the Young modulus for steel in the form of a wire. What can you deduce about the nature of the forces between molecules from the facts that the wire requires a force to extend it or to compress it longitudinally and that the wire, for forces up to a certain limit, returns to its original length when the force is removed?

A device to project a toy rocket vertically makes use of the energy stored in a stretched rubber cord. Assuming that the cord obeys Hooke's law, find the length by which the cord must be extended if the rocket is to be projected to a height of 20.0 m.

(Mass of rocket = 0.30 kg.
the Young modulus for rubber = 8.0×10^8 N m^{-2},
cross-sectional area of cord = 2.5×10^{-5} m^2,
unstretched length of cord = 0.20 m,
acceleration of free fall, $g = 10$ m s^{-2}) [L]

5 The breaking stress for copper wire under tension is 3×10^8 Pa (3×10^8 N m^{-2}) and Young's modulus for this material is 1.2×10^{11} Pa. Supposing that Hooke's law holds during the whole of the extension, estimate the percentage elongation of a copper wire when it breaks.

The following readings were obtained in an experiment on the loading and unloading of a length of new copper wire:

Loading:

Load F/N	0	75	150	195	220	250	244
Extension e/mm	0	2.0	4.0	6.0	8.0	10	12

Unloading

Load F/N	150	75	0
Extension e/mm	9.2	7.0	5.0

(a) Plot a graph of F(y-axis) against e. Find from the graph the external work done on the wire in extending it 12 mm, the change in internal potential energy, and (assuming that the temperature remains constant) the energy which escapes as heat during the loading.

(b) State and explain what you would expect to observe if the wire were again loaded in steps up to about 240 N. [O]

17 Surface tension

The phenomena of surface tension are due to the Van der Waals' forces (p. 138), which are attractions between neighbouring atoms or molecules. An atom within the body of a liquid is attracted by its neighbours on all sides, as shown in Fig. 78, but an atom on the surface has neighbours only on one side, and consequently experiences a resultant force inwards. This keeps the bulk of the liquid together, and only a few energetic atoms can escape by evaporation.

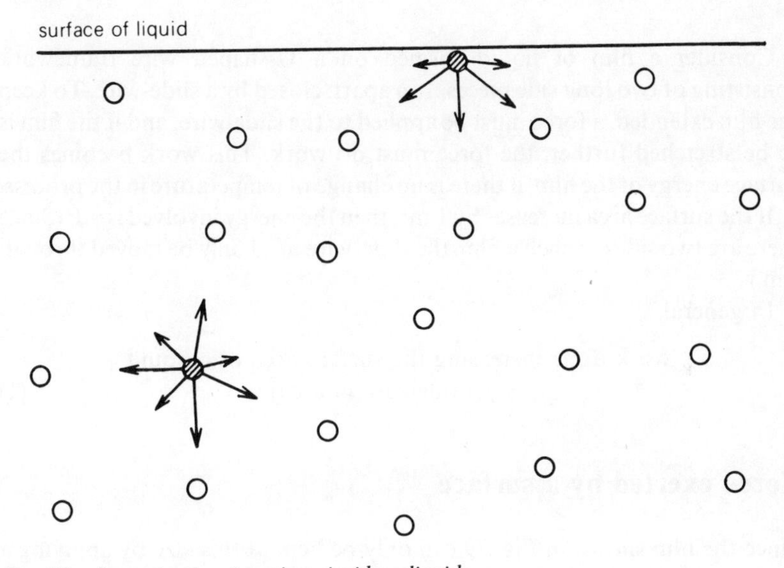

Fig. 78 Interatomic attractions inside a liquid

Energy of a surface

The fact that a liquid forms into spherical drops (except for the distortion produced by the force of gravity) demonstrates that energy is associated with a surface. A spherical shape has least surface area for a given volume, so that it would seem that the liquid is most stable and has least potential energy when its surface area is a minimum.

The surface energy (γ) of a liquid is defined as the potential energy per unit area of the surface, and measured in joules per square metre. Figure 79 shows how this definition is interpreted.

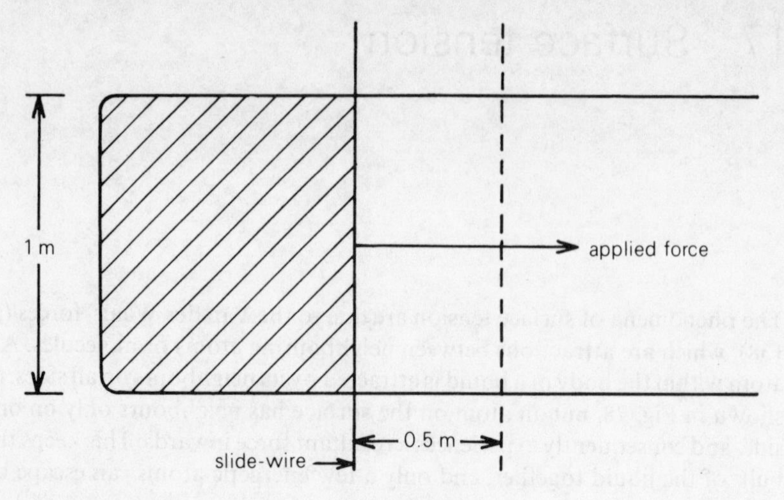

Fig. 79 Work done in stretching a film of liquid

Consider a film of liquid formed on a U-shaped wire framework consisting of two long side pieces, 1 m apart, closed by a slide-wire. To keep the film extended, a force must be applied to the slide-wire, and if the film is to be stretched further, the force must do work. This work becomes the surface energy of the film, if there is no change of temperature in the process.

If the surface area increases by 1 m^2, then the energy involved is γ J. (Since there are two sides to such a film, the slide-wire need only be moved through $\frac{1}{2}$ m.)

In general,

$$\text{work done increasing the surface area of a liquid} \\ = \gamma \times (\text{increase in area}) \qquad (1)$$

Force exerted by a surface

Since the film shown in Fig. 79 can only be kept at this size by applying a force to the slide-wire, the film itself must be exerting an equal and opposite force on the wire. This is due to the attraction between the atoms of the film and those of the wire.

The value of this force is γ N m^{-1}; this can be used as an alternative definition of the quantity γ. It agrees with equation (1) since if the applied force is 2γ (γ acting on each surface of the film) and the slide-wire is moved through $\frac{1}{2}$ m, then the work done is γ J, as before.

The surface tension of a liquid

(a) depends on the medium on the other side of the surface (air, glass, another liquid, etc.), and

(b) decreases as its temperature rises.

For these reasons, the values for the surface tension of various liquids given in Table 6 are quoted only to one significant figure.

Table 6

Liquid	Surface tension/N m^{-1} or J m^{-2}
Water	7×10^{-2}
Soap solution (depending on strength)	$2\text{–}3 \times 10^{-2}$
Alcohol	2×10^{-2}
Mercury	50×10^{-2}

Angle of contact

When a drop of liquid is placed on a horizontal surface, it may spread out as shown in (a) of Fig. 80, or sit up as in (b). The degree to which the liquid 'wets' the surface depends upon the strength of the Van der Waals' attractions between their atoms, relative to the strength of the internal attractions within the liquid.

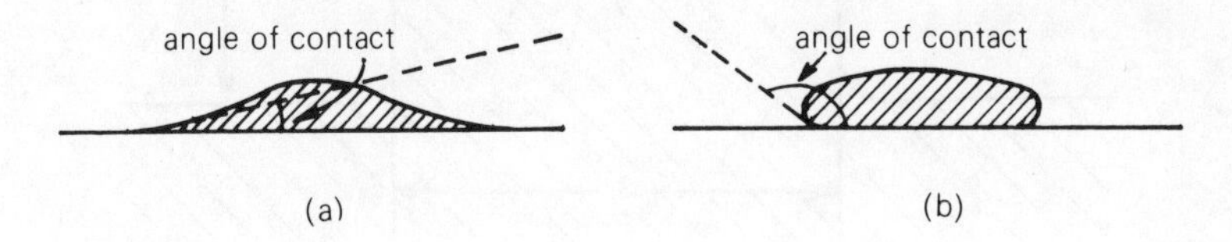

Fig. 80 The angle of contact between a liquid and a surface

The angle at which the two surfaces meet is called the *angle of contact*, for those particular substances. As marked in Fig. 80, this angle is always measured *through* the liquid. Liquids which wet surfaces have a small or zero angle of contact, while the others make an obtuse angle. The state of cleanliness of the surface makes a great deal of difference to the resulting angle of contact.

Capillary rise

If a capillary tube is dipped into a liquid which wets it, the liquid is observed to rise up inside the tube; this effect is greater for narrower tubes. In the equilibrium position, the weight of the liquid drawn up is balanced by the surface tension attraction between the atoms of the liquid and those of the material of the tube. This is illustrated in Fig. 81.

If the radius of the tube is r, then the force of surface tension acts around a circle of circumference $2\pi r$, at an angle of θ, the angle of contact.

$$\text{Total upward force} = \gamma \cos\theta \,.\, 2\pi r.$$

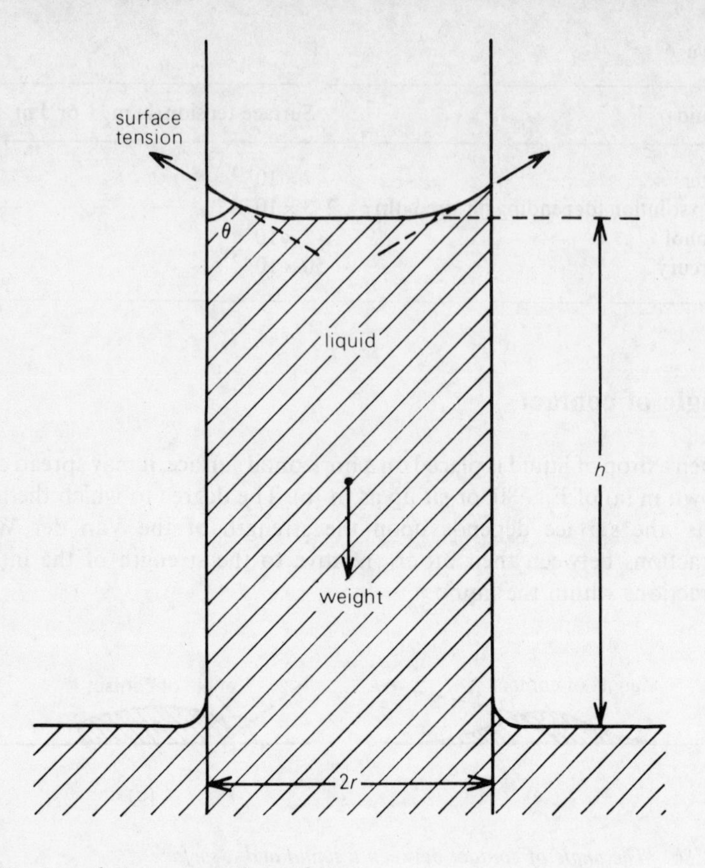

Fig. 81   Liquid rise in a capillary tube

If the liquid is drawn up to a height h, then the weight of this column of liquid is $\pi r^2 h\rho g$, where ρ is the density of the liquid. Hence

$$\gamma \cos\theta \,.\, 2\pi r = \pi r^2 h\rho g$$

$$\Rightarrow \boldsymbol{\gamma \cos\theta = \frac{rh\rho g}{2}}.$$

For water and glass, the angle of contact is zero and the relation becomes

$$\gamma = \frac{rh\rho g}{2}.$$

For mercury and glass, the angle of contact is about 140°, so that $\cos\theta$ is negative. This corresponds to the observed effect that the mercury is depressed inside the tube relative to its outside level.

It should be noticed that the weight of the small amount of liquid forming the meniscus has been ignored. The height h must be measured between the bottom of the meniscus and the horizontal part of the surface of the liquid outside the tube.

EXERCISES

1 Calculate the work that must be done against surface tension to blow a soap bubble of radius 0.1 m, assuming a value of 3×10^{-2} J m^{-2} for the surface tension of the soap solution.

2 A glass capillary tube of internal radius 0.4 mm is placed vertically into a beaker of water. If the water rises to a height of 34 mm inside the capillary, deduce a value for the surface tension of water.

3 A capillary tube of internal radius 0.3 mm is placed vertically into a beaker of liquid whose density is 700 kg m^{-3}. Taking the surface tension of the liquid as 6.0×10^{-2} N m^{-1} and its angle of contact as 15°, calculate the height to which the liquid rises inside the tube.

Pressure difference across a curved liquid surface

Inside a bubble, as inside a balloon, the pressure must be greater than it is outside, in order to maintain the curvature of the surface against the pull of surface tension.

Figure 82 shows how the pressure difference may be calculated by considering the equilibrium of one half of a spherical air bubble inside a liquid. The bubble is divided into two by a plane circular section of which YOY′ is a diameter, and OX is the line of symmetry of the hemisphere being considered.

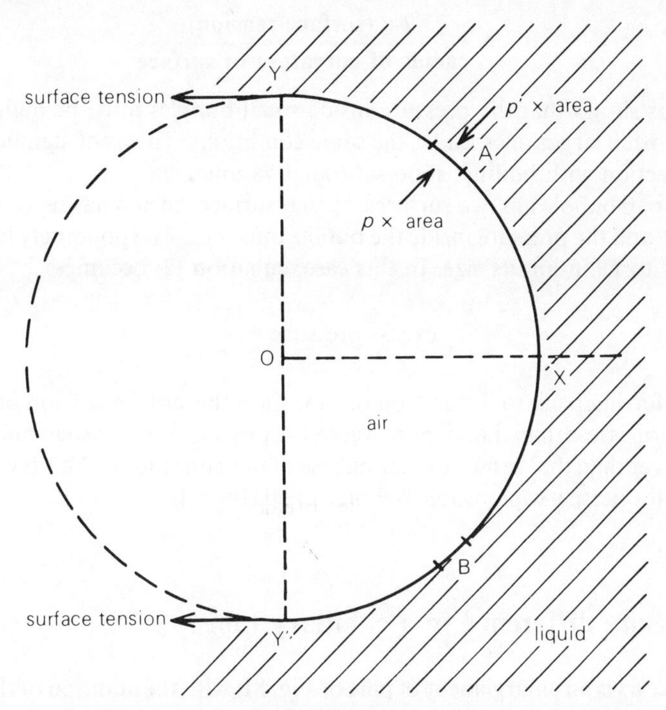

Fig. 82 Calculation of the pressure difference across a curved surface

Let the pressure inside the bubble be p, and that in the liquid outside be p'. Then $p > p'$.

Consider a small area A at some point on the curved surface of the bubble. Because of the pressure difference on its two sides, there will be a resultant force on the area, equal to $(p - p') \times A$ acting outwards normally to the surface. This force may be split into two components, one parallel to OX and the other parallel to OY.

Since the hemisphere is symmetrical about OX, the second component will be balanced by an equal and opposite force on the symmetrically placed area B on the other side of OX. The remaining components due to all the small areas of the bubble will combine to give a resultant force along OX, which is balanced by the surface tension between the two halves of the bubble, acting across their circle of contact.

Total force due to difference of pressure

$$= (p - p') \times (\text{area of circle of contact, radius OY}).$$

Force due to surface tension $= \gamma \times$ (circumference of circle of contact).

If r is the radius of the bubble,

$$(p - p')\pi r^2 = \gamma 2\pi r$$

$$\Rightarrow (p - p') = \frac{2\gamma}{r} \qquad (2)$$

i.e. **pressure difference across a curved surface**

$$= \frac{\mathbf{2 \times (surface\ tension)}}{\mathbf{radius\ of\ curvature\ of\ surface}}.$$

This shows that the pressure inside small bubbles must be higher than that inside larger ones under the same conditions. This is of significance in connection with boiling, supersaturated vapours, etc.

A soap bubble has *two* surfaces, so that surface tension has twice as much effect, and the pressure inside the bubble must be correspondingly higher in order to maintain its size. In this case, equation (2) becomes

$$\text{excess pressure} = \frac{4\gamma}{r}.$$

Referring back to Table 6 on p. 143, since the surface tension of a soap solution is less than that of pure water, the pressure inside a soap bubble will be lower than that inside a water bubble of the same radius. This is why soap bubbles persist while water bubbles break quickly.

Pressure difference in a capillary tube

Figure 83 is an enlargement of part of Fig. 81, with the addition of the point O which is the centre of curvature of the liquid meniscus, assumed to be spherical.

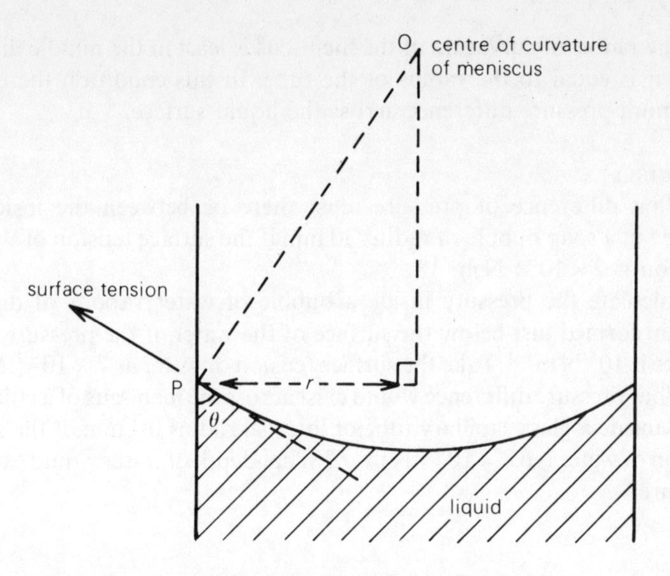

Fig. 83 Calculation of curvature of a meniscus in a tube

Since the surface tension at the point of contact P acts along the tangent to the curve, it can be seen that the radius of curvature of the meniscus is $OP = \dfrac{r}{\cos\theta}$.

Hence from equation (2),

$$\text{pressure difference across the meniscus} = \frac{2\gamma\cos\theta}{r}.$$

The pressure will be less *below* a surface of the shape illustrated in Fig. 83, and liquid will be drawn up into the tube until the pressures are equal at the same liquid level both inside and outside the tube.

If the liquid rises right to the top of the tube but does not overflow, its surface will take one of the three shapes illustrated in Fig. 84. It can be seen

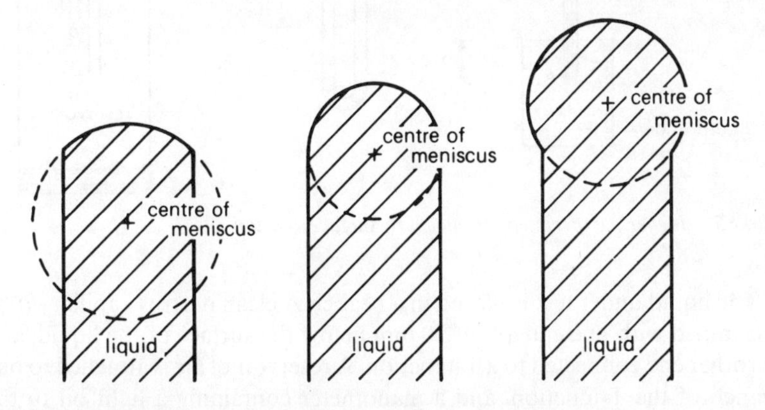

Fig. 84 Possible shapes of a meniscus at the top of a tube

that the radius of curvature of the meniscus is least in the middle diagram, when it is equal to the radius of the tube. In this condition there is the maximum pressure difference across the liquid surface.

EXERCISES

4 What difference of pressure must there be between the inside and outside of a soap bubble of radius 30 mm if the surface tension of the soap solution is 2×10^{-2} N m^{-1}?

5 Calculate the pressure inside a bubble of water vapour of diameter 0.1 mm formed just below the surface of the water, if the pressure on the surface is 10^5 N m^{-2}. Take the surface tension of water as 7×10^{-2} N m^{-1}.

6 What pressure difference would exist across the meniscus of a column of water inside a glass capillary tube of internal radius 0.4 mm, if the surface tension of water is 6.8×10^{-2} N m^{-1}? What depth of water would exert this pressure?

Jaeger's experiment to measure surface tension

This is a simple experiment which can give quite accurate results. The apparatus is as shown in Fig. 85.

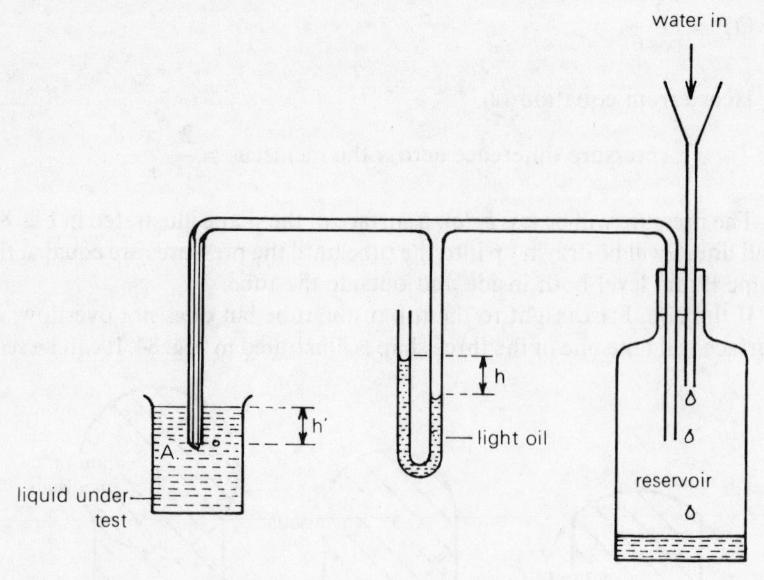

Fig. 85 Jaeger's experiment to measure surface tension

The liquid under test is placed in a beaker. A clean narrow capillary tube is clamped with one end about 20 mm below the surface of the liquid, and the other end connected to a T-junction. A reservoir of air is attached to one branch of the T-junction, and a manometer containing a light oil to the other branch.

Water drips slowly into a funnel at the top of the reservoir vessel, and drives air through the apparatus so that it bubbles out through the capillary tube. The rate of flow is adjusted to give a steady slow sequence of bubbles.

As each bubble grows, the pressure builds up in the apparatus and the levels of the manometer alter; when the bubble breaks away, the pressure drops and then begins again to build up. The extreme positions of the oil surfaces in the manometer can be located using a travelling microscope, and their difference in height, h, is then recorded.

The maximum pressure inside the apparatus is $[hg \times (\text{density of oil})]$ above atmospheric pressure. Allowance must be made for the fact that the aperture A of the capillary tube is below the surface of the liquid in the beaker, and so is at a pressure above atmospheric. The depth of liquid h' must be measured using the travelling microscope—it should not be too small, or a large percentage error may be introduced into the result.

This gives a value for the maximum pressure difference across the meniscus at A, which will occur when the meniscus is hemispherical. The radius of the capillary at A must be measured using the travelling microscope. Equation (2) is then applied to the results.

Other results can be obtained by varying the depth of A below the surface and repeating the experiment.

The experiment may be used to study the variation of surface tension with temperature. The results should lie on a smooth curve showing a gentle fall as the temperature rises.

There is no satisfactory simple equation to fit this curve, but since the atoms of the liquid will be slightly farther apart at a higher temperature, the Van der Waals' attraction would then be lower.

Examination questions 17

1 (a) Draw diagrams to illustrate the appearance of water drops of various sizes resting on a horizontal wax surface.

Explain how molecular theory accounts for the shape of very small drops.

Account for the shape of the larger drops.

(b) Derive an expression for the excess pressure inside a spherical drop of liquid in terms of the surface tension of the liquid and the radius of the drop.

(c) Describe an experiment to determine the surface tension of a liquid. [AEB, June 1978]

2 A thin film of liquid is trapped between two plates. The force required to pull the plates apart will increase if

A the surface tension of the liquid is reduced.

B the separation of the plates is increased.

C the area of the liquid surface in contact with the plates is increased.

D the pressure of the air decreases.

E the angle of contact is increased. [C]

3 A vertical capillary tube is placed in a liquid with 100 mm projecting and the liquid rises to a height of 70 mm in the tube. The tube is now lowered a further 50 mm into the liquid. Which one of the following groups of three statements applies?

A The liquid level rises in the tube to the top; the radius of curvature of the meniscus increases; the liquid that occupied 20 mm of the tube is ejected.

B The liquid level rises in the tube to the top; the radius of curvature of the meniscus is unchanged; no liquid is ejected.

C The liquid level rises in the tube to the top; the radius of curvature of the meniscus increases; no liquid is ejected.

D The liquid level rises in the tube to the top; the radius of curvature of the meniscus decreases; no liquid is ejected.

E The liquid level stays at the same point in the tube; the radius of curvature of the meniscus is unchanged; no liquid is ejected. [C]

4 (a) The energy to create unit surface area of a liquid, γ, may be expressed by the relation

$$\gamma \approx \tfrac{1}{4}nA\,\varepsilon.$$

Derive this expression, explaining the meaning of each term.

(b) A force per unit length σ may be considered to act in a liquid surface on one side of a line in the surface, and at right-angles to the line. Deduce the relation between σ and γ, and show that they have the same dimensions. [JMB]

5 Define surface tension and use your definition to obtain an expression for the work done per unit area in changing the area of a surface.

Explain the following phenomena:

(a) water forms globules on a greasy plate but not on a clean one,

(b) two flat glass blocks with a thin layer of water between them are difficult to pull apart, (c) small bubbles on the surface of water cluster together.

Two soap bubbles, one of radius 50 mm and the other of radius 80 mm, are brought together so that they have a common interface. Calculate the radius of curvature of this interface and explain whether it is convex towards the larger or smaller bubble. [L]

Sound

18 Frequency and pitch

Musical sound is distinguished from *noise* by having a recognisable frequency of vibration. It is produced by some vibrating source, carried by a material medium, and sets up sympathetic vibrations in the stretched membrane of our eardrum, so that we hear the sound.

Sources of musical sound fall broadly into three groups:
(a) percussion instruments, e.g. drums, the xylophone;
(b) wind instruments, e.g. the clarinet, the organ;
(c) stringed instruments, e.g. the violin, the piano.

All these instruments can produce notes of the same *pitch*, which is the musical term corresponding to *frequency*. A note of low pitch is produced by a low frequency of vibrations of the source, and similarly a higher-pitched note results from more rapid vibrations. Doubling the frequency of the vibrations produces a note one octave higher; the ear recognises the relation between the frequencies of the two notes.

The *loudness* of the note depends upon the amount of energy given out by the vibrating source. In musical instruments, the original vibration is usually amplified in some way, e.g. by the sounding-board of a piano, or the resonating air-space inside the body of a drum.

Most listeners can distinguish fairly easily between the same note played on a variety of instruments. This is generally because of the different *harmonics* which are produced along with the main note and give it a particular *quality*.

The tuning fork is the simplest of musical instruments, giving a pure note free from harmonics. For this reason, it is used in the laboratory as a standard of frequency with which other sources of sound are compared.

Determination of the frequency of a tuning fork

The tuning fork must be electrically maintained in vibration during the experiment. Two light pieces of card are fastened to the prongs of the fork, and a longitudinal slit is cut in each card so that the slits are in line when the fork is not vibrating. A white circular disc painted with a large number of evenly-spaced black dots is set up behind the fork so that one dot can be viewed through the slits. (Fig. 86.)

If the fork is now made to vibrate, as the cards move up and down the slits will pass opposite each other again twice during each vibration, and the time

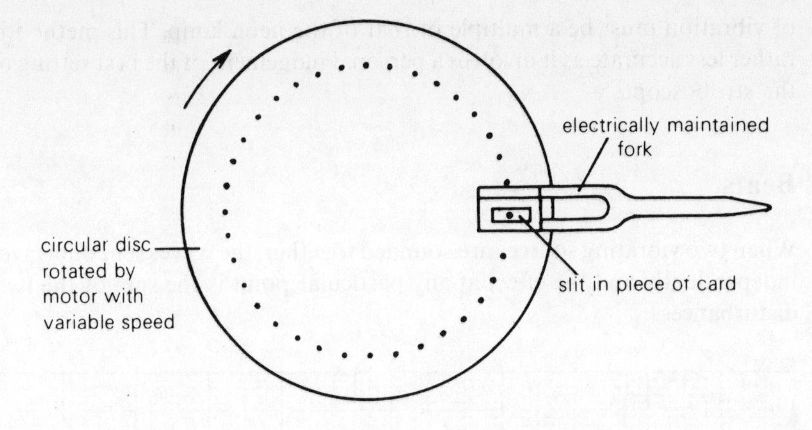

Fig. 86 Determination of the frequency of a tuning fork

interval between these will be $1/2f$ second, where f is the frequency of vibration of the tuning fork.

The circular disc is now made to rotate slowly and its speed increased until, viewed through the slits, the black dot appears to be stationary. In fact, one dot has exactly taken the place of the preceding dot, in the same time that the slits took to pass each other and return into line.

If there are m dots around the disc, and this is rotating n times a second, the time taken for one rotation of the disc is $1/n$ seconds, and so the time for one dot to replace the next is $(1/m) \times (1/n)$ seconds. Hence

$$\frac{1}{2f} = \frac{1}{mn}$$

$$\Rightarrow \quad f = \tfrac{1}{2}mn.$$

The rate of rotation of the disc, n, must be measured using a revolution-counter attached to the axle of the disc, if there is no means provided to give the speed of the motor directly.

If the speed of rotation is now increased, another apparently stationary state will occur, when $1/2f = 2/mn$, as the first dot has now been replaced by the *third*, after a time interval of $2 \times 1/mn$, and other similar situations will occur for still faster rotations.

The result obtained in this way for the frequency of the tuning fork must be corrected to allow for the fact that it vibrates more slowly than normal with the added load of the two pieces of card. This correction can only be made if a second identical unloaded fork is available, and the difference in frequency of the two forks can be found by the method of 'beats', as described below.

An alternative to this experiment is to view the vibrating tuning fork using a neon stroboscope, in a darkened room. This apparatus causes flashes of the neon lamp at a controlled rate, the frequency being shown on a meter. If the vibrating tuning fork appears to be stationary, its frequency of vibration must be a multiple of that of the neon lamp. This method is rather less accurate as it involves a personal judgement of the best setting of the stroboscope.

Beats

When two vibrating sources are sounded together, the waves sent out travel independently and the effect at any particular point is the sum of the two disturbances.

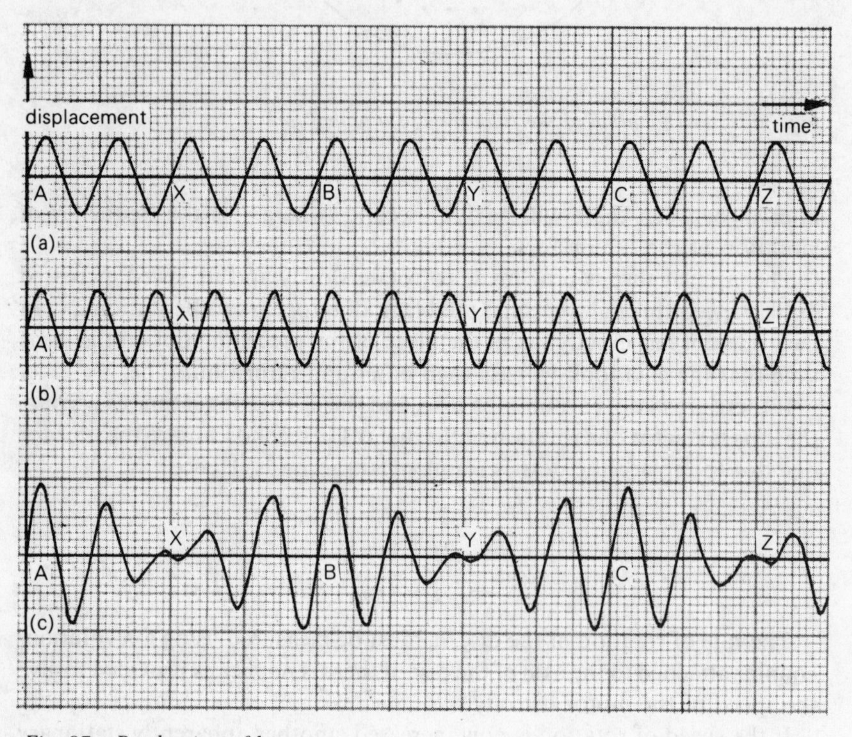

Fig. 87 Production of beats

In Fig. 87, (a) shows the displacement of a particular particle at different times as a wave passes over it, and (b) shows the same but for a wave of higher frequency. Figure (c) then shows their combined effect, found by adding together (algebraically) the displacements at each moment: it can be seen that the particle itself vibrates with the higher frequency, but the *amplitude* of its vibration varies throughout the cycle.

This variation in intensity is called 'beats', and it is heard as a throbbing of the note. Beats arise because the maximum amplitude of the combined oscillations occurs when the two components are in phase (at times A, B, C . . . in Fig. 87), and the minimum, midway between at times X, Y, Z . . . , when the component displacements are in opposite directions.

In the time interval AB, one wave has completed one more cycle than the other, and one beat has occurred. **The frequency of the beats is always equal to the difference in frequency of the original two vibrations.**

The frequency of a loaded tuning fork can thus be compared with that of a similar unloaded one by sounding the two together and counting the beats heard in a given time.

Beats are also used when one source of sound is being tuned to give the same note as a second source, both in the laboratory and in the orchestra. As the two notes approach the same pitch, the presence of beats can be recognised, and fine adjustment will then result in the beats being 'tuned out'.

EXERCISES

1 'Middle' C has a frequency of 256 Hz. What is the frequency of 'top' C (one octave higher)?

2 What beat frequency will be produced when the notes B (240 Hz) and C (256 Hz) are sounded together?

3 An A tuning fork (frequency 427 Hz) is loaded with a small piece of Plasticine, and sounded at the same time as an identical unloaded fork. If 20 beats are heard per minute, what is the real frequency of the loaded fork?

The Doppler effect

The relation between the frequency f of a sound, its wavelength λ and its velocity V is

$$V = f\lambda. \tag{1}$$

If either or both the source of sound or the observer is moving, the note heard appears to differ in pitch from the stationary condition. This is called the *Doppler effect*. A good example of this is the drop in pitch of the note of the engine of a racing car, noticed as it passes the observer.

Calculation of this apparent change in frequency must be approached in two different ways, according to the circumstances.

(*a*) *Source of sound moving, observer stationary.* If the source is moving *towards* the observer with speed u, the number of waves f sent out in 1 s occupy only a distance $(V-u)$ in air, so the actual wavelength λ' is shortened, being equal to $(V-u)/f$. This is illustrated in Fig. 88.

Since the wavelength is shorter, a larger number of waves will reach the observer per second. Applying equation (1),

$$\text{Observed frequency of waves} = \frac{V}{\lambda'}$$

$$= \frac{V}{(V-u)} f.$$

This will be a higher note than the original.

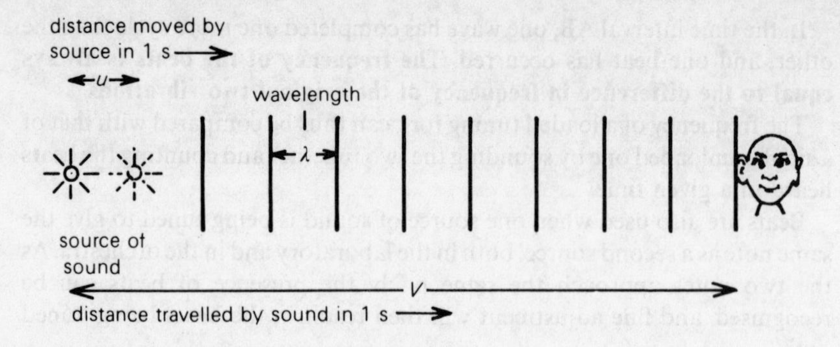

Fig. 88 Doppler effect with moving source of sound

If the source moves at the same speed *away* from the observer, the sign of u must be changed

$$\text{observed frequency} = \frac{V}{(V+u)}f \qquad \text{(a lower note)} \qquad (2)$$

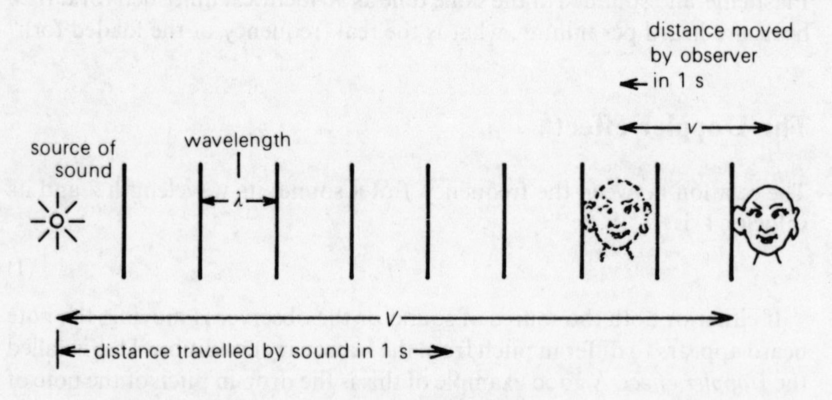

Fig. 89 Doppler effect with observer moving

(*b*) *Source of sound stationary, moving observer*. In this case, the waves are travelling through the air quite normally, but the observer, moving *towards* the source, receives more of them in 1 s. The apparent velocity of the waves relative to the observer is now $(V+v)$, so that since the wavelength is unaltered, the apparent frequency is

$$\frac{(V+v)}{V}f.$$

This means that a higher note is heard.

If the observer moves *away* from the source of sound, the relative velocity of the waves will be $(V-v)$ and their apparent frequency falls to

$$\frac{(V-v)}{V}f \qquad \text{(a lower note).}$$

The effect of *wind* must be considered separately from either of the above, for it will alter the speed with which the waves travel through the air, and the term V must be modified to include the component velocity of the wind.

Other examples of the Doppler effect

Light waves

If a source of light waves is moving towards or away from an observer, the frequency of the light can appear to alter in accordance with the Doppler effect.

This is noticeable when working with a line spectrum, in which the lines characteristic of a particular hot gas are accurately known. An apparent change in the frequency of the lines will appear as a shift of the whole spectrum. If the spectrum is displaced towards the *red* end, this means that every line appears to be of lower frequency than its normal value, hence the source must be in relative motion away from the observer. Equation (2) may be applied to the observed change in frequency to obtain a value for this relative velocity.

A similar argument applies if the lines of the spectrum are observed to be displaced towards the blue end, in which case the source of the radiation must be moving towards the observer.

Such an effect is, in fact, noticed when making astronomical observations of the distant nebulae. In all cases, a reddening of the light is found, which seems to indicate that all these nebulae are moving away from ours at high speeds.

A Doppler shift has also been observed in the laboratory, in the spectrum of the light emitted from very hot, incandescent gases (plasma). Since the individual atoms, which are the source of the light, are in random motion in all directions, the Doppler effect shows as a broadening of the spectral lines rather than a shift. The velocity of the atoms, which can be deduced from this broadening, is related by the kinetic theory to the temperature of the gas. Observations made in this way have given values for the temperature of the outer layers of the sun, and are at present being used in connection with the plasma employed in research into thermonuclear fusion.

Radio waves

When an aeroplane is in radio communication with some ground centre, a slight difference in frequency is observed between the waves sent out and those actually received. This is a Doppler effect, and it is possible to calculate the speed of the aeroplane from the shift in frequency.

An extension of this phenomenon is used with ballistic missiles and space rockets, which are tracked by radar after launch. The radar echo shows a Doppler shift in frequency which gives a value for the speed of the rocket as it moves away.

EXERCISES

Take the speed of sound in air as 330 m s^{-1}.

4 A bugler is blowing the note of top C (512 Hz) as a file of soldiers march towards him at 2 m s^{-1}. What is the frequency of the note heard by the soldiers?

5 A racing car is approaching a spectator at a speed of 200 km h^{-1}. If the actual note of its engine is of frequency 360 Hz, what drop in frequency will be noticed as the car passes the spectator?

Examination questions 18

1 (a) Explain the terms *pitch*, *frequency* and *noise.*
(b) Describe how you would measure accurately the frequency of vibration of a tuning fork. [AEB, June 1976]

2 A whistle is whirled rapidly round in a horizontal circle on the end of a piece of flexible tubing through which air is blown steadily. A distant observer hears a note which fluctuates in pitch over a range of 20 Hz once a second. The effect of doubling the rate of rotation, all other aspects remaining constant, would be that the note would fluctuate over

A less than 20 Hz once a second.
B less than 20 Hz twice a second.
C more than 20 Hz once every two seconds.
D more than 20 Hz once a second.
E more than 20 Hz twice a second. [L]

3 Derive from first principles an expression for the frequency f' of sound in still air heard by a stationary observer as a source of sound of frequency f approaches the observer with a velocity v. The velocity of sound is c.

In the case where f is 1000 Hz and c is 300 m s^{-1}, estimate how accurately f' must be measured to distinguish between source velocities of 30 and 35 m s^{-1}. [OC]

19 Vibrations of stretched springs

If a stretched wire fixed at both ends, or the string of a musical instrument, is plucked at the centre, it will vibrate in the mode shown in Fig. 90. This is the simplest form of vibration, and consists of a single loop.

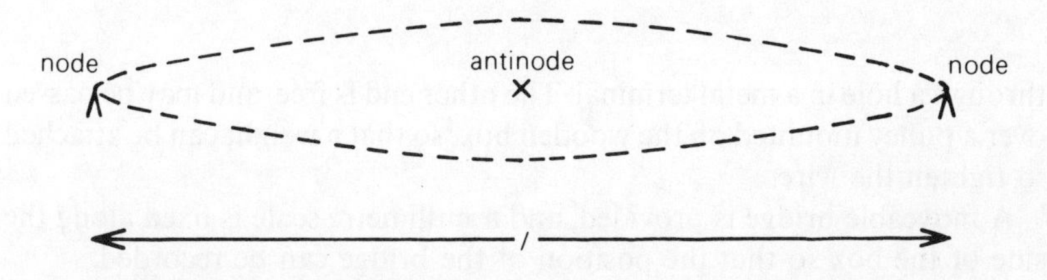

Fig. 90 Fundamental mode of vibration of a stretched string

The mid-point of the wire vibrates through the greatest distance; such a point is called an *antinode*. Points of the wire which do not move at all during the vibration are called *nodes*; since the wire is fixed at the ends, these are nodes.

The sound given out by the vibrating wire is called its *fundamental* note. The pitch of this note depends on:
(a) the length of the wire: long wires give low notes,
(b) how taut it is: taut wires give high notes, and
(c) how thick and heavy it is: thick wires give low notes.
These three relations are expressed by the equation

$$f_0 = \frac{1}{2l}\sqrt{\frac{T}{m}} \qquad (1)$$

in which f_0 is the frequency of the fundamental note, l is the length of the wire, T is its tension, in newtons, and m is the mass in kilograms per metre length of the wire.

The mathematical basis of this equation is given later (Chap. 38) but it may be verified experimentally using the sonometer, described below.

The sonometer

The sonometer consists of a stretched wire mounted on the top of a long wooden box which acts as an amplifier. One end of the wire is fixed securely

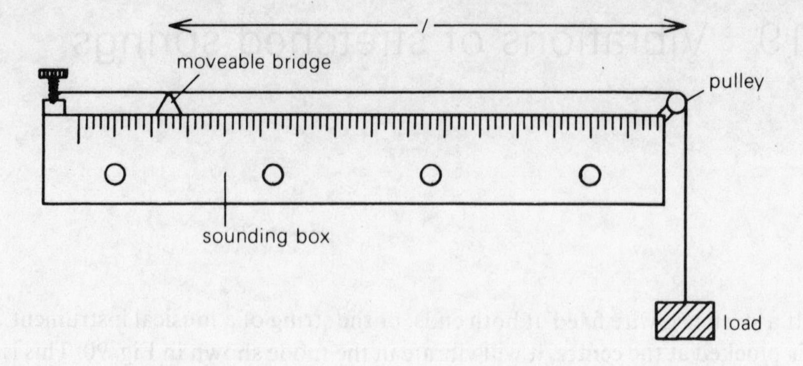

Fig. 91 The sonometer

through a hole in a metal terminal. The other end is free, and may be passed over a pulley mounted on the wooden box, so that a weight can be attached to tighten the wire.

A moveable bridge is provided, and a millimetre scale is fixed along the side of the box so that the position of the bridge can be recorded.

This apparatus can be used for three separate experiments concerned with equation (1).

Variation of fundamental frequency with length of wire

The bridge should be positioned close to the fixed end of the wire at the beginning of the experiment, and the lowest frequency available chosen from a set of tuning forks. Quite a large mass, probably several kilograms, should be hung from the free end of the wire.

The wire is then plucked at the centre and its note compared by ear with that of the chosen tuning fork. It may be necessary to alter the value of the hanging mass at this stage so that the two notes are reasonably close.

Final tuning is done by adjusting the position of the moveable bridge to vary the length of wire in use. There are two ways of judging the best position. If the experimenter has a musical ear, agreement between the two notes can be found by using the method of beats, as described in the previous chapter (p. 155). Alternatively, the vibration of the wire may be studied by using a small paper rider, i.e. a piece of paper measuring about 10 mm × 3 mm, folded in half and placed on the wire at its mid-point. If the tuning fork is sounded and then held with its stem touching the wire where it crosses the bridge, the wire will be set into vibration sympathetically with the fork, and the movement of the paper rider will show the amplitude of the vibrations. When the length of the wire has been correctly adjusted, it will be in resonance with the tuning fork and the vibrations will be clearly seen.

The length l of the vibrating portion of the wire should be recorded against the frequency f_0 of the tuning fork used. The experiment is then repeated using the higher frequency tuning forks in turn, but the mass suspended from the wire must not be altered.

The graph of l against $1/f_0$ should be a straight line.

Variation of fundamental frequency with tension in wire

If the pulley is smooth, the tension in the wire will be equal to the weight of the load hanging from it.

The experiment will again use the whole range of tuning forks, so it is convenient to start with the length l of wire corresponding to the lowest frequency fork, as found in the previous experiment. Higher frequency notes will then be obtained from the wire by increasing its tension, keeping the same length in use throughout.

A simple way to make the final adjustment to the hanging mass when the wire and a particular tuning fork are nearly matched is to add a scale-pan containing sand. The quantity of sand can be varied until satisfactory agreement between the notes is obtained, and the scale-pan and sand should then be weighed together on chemical scales; a reading to the nearest gram is amply precise.

The graph of the total hanging mass against $(f_0)^2$ should be a straight line.

Use of different sonometer wires

Investigation of the factor m in equation (1) (mass per metre length of the wire) is most easily carried out by performing one of the two experiments just described, using each wire in turn. The slopes of the resulting graphs will differ because of the difference in mass of the wires, and the ratio of the values of m may be deduced from the two graphs if desired.

Harmonics

It is possible for a stretched wire to vibrate in a more complicated pattern than the single loop so far considered. Figure 92 shows some of the possible modes of vibration.

The first two modes illustrated can be produced by plucking or bowing the wire at its mid-point and then 'stopping' the vibration at one of the points marked N, by touching the wire gently with a finger. This creates a node at N and so suppresses the fundamental mode. Antinodes will occur at the points marked A, halfway between the nodes.

The modes of vibration shown in the third and fourth diagrams of Fig. 92 cannot occur if the wire is plucked at the centre. It must be set into vibration at one of the points required to be antinodes, e.g. to set up a vibration pattern of two equal loops, the wire must be plucked one-quarter of the way along and not at the mid-point.

The notes produced by these more complicated modes of vibration are

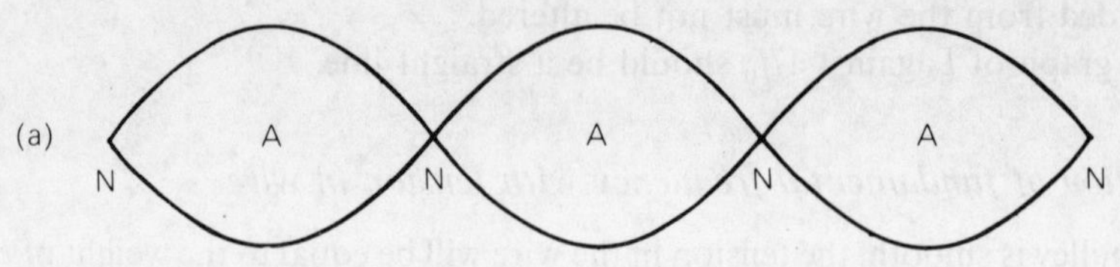

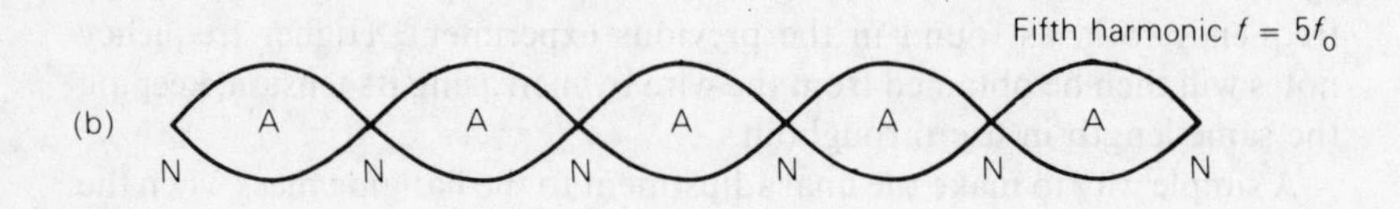

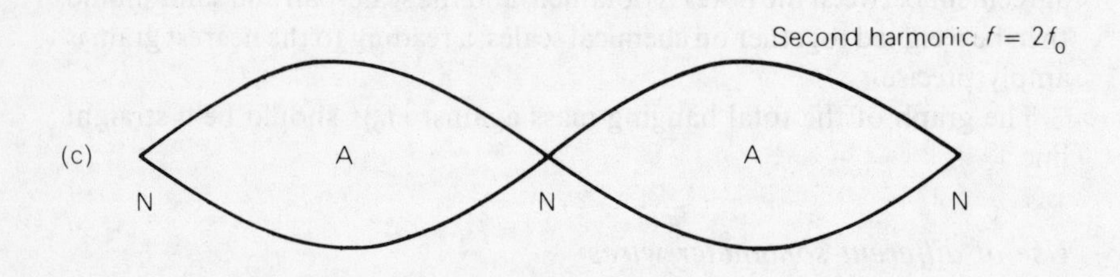

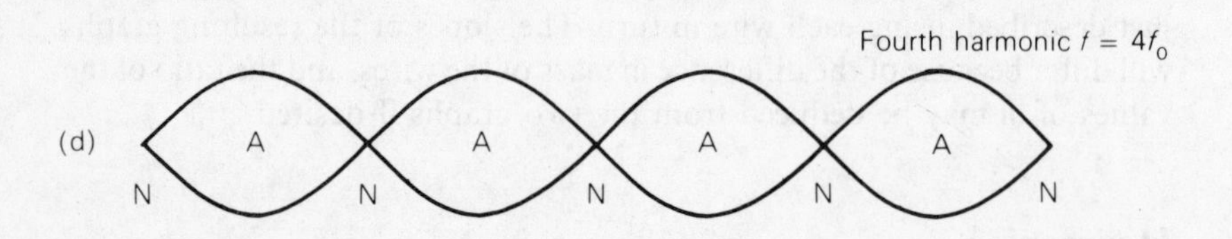

Fig. 92 Higher modes of vibration of a stretched string

called the *harmonics* of the fundamental note, and their frequencies are in simple ratio to the fundamental frequency.

Those harmonics in which an antinode occurs at the mid-point of the wire are sometimes called *overtones* of the fundamental note. However, the musician uses the term 'overtones' to refer to the extra frequencies added to the note by the amplification system, e.g. the resonance box of the violin or the sounding-board of the piano. It is these overtones which give the particular *quality* associated with the music of an instrument.

Figure 93 shows the actual waveform of a violin note, which can be studied if the plates of an oscilloscope are connected across the output of a microphone. The difference in shape of this waveform from the simple sine curve produced by a tuning fork demonstrates the presence of these overtones.

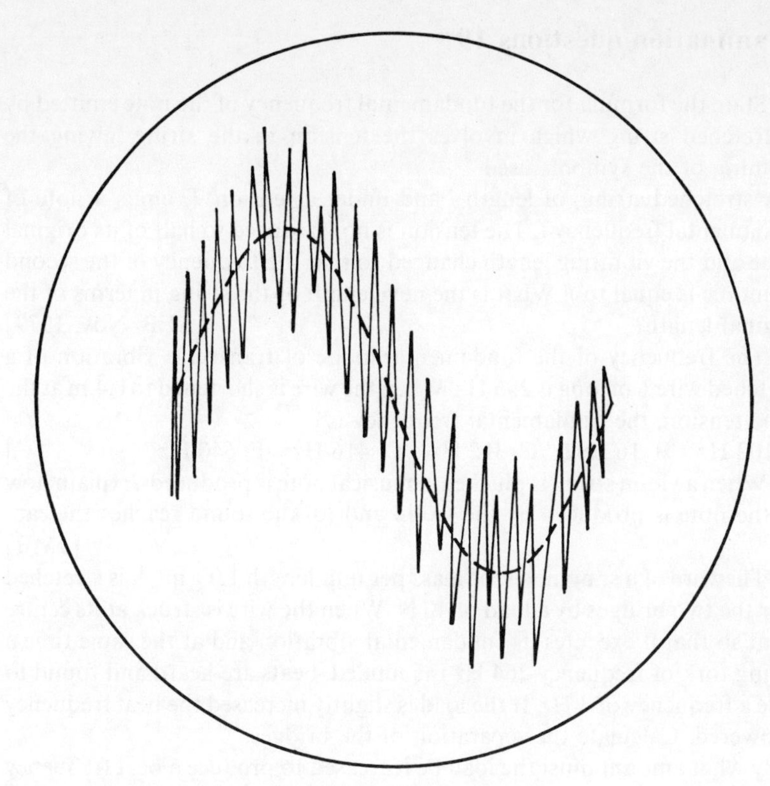

Fig. 93 Waveform of the note produced by a violin

EXERCISES

1 A mass of 10 kg is suspended from the end of a vertical wire 1 m long, whose mass per metre is 10^{-2} kg. If the wire is plucked, find the frequency of its fundamental note.

2 A vibrating stretched wire gives out a fundamental note of 240 Hz. If the tension in the wire is doubled, what will be the new frequency of its note?

3 Calculate the frequencies of the first and second overtones of middle C (256 Hz).

4 A stretched wire of length 1 m gives out a fundamental note of frequency 200 Hz. If the vibration is stopped at a point 0.25 m from one of the ends, what will then be the frequency of the predominating note?

Stationary waves

The modes of vibration which have been considered in this chapter are examples of *stationary* (or *standing*) *waves*. They are actually the result of the disturbance caused by the plucking of the wire travelling in both directions down its length and being reflected back at the fixed ends. Interference occurs between the two reflected waves, and the vibrating loops which result are the interference pattern.

Examination questions 19

1 State the formula for the fundamental frequency of the note emitted by a stretched string which involves the tension in the string, giving the meaning of the symbols used.

A stretched string, of length l and under a tension T, emits a note of fundamental frequency f. The tension is now reduced to half of its original value and the vibrating length changed so that the frequency of the second harmonic is equal to f. What is the new length of the string in terms of the original length? [AEB, Nov. 1979]

2 The frequency of the fundamental mode of transverse vibration of a stretched wire 1 m long is 256 Hz. When the wire is shortened to 0.4 m at the same tension, the fundamental frequency is

A 102 Hz B 162 Hz C 312 Hz D 416 Hz E 640 Hz [C]

3 When a violin string is plucked, a musical note is produced. Explain how (a) the note is produced by the violin and (b) the sound reaches the ear. [JMB]

4 The wire of a sonometer, of mass per unit length 1.0 g m^{-1}, is stretched over the two bridges by a load of 40 N. When the wire is struck at its centre point so that it executes its fundamental vibration and at the same time a tuning fork of frequency 264 Hz is sounded, beats are heard and found to have a frequency of 3 Hz. If the load is slightly increased the beat frequency is lowered. Calculate the separation of the bridges.

By what amount must the load be increased to produce a beat frequency of 10 Hz if the same tuning fork is used? [L]

5 (a) The four strings of a violin are of different gauge, and possibly of different material; they are under approximately the same tension. Their vibrations are transferred through the bridge and sound-post to the body of the instrument.

(i) Explain the mechanisms by which, when a given string is excited, a sound of definite pitch reaches the ear.

(ii) How does the associated frequency depend on the vibrating length of the string which is excited?

(iii) How does the quality of the note heard depend on the place at which the string is excited, and whether it is plucked or bowed?

(b) The fundamental frequency of a sonometer wire is 50 Hz. Its oscillations are maintained by the output of a signal generator at 100 Hz, and it is observed in the light from a stroboscopic flashing lamp. With the help of diagrams, describe and explain briefly what you would expect to see if the flash frequency of the lamp were (i) 25 Hz, (ii) 50 Hz, (iii) 100 Hz, (iv) 200 Hz.

(c) If the wire is now maintained in oscillation at a frequency of exactly 50 Hz and the lamp is adjusted to flash at 50.1 Hz, what would you expect to see? Explain your answer. [O]

20 Vibrations of air columns

Sound waves consist of regular variations of pressure transmitted from particle to particle through a medium. These changes of pressure are set up by the vibrations of the source. The regularity of the variations (considering a musical note as distinct from *noise*) means that a condition of high pressure, called a *compression,* will be followed by a condition of low pressure, called a *rarefaction,* as the wave travels along.

This type of wave is called a *longitudinal* wave, since the particles involved in it actually vibrate in the direction in which the wave is travelling (and not transversely, as in the case of a stretched string).

The distance between the centres of two compressions, or of two rarefactions, is defined as the wavelength of a compression wave.

Variation of pressure in a compression wave

Figure 94 shows the variation of pressure with distance along a compression wave. The line of dots above the curve shows the positions of the particles carrying the wave.

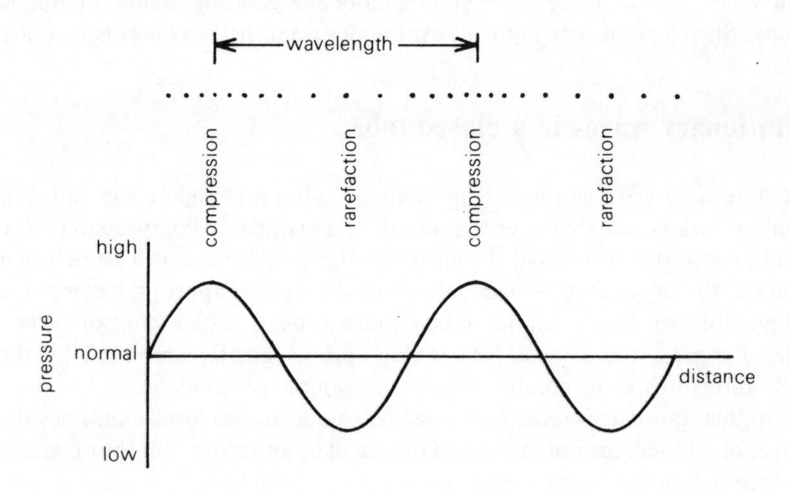

Fig. 94 Graph of pressure variation along a compression wave

If this type of wave is reflected by a fixed barrier, a compression is reflected back as a compression and a rarefaction is reflected as a rarefaction. Interference between the incident and reflected waves sets up a *stationary* wave.

Nodes

At the particular point lying *half a wavelength* from the reflector, a compression which has been reflected back will arrive at the same moment as the next compression of the incident wave; at this moment the pressure at this point will be twice as great as in a single compression. Similarly when two rarefactions arrive simultaneously, the combined pressure will be twice as low as in a single rarefaction. The forces acting on a particle at this point (due to its neighbours) will always be equal and opposite, so that the particle will never move. Such a point in a stationary wave is a *node*. Further nodes occur at a spacing of half a wavelength.

Antinodes

At a different point, only *one-quarter of a wavelength* from the reflector, a compression due to the reflected wave will arrive at the same moment as a rarefaction due to the incident wave. The two pressure variations will cancel, and the point will be at *normal atmospheric pressure*. This will apply at all times since the two waves always reach that particular point half a cycle out of step.

A particle at this point will experience two forces in the same direction as the waves pass it, hence it will vibrate more energetically than in the single wave. Such a point is an *antinode*. Antinodes occur halfway between nodes.

Stationary waves in a closed tube

A stationary wave can be set up in an air column closed at one end. If a tuning fork is sounded over the mouth of a cylinder, a compression wave will be sent down through the air inside the cylinder and will be reflected back at the closed end. Provided that the tube is of the appropriate length, it is possible for the air column to be set into a steady state of motion. When this happens, the note of the tuning fork is greatly amplified by the vibrations of the air column; this is an example of *resonance*.

In this stationary wave, there must be a node at the closed end, since the layer of air there cannot move, and there will be an antinode at the open end where the air can move freely.

The fundamental stationary wave will have only this one node and antinode, as this is the simplest possible mode of vibration. The dotted lines shown in Fig. 95(a) are a conventional way of representing this condition. It

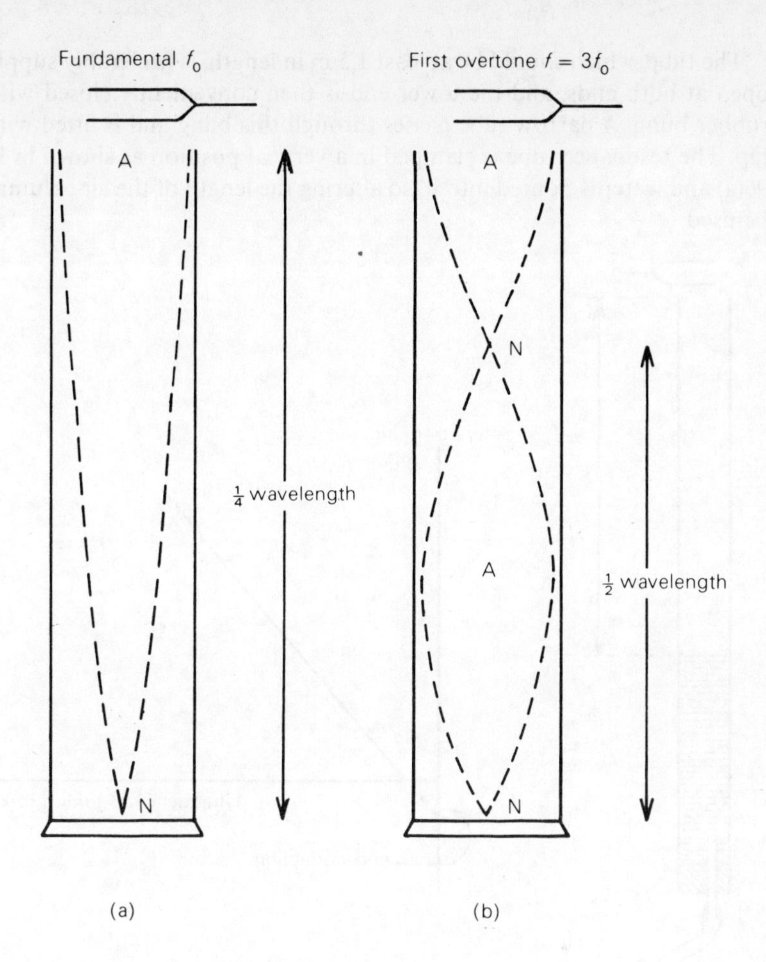

Fig. 95 Stationary waves in an air column closed at one end

should be noted that **the length of the air column is equal to $\frac{1}{4}$ wavelength of the fundamental note.**

Figure 95(b) illustrates the condition of the same air column when it is vibrating with the frequency of the harmonic called the first overtone. It can be seen from the figure that the air column now contains $\frac{3}{4}$ wavelength of the stationary wave, hence the frequency of this note is 3 × (frequency of the fundamental).

Measurement of the velocity of sound using a resonance tube

Using the relation $v = f\lambda$, it is possible to deduce a value for the velocity of sound in air by measuring the wavelength λ of the stationary wave corresponding to a particular frequency f. This is done by varying the length of a closed air column until it shows resonance with a standard tuning fork.

The tube, which should be at least 1.5 m in length, is preferably supplied open at both ends, and the lower end is then conveniently closed with a rubber bung. A narrow tube passes through this bung and is fitted with a tap. The resonance tube is clamped in a vertical position as shown in Fig. 96(a) and water is poured into it, so altering the length of the air column to be used.

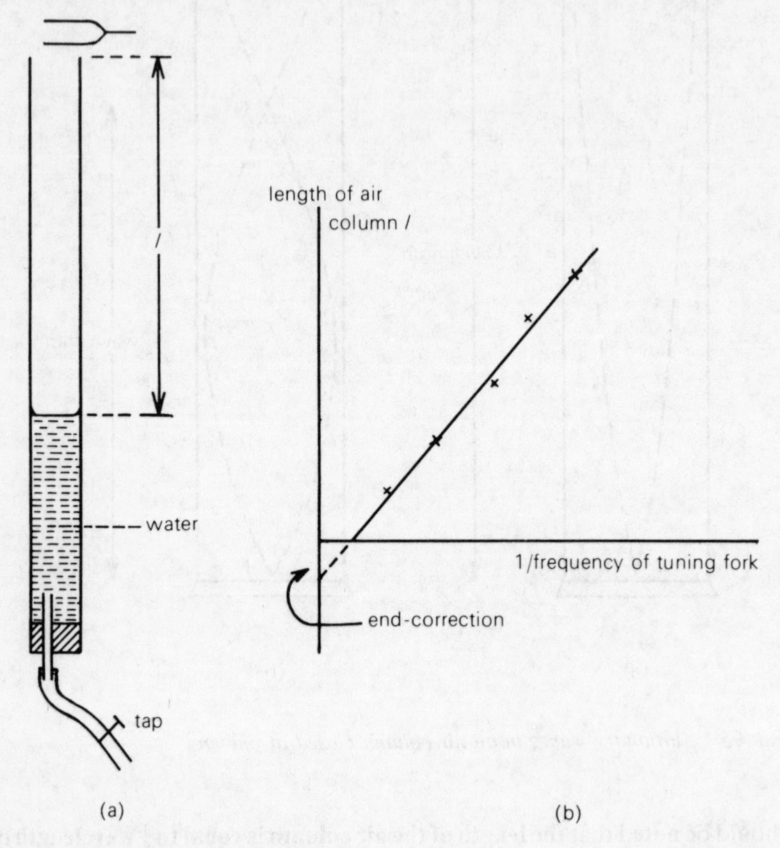

Fig. 96 Determination of the velocity of sound using a resonance tube (a) apparatus used, (b) graph of results

One of the tuning forks is now struck on a rubber pad and held over the mouth of the resonance tube. The tap is opened so that water runs out slowly, and it will be found that the note of the tuning fork is amplified clearly as the water falls to a particular level, L. This is the resonance position required. If the water level is adjusted carefully, it should be possible to determine the optimum position of L to about a millimetre (the use of a pipette may even be recommended in the final stages).

The length l of the air column is measured from the mouth of the tube to the surface of the water. This length will be the distance between a node and the nearest antinode of the stationary wave, and is equal to $\frac{1}{4}$ wavelength. The frequency of the tuning fork used must be recorded.

The experiment should be repeated using different tuning forks so that a table of corresponding values of l and f is obtained. The graph of l against $1/f$ should be a straight line as shown in Fig. 96(b).

The line will not, in practice, pass through the origin of the two scales. This is because the antinode at the open end occurs actually a little distance above the mouth of the tube, from which l was measured. This distance is known as the *end-correction* of the tube, and may be found from the intercept on the l-axis.

Since $\lambda = 4l$, the slope of the line is $fl = \frac{1}{4}f\lambda = \frac{1}{4}v$. This gives a result for the velocity of sound which is usually surprisingly accurate.

If the resonance tube is sufficiently long, it may be possible to find a second resonance position L′ for a particular tuning fork, when the air column is vibrating in the mode shown in Fig. 95(b). The distance LL′ between these two positions will be $\frac{1}{2}\lambda$, and the value for the velocity of sound can be calculated from this result, perhaps more easily than by the use of a graph.

Although the experiment can be carried out using a small loudspeaker attached to an a.c. oscillator, the difficulty of reading the scale of the oscillator leads to a general loss of accuracy compared to the results obtainable using standard tuning forks.

Stationary waves in an open tube

Stationary waves can also be set up in an air column open at both ends. In the fundamental mode, antinodes will occur at the two open ends and a single node at the centre as shown in Fig. 97(a), and **the wavelength of the**

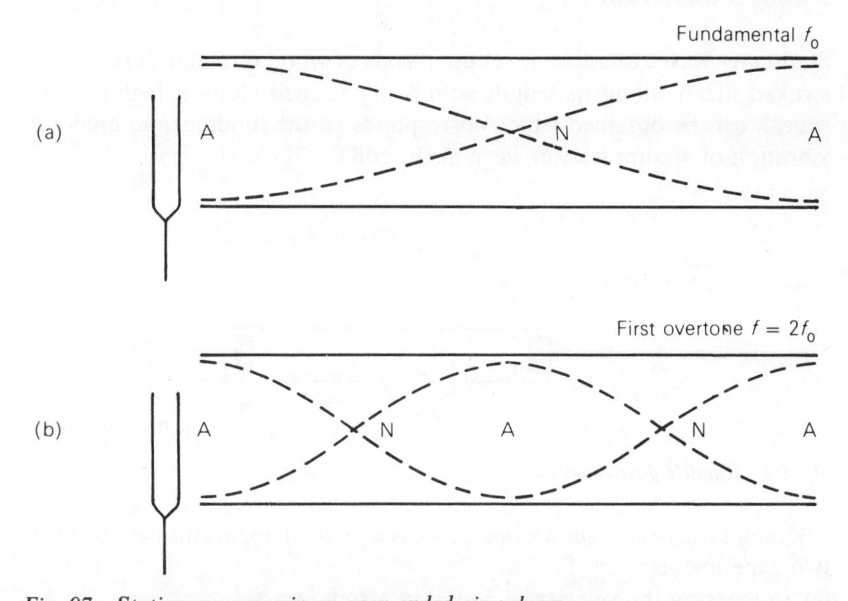

Fig. 97 Stationary waves in an open-ended air column

fundamental note will be 2 × length of the tube (if the end-corrections are included). An open-ended air column has a fundamental frequency twice as high as a closed air column of the same length.

Figure 97(b) shows the mode of vibration of the harmonic called the first overtone. In this case, the frequency of the note is 2 × fundamental frequency and is one octave higher.

Experimental work with open-ended air columns is carried out by attaching a stiff cylindrical paper collar over one end of the tube, using elastic bands. The position of this collar is then adjusted to vary the length of the tube when searching for resonance. It is not quite such a simple technique as that used with the closed tube, which is therefore preferred as a means of determining the velocity of sound.

EXERCISES

Take the velocity of sound in air as 330 m s^{-1} where necessary in these questions.

1 What is the length of a closed air column which will give a fundamental note of 330 Hz?

2 A closed air column 0.20 m long is in resonance with a tuning fork of frequency 384 Hz. Assuming that the air column is vibrating in its fundamental mode, deduce the value of the end-correction involved.

3 What is the length of an open-ended air column which will give a fundamental note one octave higher than a closed air column 0.50 m long?

4 Calculate the separation of the nodes in a closed air column which is sounding its first overtone, a note of frequency 450 Hz.

Kundt's dust-tube

Stationary waves can also be set up in rods of wood or metal. If the rod is stroked firmly along its length with a dry resined cloth, a high-pitched squeak can be obtained which corresponds to the fundamental mode of vibration of a compression wave in the rod.

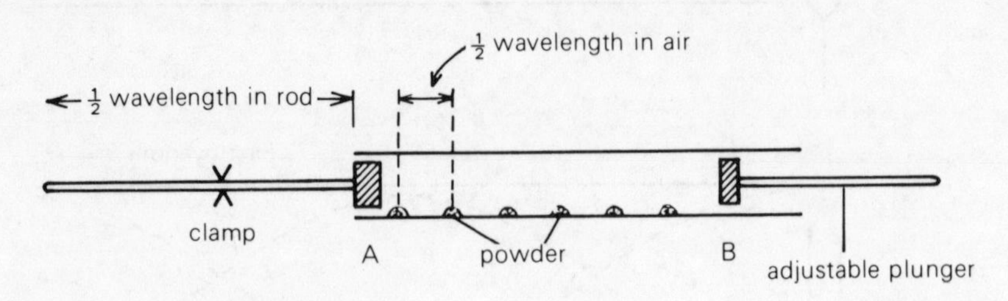

Fig. 98 Kundt's dust-tube

Kundt's dust-tube, shown in Fig. 98, is a piece of apparatus designed for two experiments:

(a) to measure the velocity of sound in a rod, and

(b) to measure the velocity of sound in a gas other than air.

The tube itself is made of glass, about 1.5 m long and 50 mm in diameter. The sounding-rod, usually supplied with it, is about 1 m long and has a stout cork disc A at one end, which just fits inside the glass tube. A second similar disc B, on the end of a plunger, may be moved along inside the other end of the tube, to vary the length of the air column in use.

Before the experiment starts, the glass tube should be carefully cleaned and dried. A thin layer of some light powder (such as lycopodium powder) is then laid down the length of the tube; this is simply done by scattering the powder evenly along a metre rule which is passed into the tube and then turned over. The dust-tube is held firmly in a horizontal position.

The sounding-rod should be clamped exactly at its mid-point (thus creating a node there) with the disc A a short distance inside the tube. When the rod is rubbed hard enough to squeak, the vibrations pass down its length to A and cause sympathetic vibrations in the air column inside the tube. The position of disc B is adjusted until resonance is obtained. Since the note of the rod is of high frequency, its wavelength is fairly short and B should not need to be moved through more than about 50 mm.

At resonance, the movement of the air particles inside the tube is quite large, and the fine powder rises in swirls which settle at the nodes of the stationary wave, producing a series of ridges. The distance between these ridges is measured with a metre rule, and is equal to $\frac{1}{2}\times$ (wavelength of the note in air).

Since the frequency of the vibrations in the air column is the same as that of the vibrating rod,

$$\frac{\text{velocity of sound in air}}{\text{wavelength in air}} = \frac{\text{velocity of sound in rod}}{\text{wavelength in rod}}. \qquad (1)$$

The rod is assumed to be vibrating in its fundamental mode, and the wavelength of this vibration is taken as $2\times$ length of rod (as in the case of an open-ended air column). Equation (1) may therefore be used either (a) to give the velocity of sound in the material of the rod, assuming its value in air, or (b) to deduce the velocity of sound in a gas other than air, by repeating the experiment using the same rod but filling the tube with the required gas.

Examination questions 20

1 What are (i) progressive waves, (ii) stationary waves, (iii) nodes and (iv) antinodes?

Describe a stationary wave method of measuring the speed of sound in air. [AEB, Nov. 1974]

2 An organ pipe of effective length 0.6 m is closed at one end. Given that the speed of sound in air is 300 m s^{-1}, the two lowest resonant frequencies are

A 125,250 Hz B 125,375 Hz C 250,500 Hz
D 250,750 Hz E 500,1000 Hz [C]

3 A resonance tube open at both ends and responding to a tuning fork
A always has a central node.
B always has a central antinode.
C always has an odd number of nodes.
D always has an even number of nodes.
E always has an odd number of nodes + antinodes. [C]

4 (a) Distinguish between a progressive wave and a stationary wave. Your answer should refer to energy, amplitude and phase.
(b) A small loudspeaker emitting a pure note is placed just above the open end of a vertical tube, 1.0 m long and about 40 mm in diameter, containing air. The lower end of the tube is closed. Describe in detail and explain what is heard as the frequency of the note emitted by the loudspeaker is gradually raised from 50 Hz to 500 Hz. (You may assume that the speed of sound in air is 340 $\mathrm{m\,s^{-1}}$ and need make only approximate calculations). [L]

5 A brass rod clamped at its mid-point excites an air resonance column of the 'Kundt's tube' type. The length of the rod is 1.5 m and the distance between successive antinodes in the air column is 0.125 m.

The ratio $\dfrac{\text{speed of sound in brass}}{\text{speed of sound in air}}$ is

A 0.8 B 1.2 C 5.6 D 12.0 E 16.4 [O]

Alternating Current

21 Alternating current

Measurement of a.c.

If an alternating current is passed through a d.c. moving-coil ammeter, the pointer remains on zero although it can be seen to be trembling as the coil attempts to follow the changes of direction of the current.

Specially designed a.c. meters are available, based either on magnetic repulsion (moving-iron) or on the heating effect of the current. These are cheap, but not very sensitive. They have a 'cramped' (non-linear) scale, as illustrated in Fig. 99, since both effects depend on the *square* of the current. This can be an advantage for certain uses, i.e. if readings will be restricted to a particular part of the scale.

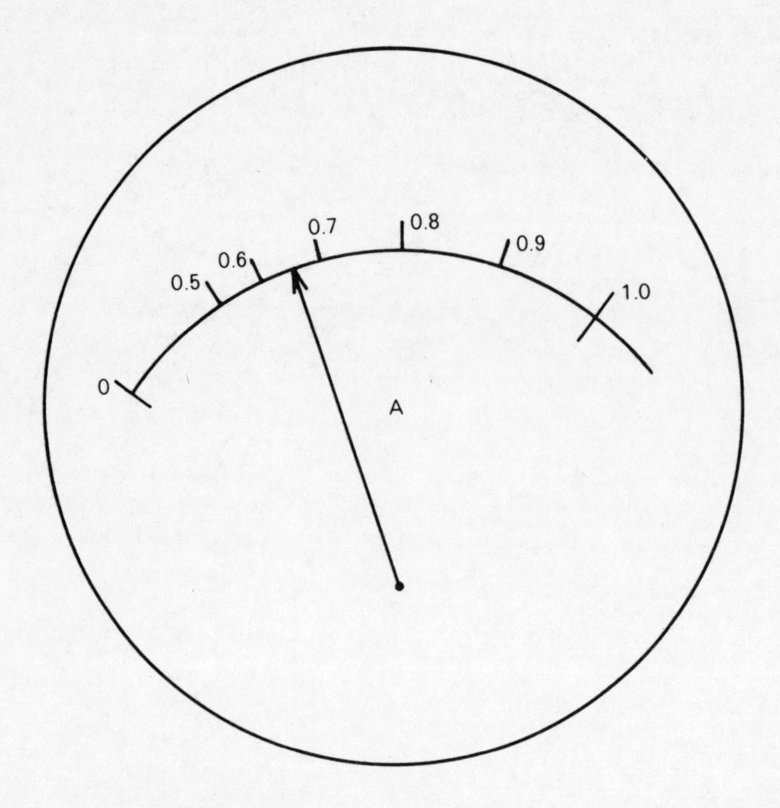

Fig. 99 The scale of an a.c. ammeter

For accurate work, it is best to rectify the a.c. and then measure it with a moving-coil meter. Rectification will be considered in Chap. 29.

Root-mean-square value

This 'average' value of an alternating current is defined as **the value of the direct current which would have the same heating effect in a resistance.**

Since the heating effect depends on the square of the current, this definition means

$$(\text{equivalent d.c.})^2 = \text{average value of (a.c.)}^2$$
$$\text{or } \mathbf{equivalent\ d.c.} = \sqrt{\mathbf{average\ of\ (a.c.)^2}}.$$

This is known as the root-mean-square or r.m.s. value.

Direct-reading a.c. meters are calibrated on the basis of this relation, and show the r.m.s. value of the current.

The mains supply

The alternating current supplied by the Central Electricity Generating Board is generated by alternators as described in Chap. 8. Its wave form is *sinusoidal*, following an equation of the form deduced earlier (p. 64):

$$E = C\omega \sin \omega t$$

where C is a constant, deducible from the design of the generator, E is the e.m.f. generated, and ω is the angular velocity of the coil of the generator, in radians per second.

The frequency of the mains supply is stated as 50 Hz, though in practice this varies a little according to the load. Any drop in frequency in peak hours is made up during the night so that electric clocks and timers are maintained correctly.

Thus ω may be taken as $50 \times 2\pi = 100\pi$ rad s^{-1}.

The voltage of the mains is stated as 240 V; this is the root-mean-square value.

Calculation of the r.m.s. of sinusoidal voltage is a simple application of calculus, and gives the result

$$V_{\text{r.m.s.}} = \frac{1}{\sqrt{2}} \times (\text{peak value of voltage}).$$

The peak of the mains supply is therefore about 340 V (Fig. 100).

The national grid system

Mains electricity is distributed throughout the country by means of the National Grid. This was designed so that, should a breakdown occur

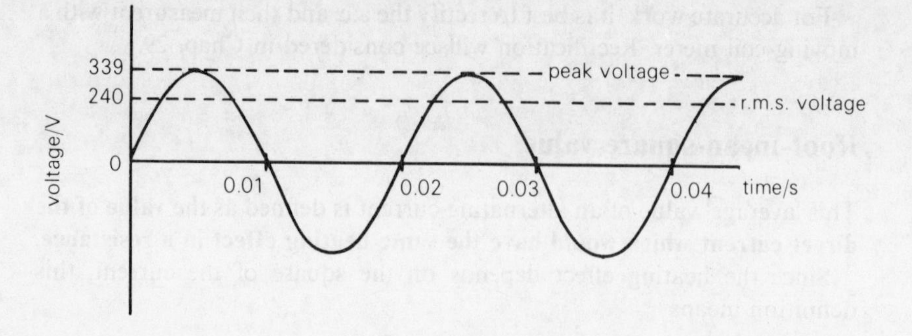

Fig. 100 Graph showing the variation of voltage with time for the C.E.G.B. mains supply

anywhere, the supply could be maintained by drawing electricity from other areas. All generating stations of the C.E.G.B. feed into the grid system, and it is also linked with France.

Most power stations generate electricity at about 450 V (r.m.s.). It is not economic to carry it over long distances at this voltage, however, for the following reasons.

Power is lost during transmission due to the production of heat in the wires. These losses are proportional to the resistance of the wires, but also to the square of the current. The resistance can be reduced by using copper wires, but this is an expensive metal and the wires must be kept thin.

Power losses are therefore minimised by transmitting at low current. To provide adequate power for the consumer, this means that the voltage of the supply must be high, and in fact over most of the distance it is transmitted at 132 000 V. Even higher voltages are being considered for the new supergrid network.

The use of such high voltages means that the supply cannot be carried by underground cables as the cost of adequate insulation for these would be too great. Underground cables are used at the beginning of the distribution network, and at the end when the voltage has been stepped down at the various sub-stations, but over most of the distance the current is carried by high-tension wires supported on pylons. Much opposition to this scheme is raised by people who do not appreciate the economics behind it.

EXERCISES

1 A bicycle dynamo operates a 6 V, 3 A lamp bulb. What is the peak value of the voltage supplied?

2 The U.S. electricity supply is quoted as 115 V, 60 Hz. Illustrate this by drawing a sketch graph of the type shown in Fig. 100.
What will be (i) the peak voltage, and (ii) the period of the wave cycle?

Examination questions 21

1 Half-wave rectification of an alternating sinusoidal voltage of amplitude 200 V gives a waveform as shown in Fig. 146(b) (p. 251). The r.m.s. value of the rectified voltage is
A 0 V B 70.7 V C 100 V D 141.4 V E 200 V. [C]

2 The r.m.s. value of a sinusoidal alternating e.m.f. of peak value 4 V is
A 0 B 2 V C $(2\sqrt{2})$ V D 4 V E $(4\sqrt{2})$ V [O]

22 A.C. and inductance

If an alternating current passes through a coil of wire, its self-inductance will give rise to a back e.m.f. as described in Chap. 8. This e.m.f. opposes the current and so the coil has much the same effect as a resistance. The opposition due to an inductance is called its *reactance*, and it is measured in ohms.

Supposing the current I to be sinusoidal, it will have a waveform following an equation

$$I = C\omega \sin \omega t$$

as discussed in the preceding chapter (p. 175).

It is now more appropriate to replace the term ω, the angular velocity of the generator rotor, by $2\pi f$, where f is the frequency of alternation of the supply. This will apply whenever the *use* of the current is being considered rather than its production.

Since the maximum value of the current, I_{max}, will be the value of I when the variable $\sin 2\pi ft = 1$,

$$I_{max} = C\,.\,2\pi f$$

and the equation for the current may be written

$$I = I_{max} \sin 2\pi ft. \qquad (1)$$

Now, from equation (2) of Chap. 8 (p. 66),

$$\text{back e.m.f.} = -L\,.\frac{dI}{dt}$$

where L is the self-inductance of the circuit, so that in this case,

$$\text{back e.m.f.} = -L\,.\,2\pi f I_{max} \cos 2\pi ft.$$

An applied voltage V, equal and opposite to this, will be required to maintain the current I through the inductance;

$$V = 2\pi fL\,.\,I_{max} \cos 2\pi ft. \qquad (2)$$

Figure 101 shows the variation with time of the current through the coil (broken curve), and the applied voltage needed to produce this current (full curve). The two graphs are out of step, because
(i) when the current is growing most rapidly, as at point A, the maximum voltage will be required to overcome the back e.m.f.,
(ii) when the value of the current is stationary, as at B, the back e.m.f. will

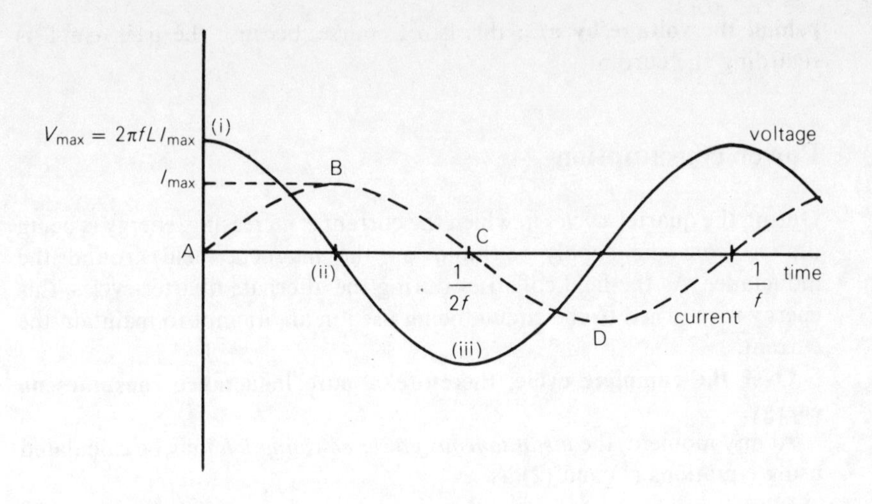

Fig. 101 Graphs showing the effect of inductance on an a.c. supply

be zero and no applied voltage will be necessary,
(iii) when the current is falling, between B and D, the back e.m.f. will try to maintain it, and an applied voltage in the reverse direction will be required to overcome this.

Phase difference between current and voltage

Since, for any angle θ,

$$\cos\theta = \sin\left(\frac{\pi}{2} - \theta\right) \qquad \text{(complementary angles)}$$

$$= \sin\left[\pi - \left(\frac{\pi}{2} - \theta\right)\right] \qquad \text{(supplementary angles)}$$

$$= \sin\left(\theta + \frac{\pi}{2}\right),$$

equation (2) may be re-written as

$$V = 2\pi f L . I_{max} \sin\left(2\pi f t + \frac{\pi}{2}\right).$$

Comparing this with equation (1)

$$I = I_{max} \sin 2\pi f t,,$$

emphasises the similarity between the two curves of Fig. 101. The term $\pi/2$ occurring inside the bracket shows the amount by which the two curves are out of step, which is called their *phase difference*. **The current is said to lag**

behind the voltage by $\pi/2$; this is, of course, because the back e.m.f. is retarding the current.

Power consumption

During the quarter-cycles in which the current is increasing, energy is being drawn from the supply to build up the magnetic field around the inductance. As the field collapses during the alternate quarter-cycles, this energy is returned to the circuit, being used in an attempt to maintain the current.

Over the complete cycle, therefore, **a pure inductance consumes no energy**.

At any moment, the *instantaneous power consumption* may be calculated using equations (1) and (2) as

$$\begin{aligned} \text{current} \times \text{voltage} &= I_{max} \sin 2\pi f t \, . \, V_{max} \cos 2\pi f t \\ &= \tfrac{1}{2} I_{max} \, V_{max} \sin 4\pi f t. \end{aligned}$$

Reactance

The *reactance* of the inductance is defined as

$$\frac{\textbf{maximum value of voltage across coil}}{\textbf{maximum value of current through coil}}$$

or

$$X_L = \frac{V_{max}}{I_{max}}.$$

Considering equation (2),

$$V_{max} = 2\pi f L I_{max}.$$

Hence **reactance** $\boldsymbol{X_L = 2\pi f L.}$

High-frequency currents have great difficulty in passing through an inductance. A coil made up of a large number of turns wound on a soft-iron core, having high inductance L, is called a *choke*. It can be used in communication circuits to separate out the comparatively low-frequency speech vibrations from the higher-frequency carrier wave.

EXERCISES

1 A.C. of frequency 60 Hz flows through a pure inductance. If time is to be reckoned from a moment at which the current is zero, give (i) the times at which the current through the coil will be a maximum, and (ii) the times at which the voltage across the coil will be a maximum.

2 A.C. of frequency 50 Hz flows through a circuit containing a coil of self-inductance 2 henry. If the maximum value of the current is 2 A, calculate (i) the maximum voltage developed across the coil, and (ii) the voltage across the coil when the current through it has fallen to 1 A.

3 An inductance of 2 henry is supplied with 50 Hz alternating current whose maximum value is 3 A. Find (i) the points in each cycle at which maximum power is being absorbed by the inductance, and (ii) the value of this power consumption.

4 What is the reactance of a coil whose inductance is 1 millihenry, when supplied with a.c. of frequency (i) 50 Hz, (ii) 1000 kHz?

Mutual inductance

Mutual inductance is the term used when *two* coils are magnetically linked, so that a changing current in one coil induces an e.m.f. in the other. The mutual inductance of the two coils is defined in the same way as self-inductance (p. 66), and is similarly measured in henrys.

The transformer

A transformer consists generally of two coils wound one on top of the other on a soft-iron yoke, so that their mutual inductance is high. Since the purpose of a transformer is to step up (or down) an alternating voltage, the coils have very different numbers of turns.

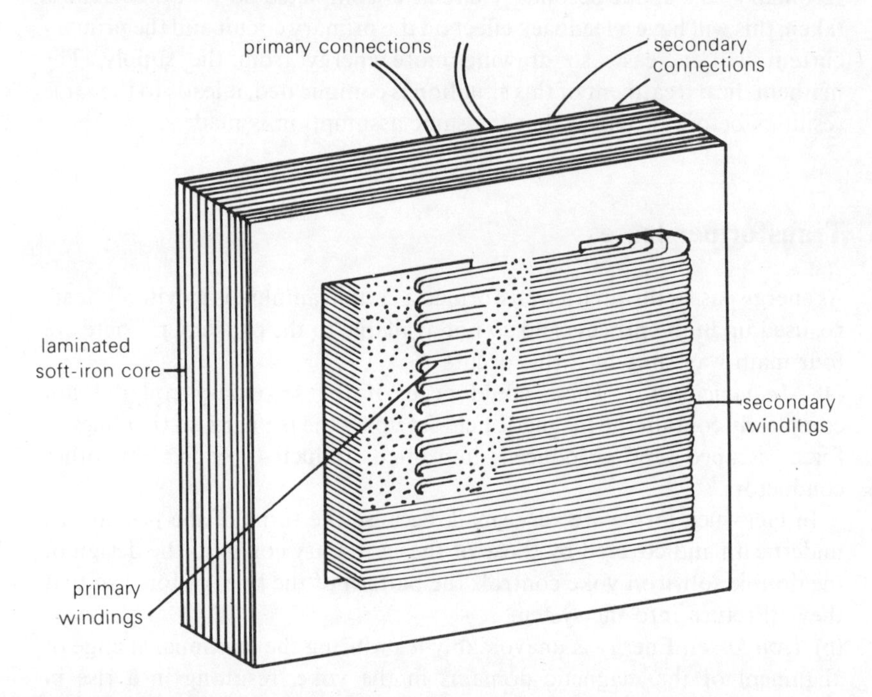

Fig. 102 Cut-away view of a small transformer

Suppose that the a.c. supplied to the primary coil sets up a magnetic flux B in the core, which will, of course, vary with time. This changing flux will induce an e.m.f. E_2 in the secondary coil, and at the same time, a back e.m.f. E_1 in the primary.

If there are N_1 turns in the primary coil,

$$E_1 \propto N_1 \frac{dB}{dt}$$

and similarly

$$E_2 \propto N_2 \frac{dB}{dt}$$

if there are N_2 turns in the secondary coil. Hence

$$\frac{E_1}{E_2} = \frac{N_1}{N_2}.$$

If the resistance of the primary coil is low, then the applied voltage will only be used to overcome this back e.m.f. and may be taken as equal to it. The familiar transformer law results:

$$\frac{\text{induced e.m.f. in secondary}}{\text{applied e.m.f. in primary}} = \frac{\text{number of turns of secondary winding}}{\text{number of turns of primary winding}}.$$

This argument is acceptable so long as no current is drawn from the secondary coil. If the secondary circuit is completed so that a current is taken, this will have a feedback effect on the primary circuit and the primary current will increase, so drawing more energy from the supply. The mathematical treatment of this situation is complicated; it leads to the same result as before provided that the same assumption is made.

Transformer losses

As energy passes through a transformer, a certain amount is inevitably 'lost', i.e. used up in the process and so not available to the consumer. There are four main ways in which this occurs.

(a) *Magnetic losses.* These could occur if the secondary coil did not completely contain the magnetic field set up by the primary, so that lines of force escaped and produced unwanted inductive effects in other conductors.

In fact, such losses are very small, because the turns of the primary lie underneath and covered by those of the secondary coil; also the design of the double soft-iron yoke controls the pattern of the lines of force so that they all return into the system.

(b) *Iron losses.* Energy is unavoidably lost during the continual change of alignment of the magnetic domains in the yoke, resulting in a rise in temperature. These are hysteresis losses which cannot be entirely avoided, but only minimised by a wise choice of material, as described in Chap. 7.

This loss of energy is the basic cause of the 'mains hum', the low note heard when a transformer is in use.

(c) *Eddy currents.* The iron yoke is itself a conductor in a changing magnetic field, and consequently an e.m.f. is induced in the iron and tends to set up *eddy currents.* These are small alternating currents which flow between the opposite faces of the yoke. Although iron is a good conductor of electricity, heat will be produced by these currents, causing a loss of energy.

Eddy currents are reduced by dividing the iron yoke into thin layers called laminations, separated by even thinner layers of insulating material. This cuts down the current while the magnetic flux remains undisturbed.

(d) *Copper losses.* Some energy is lost in the form of heat as the currents flow through the actual windings of the coils.

These are reduced by using *copper* wire, as being the best conductor available, and by using thicker wire for the primary coil, which carries the larger current.

The total energy loss can be made very small if all these points are observed, and generally the transformer is considered to be about the most efficient of machines, its efficiency of about 98% contrasting with only 30–40% from the best mechanical systems.

Examination questions 22

1 (a) Explain the terms *peak voltage*, *r.m.s. voltage* and *phase difference* as applied to sinusoidal a.c. waveforms.

(b) What is the reactance of a 10 H inductor when it is connected across a supply of frequency 50 Hz? What would be the power dissipated in a resistor which offers the same opposition to current flow as this inductor when it is connected across a 250 V (r.m.s.) supply? [AEB, Nov. 1975]

2 (a) Draw a fully labelled diagram showing the structure of a transformer capable of giving an output of 12 V from a 240 V a.c. supply. State *two* sources of power loss in the transformer and describe how they may be minimised.

(b) Explain why, when the secondary is not delivering a current, the transformer consumes very little power from the supply.

(c) Why will the transformer not work with a direct current?

(d) A factory requires power of 144 kW at 400 V. It is supplied by a power station through cables having a total resistance of 3 Ω.

If the power station were connected directly to the factory,

(i) show that the current through the cables would be 360 A,

(ii) calculate the power loss in the cables,

(iii) calculate the generating voltage which would be required at the power station,

(iv) calculate the overall efficiency.

If the output from the power station provided a 10 000 V input to a transformer at the factory, the transformer having an efficiency of 96 %,
(v) show that the current through the cables would be 15 A,
(vi) calculate the power loss in the cables,
(vii) calculate the generating voltage which would be required at the power station,
(viii) calculate the overall efficiency. [AEB, Nov. 1979]

3 An electric lamp, enclosed in a box with a photocell, first has a direct current passed through it and then an alternating current of r.m.s. value equal to that of the direct current. Discuss whether you would expect the photoelectric current to be the same in the two cases.

A current $I_0 \cos \omega t$ flows through a pure inductor of inductance L. Starting from the definition of inductance, show that the p.d. across the inductor has a maximum value $I_0 \omega L$. Find, in terms of ω, the time interval between a current maximum and the next p.d. maximum. Calculate this time difference for a.c. of frequency 50 Hz.

Explain why no power is dissipated in the inductor. [C]

23 A.C. and capacitance

A capacitor is a device for storing electricity. If a capacitor is connected to an a.c. supply as shown in Fig. 103, the two plates will be alternately charged and discharged, and an a.c. ammeter in the circuit will register a current.

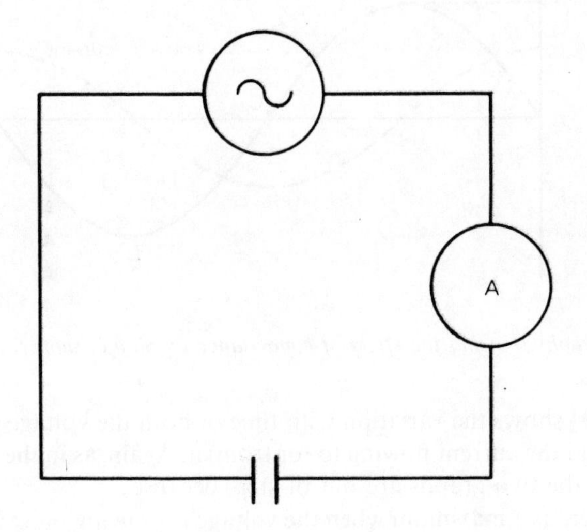

Fig. 103 A capacitor in an a.c. circuit

Although charge cannot actually pass between the plates of the capacitor, through the dielectric, the current is still able to flow around the circuit while the capacitor is charging and discharging. A large capacitance will probably not be fully charged before the a.c. supply changes direction, but charge building up on a small capacitance will cause its potential to rise quickly, and further charge would be added only with difficulty. Thus the value of the capacitance can affect the flow of current, particularly if the frequency of the alternations is low.

Suppose the supply voltage is of the form

$$V = V_{max} \sin 2\pi ft \qquad (1)$$

corresponding to equation (1) of the preceding chapter (p. 178).

If the capacitance of the circuit is C, and its (pure) resistance can be neglected, then the charge Q on the plates of the capacitor will vary with the voltage of the supply:

$$\text{charge} = \text{capacitance} \times \text{voltage}$$

i.e.

$$Q = C \, . \, V_{\text{max}} \sin 2\pi ft.$$

Now the reading of the ammeter will depend on the rate at which charge flows through it.

$$\text{Current } I = \frac{dQ}{dt} = 2\pi fC \, . \, V_{\text{max}} \cos 2\pi ft. \qquad (2)$$

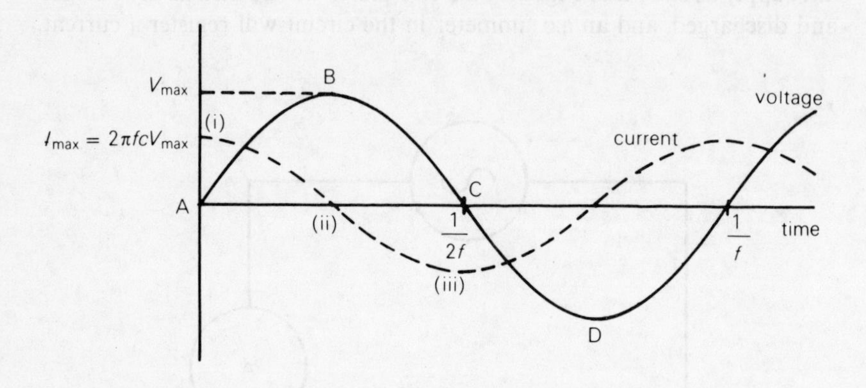

Fig. 104   Graphs showing the effect of capacitance on an a.c. supply

Figure 104 shows the variation with time of both the voltage across the capacitor and the current flowing to (or from) it. Again, as in the case of an inductance, the two graphs are out of step, because

(i) the current is a maximum when the voltage is growing most rapidly, at point A,

(ii) the current is momentarily zero when the voltage is a maximum, at point B,

(iii) the current flows in the reverse direction as the voltage falls, from B to D.

Phase difference between current and voltage

From equation (2), the maximum value of the current I is

$$I_{\text{max}} = 2\pi fC V_{\text{max}}.$$

Equation (1) may then be re-written as

$$V = \frac{1}{2\pi fC} \, . \, I_{\text{max}} \sin 2\pi ft.$$

As in the preceding chapter (p. 179), equation (2) may be re-written in the form

$$I = I_{\text{max}} \sin\left(2\pi ft + \frac{\pi}{2}\right).$$

As before, presenting the two equations in this form demonstrates the similarity of the two curves shown in Fig. 104. There is again a phase difference of $\pi/2$ between the two curves, but in this case, **it is the voltage which lags behind the current.**

Power consumption

During the quarter-cycles in which the voltage across the capacitor is increasing since charge is building up on its plates, energy is being stored in it. This is returned to the circuit in the alternate quarter-cycles when the p.d. across the plates exceeds that of the supply and the charge begins to drain away.

As in the case of the inductance, **a pure capacitative circuit consumes no power when considered over a complete cycle.**

The instantaneous power consumption is calculated, using equations (2) and (1), as

$$\begin{aligned}\text{current} \times \text{voltage} &= I_{\text{max}} \cos 2\pi ft \,.\, V_{\text{max}} \sin 2\pi ft \\ &= \tfrac{1}{2} I_{\text{max}} \,.\, V_{\text{max}} \sin 4\pi ft \text{ (again).}\end{aligned}$$

Reactance

As for an inductance, the reactance of the capacitor is defined as

$$\begin{aligned}\boldsymbol{X_C} &= \frac{\textbf{maximum value of voltage across capacitor}}{\textbf{maximum value of current around circuit}} \\ &= \frac{V_{\text{max}}}{I_{\text{max}}} \\ &= \frac{\mathbf{1}}{\boldsymbol{2\pi fC}}.\end{aligned}$$

X_C will be in ohms if C is in farads.

This expression supports the idea that the reactance should be less if the capacitance is large and the frequency of the a.c. supply is high. One application of this result is in the use of coupling and decoupling capacitors in amplifying circuits, to separate out the audio-frequency signals from the radio-frequency carriers. It is generally more convenient to use capacitors for this purpose rather than inductances, which are bulkier and only to be preferred if large currents are involved.

EXERCISES

1 A.C. of frequency 50 Hz flows through a circuit containing a 2 μF capacitor. If the maximum p.d. of the supply is 100 V, calculate (i) the maximum current around the circuit, and (ii) the voltage across the capacitor when the current has fallen to half of this maximum value.

2 A 0.1 μF capacitor is connected across the 240 V, 50 Hz, mains supply. Find (i) the points in each cycle at which the capacitor is absorbing maximum power, and (ii) the current flowing into it at those moments.

3 Calculate the reactance of an 0.01 μF capacitor to a.c. of freqency (i) 50 Hz, (ii) 1000 kHz.

Examination questions 23

1 Describe and explain the difference in the flow of charge when a capacitor is connected to (a) a d.c. source, and (b) an a.c. source. In what way is this flow affected by the frequency of the a.c. source? [L]

24 Impedance and resonance

It has been shown in the last two chapters that the voltage across a pure inductance leads the current by quarter of a cycle (p. 179) while the voltage across a pure capacitance lags quarter of a cycle behind the current (p. 187). Combination of these components in an a.c. circuit involves adding their reactances algebraically.

The effect of a combination of resistance and reactance can be considered either by the use of vectors or trigonometrically. The result is called the *impedance* of the circuit, and it is defined in the same way as reactance:

$$\textbf{impedance} = \frac{\textbf{maximum value of voltage around circuit}}{\textbf{maximum value of current through circuit}} \qquad (1)$$

or in symbols,

$$Z = \frac{V_{max}}{I_{max}}.$$

Impedance is measured in ohms.

Series circuits

Vector approach

The phase angle occurring in the equation of a waveform may be used to represent a.c. currents and voltages as vectors.

If the current through a circuit is represented as in Fig. 105(a), then the voltages developed by it across a resistor, an inductor, and a capacitor would be represented as shown in diagrams (b), (c), and (d) of the same figure. The lengths of the lines in the figure are proportional to the maximum values of the variables, and their directions represent phase differences.

The effect of combining various components is illustrated in Fig. 106. In this example, a pure resistor is combined with a pure inductor. The resulting voltage drop across the two components in series is represented by the hypotenuse of the triangle of vectors; its maximum value is $\sqrt{(R^2 + X_L{}^2)}I_{max}$ and its phase angle is $\tan^{-1}(X_L/R)$ in advance of the current.

The impedance of this combination (using equation (1)) will be $\sqrt{(R^2 + X_L{}^2)}$.

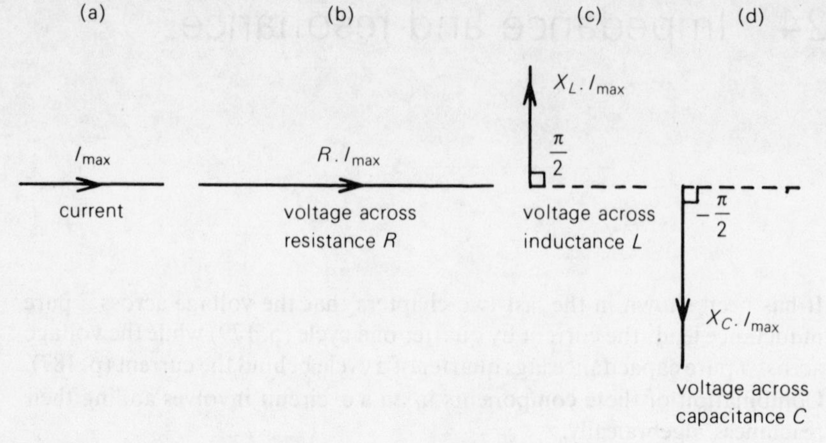

Fig. 105 Vector representation of a.c. variables

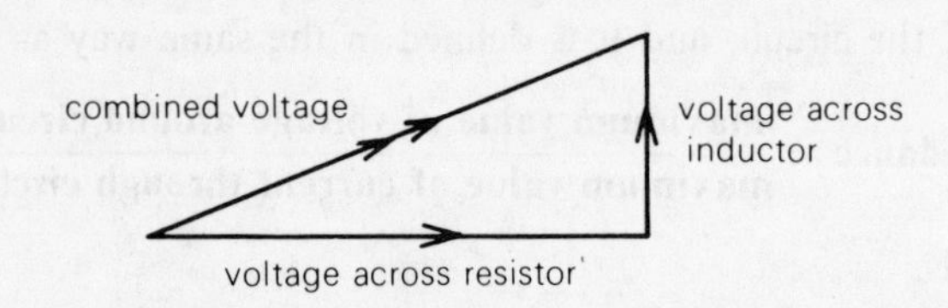

Fig. 106 Vector addition of two voltages

Similarly, combination of a pure resistor and a pure capacitor produces an impedance equal to $\sqrt{(R^2 + X_C{}^2)}$.

If resistance can be ignored, then the combination of inductance and capacitance gives simply

$$Z = (X_L - X_C).$$

Including the term for resistance, the general equation for the impedance of an a.c. series circuit is

$$Z = \sqrt{[R^2 + (X_L - X_C)^2]}.$$

This result is illustrated by Fig. 107, but it must be emphasised that this is *not* a vector polygon, since resistance and reactance cannot be considered as vectors. The use of such a diagram to calculate impedance is only justified

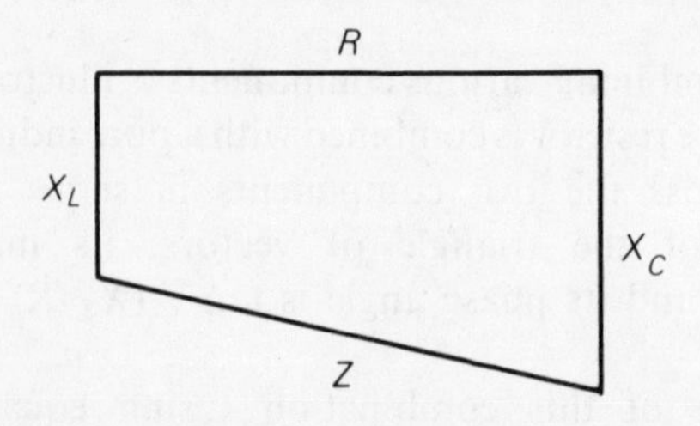

Fig. 107 A phasor diagram for an a.c. circuit

by the previous reasoning concerning voltages. Perhaps for this reason, it is often called by the alternative name of a *phasor* diagram.

Trigonometric approach

Suppose the a.c. circuit can be divided, as shown in Fig. 108, into a pure resistance R, a pure inductance L, and a pure capacitance C, in series.

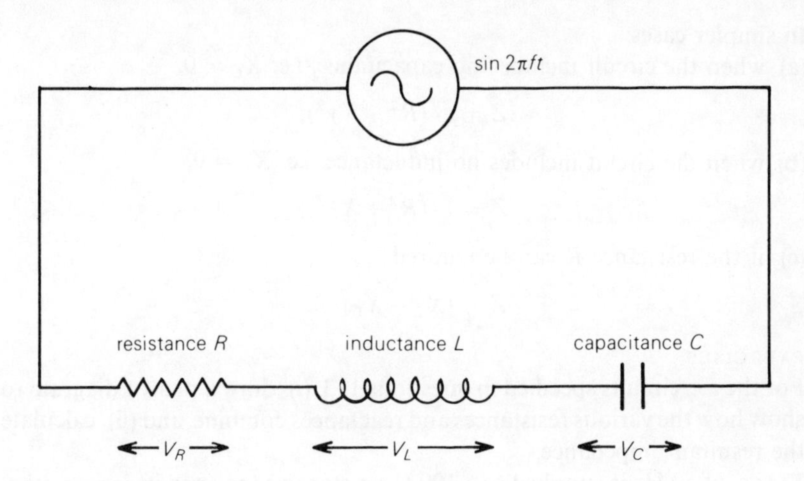

Fig. 108 An a.c. series circuit

The current at any moment must be the same through all the components; take it, as usual, as $I = I_{max} \sin 2\pi ft$.

The potential difference V_R across the resistance R will follow the equation

$$V_R = RI_{max} \sin 2\pi ft, \qquad \text{from Ohm's law.}$$

The potential difference V_L across the inductance L will follow the equation

$$V_L = X_L I_{max} \cos 2\pi ft, \qquad \text{from equation (2) on p. 178,}$$

$$\text{substituting } X_L = 2\pi fL.$$

The potential difference V_C across the capacitance C can similarly be written as

$$V_C = -X_C I_{max} \cos 2\pi ft$$

where $X_C = 1/2\pi fC$, the minus sign occurring since V_C is always half a cycle behind V_L.

Combining these three expressions, the total potential drop V around the circuit will be the result of simple addition,

$$\begin{aligned} V &= V_R + V_L + V_C \\ &= RI_{max} \sin 2\pi ft + (X_L - X_C) I_{max} \cos 2\pi ft. \end{aligned}$$

This is the general equation. The maximum value of V may be found by differentiation, which gives

$$V_{max} = \sqrt{[R^2 + (X_L - X_C)^2]}\, I_{max}.$$

The impedance Z of the circuit is then

$$Z = \frac{V_{max}}{I_{max}} = \sqrt{[R^2 + (X_L - X_C)^2]}.$$

In simpler cases:

(a) when the circuit includes no capacitance, i.e. $X_C = 0$,

$$Z = \sqrt{(R^2 + X_L{}^2)};$$

(b) when the circuit includes no inductance, i.e. $X_L = 0$,

$$Z = \sqrt{(R^2 + X_C{}^2)};$$

(c) if the resistance R can be ignored,

$$Z = (X_L - X_C).$$

EXERCISES

For the a.c. circuits specified in questions 1–3, (i) draw a phasor diagram to show how the various resistances and reactances combine, and (ii) calculate the resultant impedance.

1 a.c. of 50 Hz is supplied to a 100 Ω resistance connected in series with a pure inductance of 1 henry (1H).

2 a.c. of 60 Hz is supplied to a circuit consisting of a 10 μF capacitor in series with a 200 Ω resistor.

3 Signals of frequency 2000 Hz are received by a circuit consisting of an inductance of 2 mH whose resistance is 50 Ω, in series with a 1 μF capacitor.

4 An inductance of 1 H whose resistance is 100 Ω is connected to the 240 V, 50 Hz mains supply. Calculate (i) the r.m.s. value of the current taken, and (ii) the phase difference between the current and the voltage across the coil.

Resonance

Since reactance varies with frequency, the total impedance of an a.c. circuit will also vary. For a series circuit, it follows a curve of the type illustrated in Fig. 109.

The minimum of the curve occurs when the reactances of the component inductances and capacitances cancel out,

$$\text{i.e.}\quad X_L = X_C$$

$$2\pi f L = \frac{1}{2\pi f C}$$

$$\Rightarrow\quad f = \frac{1}{2\pi\sqrt{LC}}.$$

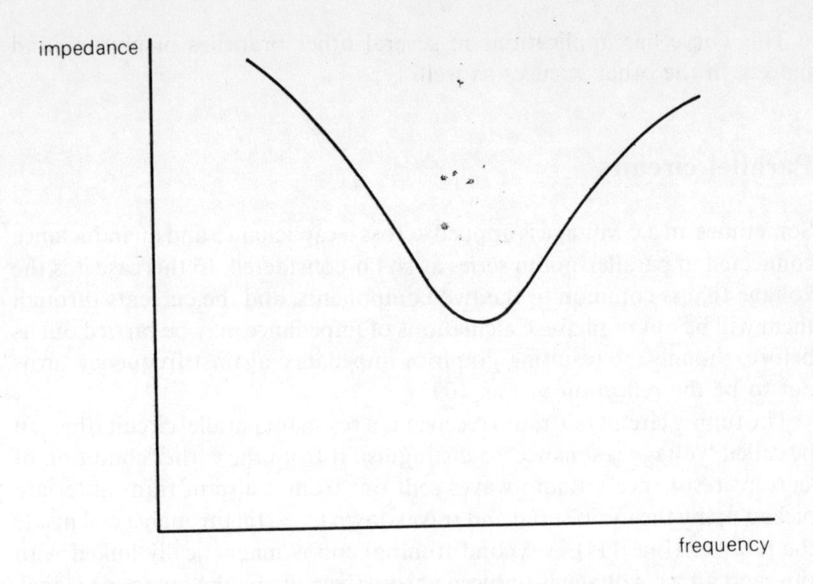

Fig. 109 Graph of frequency and impedance for an a.c. series circuit

The response of the circuit to the supply will be the inverse of this, as shown in Fig. 110. The condition in which the peak response is obtained is called *resonance*, and the frequency at which this occurs is the *resonant* (or *natural*) *frequency*.

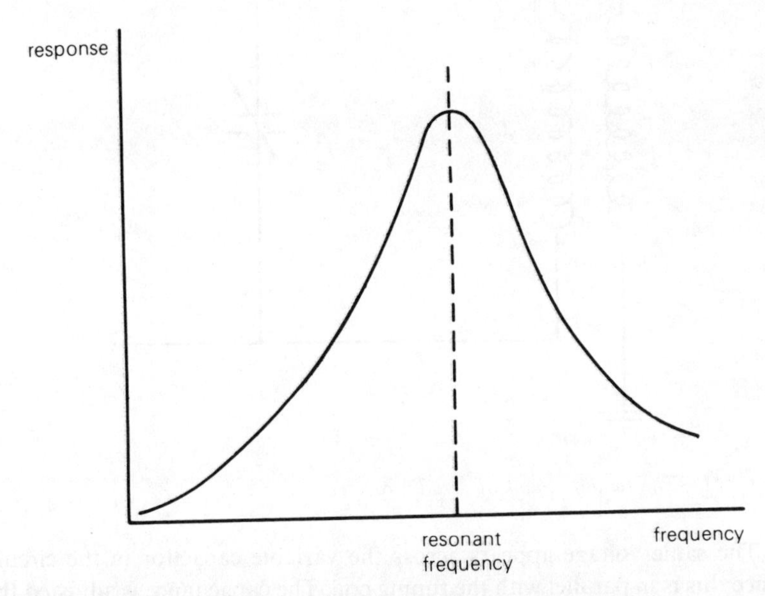

Fig. 110 Graph of frequency and response, showing resonance peak

This curve has applications in several other branches of physics, and indeed, in the other sciences as well.

Parallel circuits

Sometimes an a.c. voltage is applied across a capacitance and an inductance connected in parallel, not in series as so far considered. In this case it is the voltage that is common to the two components, and the currents through them will be out of phase. Calculations of impedance may be carried out as before, though the resulting graph of impedance against frequency turns out to be the reflection of Fig. 109.

The tuning circuit of a radio receiver is a resonant parallel circuit (this can be called 'voltage resonance' to distinguish it from the earlier condition of 'current resonance'). Radio-waves sent out from a distant transmitter are picked up by the radio aerial and travel down to earth, through a coil inside the receiver (Fig. 111). A second (tuning) coil is magnetically linked with this, and an a.c. voltage is induced across the coils by the incoming signal, which is generally a combination of many different transmissions coming simultaneously from different sources.

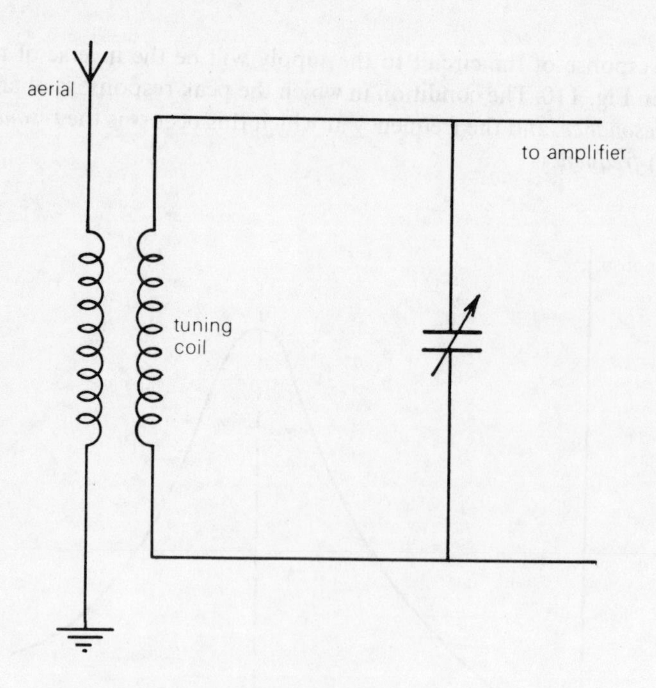

Fig. 111 The tuning circuit of a radio receiver

The same voltage appears across the variable capacitor in the circuit, since this is in parallel with the tuning coil. The capacitance is adjusted (by turning the knob of the frequency selector) until resonance is obtained with the frequency of one particular transmitter, and the circuit will respond to that while rejecting other radio broadcasts on different frequencies.

Connections from the two sides of the capacitor go to the amplifying circuits of the receiver.

In this condition, the total impedance of the circuit is a maximum (not a minimum as in the case of a series circuit), and very little current will be drawn from the external source. The circuit is oscillating at its natural frequency and little external energy is needed to sustain the oscillations. The energy in the circuit flows from the magnetic field of the inductance to the dielectric of the capacitor and back again, with losses only in the small amount of resistance in the wires. This action can be compared to the behaviour of a pendulum, in which energy is continually changed from potential to kinetic, and vice versa.

EXERCISES

5 A pure inductance of 4 mH is connected in series with a 10 μF capacitor. Calculate the resonant frequency of the circuit.

6 The aerial of a radio receiver is magnetically linked to a tuning coil of inductance 8 mH. If the circuit is to resonate to a broadcast on a frequency of 1.5 MHz, calculate the value at which the variable capacitor must be set.

Examination questions 24

1 A coil of inductance L and negligible resistance is connected in series with a non-inductive resistance $R = 150\ \Omega$ and a 1 kHz sinusoidal a.c. generator. An a.c. voltmeter reads 3.6 V when connected across the generator, 2.0 V when connected across the coil and 3.0 V when connected across R. Draw a voltage vector diagram for the circuit.

Hence, or otherwise, derive values for (i) the inductance L, and (ii) the phase angle between the current in the circuit and the potential difference across the generator. [AEB, Nov. 1974]

2 A capacitor, of capacitance 200 pF, is connected in series with a resistor and an a.c. source whose frequency is $(1/2\pi) \times 10^6$ Hz. The potential difference across the resistor and the capacitor are found to be equal. Calculate

(i) the reactance of the capacitor,
(ii) the resistance of the resistor,
(iii) the impedance of the circuit,
(iv) the phase angle between the supply voltage and the current in the circuit.

Sketch a curve to show how the circuit impedance varies over the complete frequency range. [AEB, Nov. 1975]

3 (a) Define the *impedance* of a coil carrying an alternating current. Distinguish between the *impedance* and *resistance* of a coil and explain how they are related.

Describe and explain how you would use a length of insulated wire to make a resistor having an appreciable resistance but negligible inductance.

(b) Outline how you would determine the impedance of a coil at a frequency of 50 Hz using a resistor of known resistance, a 50 Hz a.c. supply and a suitable measuring instrument. Show how to calculate the impedance from your measurements. [JMB]

4 A coil of inductance 3 mH and resistance 10 Ω is connected in series with a 2 μF capacitor and an oscillator of variable frequency but constant output voltage. The frequency of the oscillator is varied smoothly from 100 Hz to 5 kHz. The current through the circuit is observed to

A increase to a maximum and then fall.
B drop to a minimum and then rise.
C rise continuously from a lower to a higher value.
D fall continuously from a higher to a lower value.
E remain approximately constant over the whole frequency range. [L]

5 Explain the terms *reactance* and *impedance* as applied to a component of an a.c. circuit.

Describe an experiment to demonstrate the action of a *choke* in an a.c. circuit, explaining what happens.

P and Q are two boxes *each* containing one electrical component connected by two terminals on the outside of the box. It is found by experiment that:

(a) when P alone is connected to a 240 V d.c. supply, there is a current of 0.80 A.

(b) when P alone is connected to a 240 V, 50 Hz a.c. supply, there is a current of 0.48 A.

(c) When P and Q, joined in series, are connected to a 240 V, 50 Hz a.c. supply, there is a current of 0.80 A.

Use the data to identify the component in *each* box and to calculate the numerical values of its electrical properties. [L]

6 When a *sinusoidal alternating current* flows through an *inductor*, the current *lags behind* the voltage across the inductor. Explain the meaning of the italicised terms and account for the phenomena described.

An inductor and a capacitor are connected across the terminals of an a.c. supply, the frequency of which can be varied while the r.m.s. voltage across the terminals is kept constant. Draw rough sketches to show the variation with frequency of the r.m.s. current drawn from the supply when the inductor and capacitor are (a) in series, (b) in parallel.

Either: explain why it is possible for the r.m.s. current drawn from the supply by two components in parallel to be less than that drawn by either component when connected separately across the supply.

Or: give an example of a practical application of a series or parallel combination of an inductor and a capacitor. [OC]

Atomic Physics

25 Electric fields

There is a close connection between electric and magnetic fields, and theoretical treatment of the two is very similar.

Lines of force

An electric field can be mapped, as a magnetic field is, by lines of force. These show the path which would be taken by a positive charge if free to move. Lines of electric force begin at positive charges and end at negative charges.

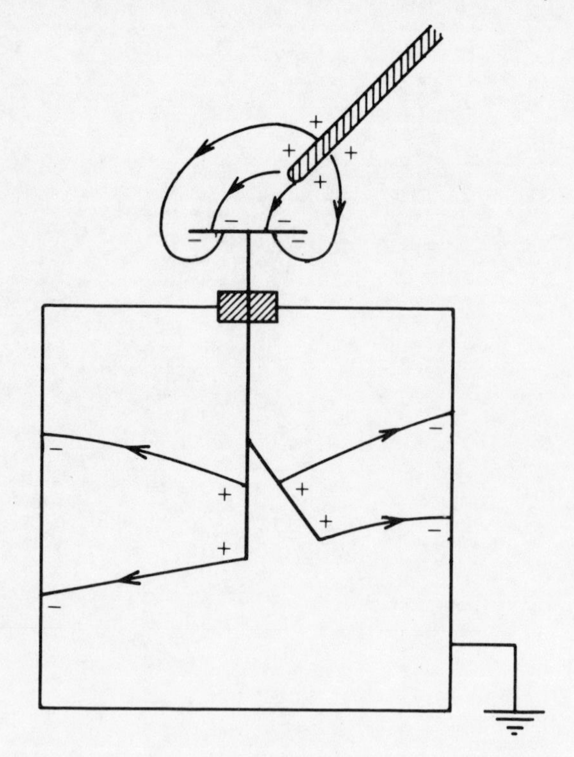

Fig. 112 Lines of electric force around a charged gold-leaf electroscope

Figure 112 shows the lines of force around a gold-leaf electroscope in the presence of a charged rod. This illustrates the forces acting on the leaves, causing them to diverge.

Equipotentials

The electric *potential* of a point is defined in terms of the work needed to bring a positive charge from infinity to that point. **If the work done is 1 joule when a positive charge of 1 coulomb is brought from infinity to a point, then the potential of that point is 1 volt.** (volts × coulombs = joules)

The electric field can therefore be mapped using another set of lines called equipotentials, drawn through points which are at the same potential. This set of lines will be everywhere at right angles to the set of lines of force (Fig. 113). A small positive charge placed at a point on an equipotential, will be repelled away along the line of force which passes through the point.

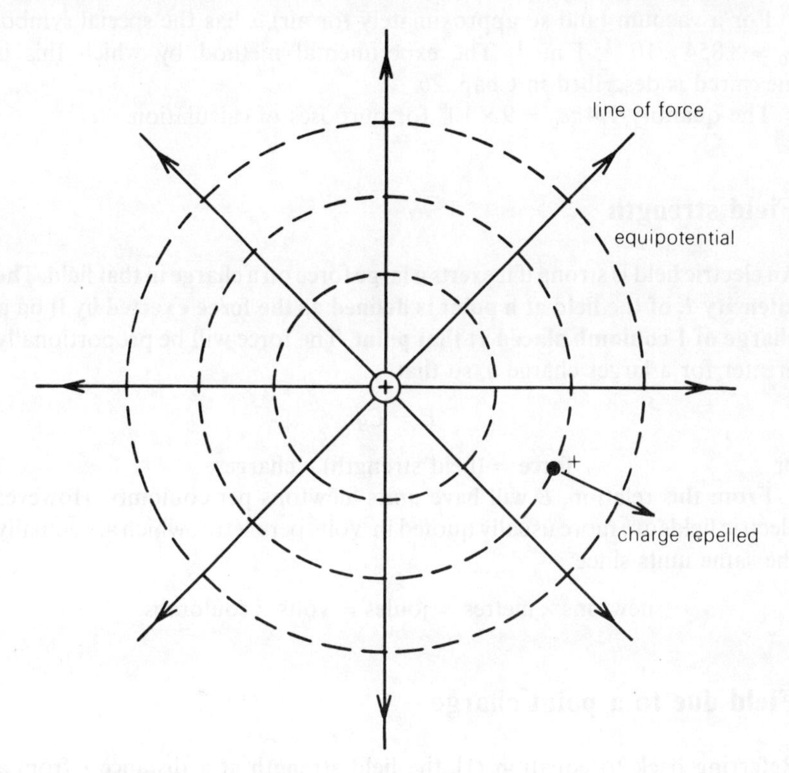

Fig. 1.13 The electric field due to a positive charge

Force between two charges

The force of attraction or repulsion between two charges depends on:
(a) the sizes of the charges, q_1 and q_2, and
(b) the distance between them, r.

The relation between force and distance is the inverse square law:

$$F \propto \frac{1}{r^2}$$

and combining both factors:

$$F \propto \frac{q_1 q_2}{r^2}.$$

If F is to be measured in newtons when q_1 and q_2 are in coulombs and r is in metres, then a numerical constant is required in order to write the relation as an equality:

$$F = \frac{1}{4\pi\varepsilon} \frac{q_1 q_2}{r^2} \tag{1}$$

where ε is called the *permittivity* of the medium between the two charges.

For a vacuum (and so approximately for air), ε has the special symbol $\varepsilon_0 = 8.854 \times 10^{-12}$ F m^{-1}. The experimental method by which this is measured is described in Chap. 26.

The quantity $1/4\pi\varepsilon_0 = 9 \times 10^9$ for purposes of calculation.

Field strength

An electric field is strong if it exerts a large force on a charge in that field. **The intensity E of the field at a point is defined as the force exerted by it on a charge of 1 coulomb placed at that point**. The force will be proportionally greater for a larger charge q, so that

$$F = Eq$$

or force = (field strength) × charge.

From this relation, E will have units 'newtons per coulomb'. However electric fields are more usually quoted in 'volts per metre', which are actually the same units since:

newtons × metres = joules = volts × coulombs.

Field due to a point charge

Referring back to equation (1), the field strength at a distance r from a charge q is equal to the force which would be exerted there on a charge of 1 coulomb, i.e. q_2 must be made equal to 1.

Then

$$E = \frac{1}{4\pi\varepsilon} \frac{q}{r^2}. \tag{2}$$

At very small distances from the charge, the electric field becomes very large. Such a field can break down and ionise the air molecules, producing an 'electric wind' or 'corona discharge'. This occurs for air at normal pressure with field strengths about 3×10^6 V m^{-1}. Such high field strength occurs particularly at the tip of a pointed charged conductor or around dust particles on its surface. Conductors which are to be used at very high potentials, such as the dome of a Van der Graaff generator, must be kept smooth and dust-free if the charge is not to leak away.

Field due to a charged spherical conductor

(a) *Outside the sphere.* A charge placed on a spherical conductor spreads out evenly all over the surface. The set of equipotentials which describe the resulting field are exactly the same as if the whole charge were concentrated at the centre of the sphere, and **expression (2) above may be used to give the field strength at all points outside the sphere.**
(b) *Inside the sphere.* Faraday's ice-pail experiments demonstrated that no charge exists on the inside of a hollow charged conducting sphere, so there can be no electric field there.

All points within the sphere must be at the same potential as the sphere itself.

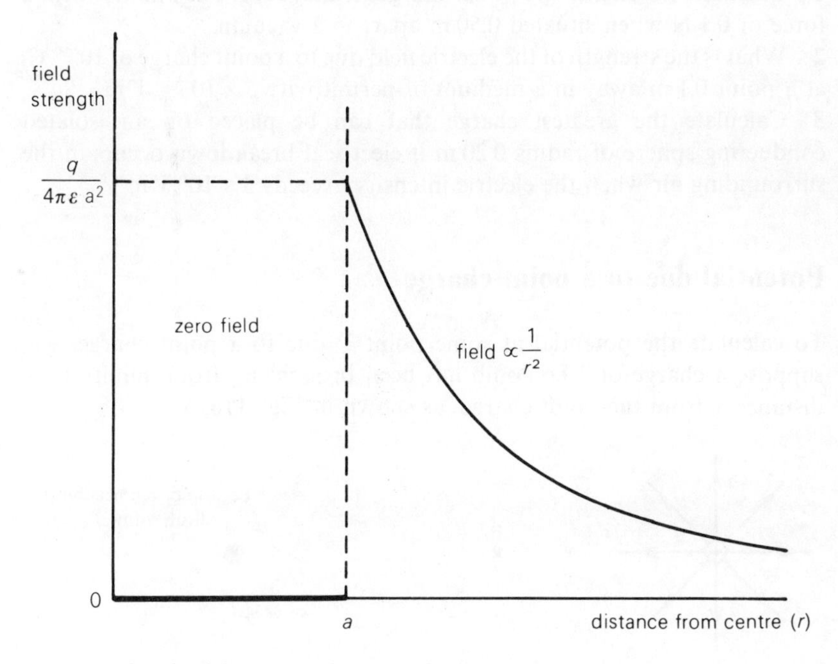

Fig. 114 Graph of distance and electric field strength for a charged sphere

Figure 114 shows how the field strength varies with distance for a charge q on a sphere of radius a. This graph should be compared with Fig. 65 on p. 116, which relates to *gravitational* field strength.

Field between two parallel plates

The electric field produced by two oppositely charged parallel plates is as shown in Fig. 115. **Over the central region between the plates, the field is**

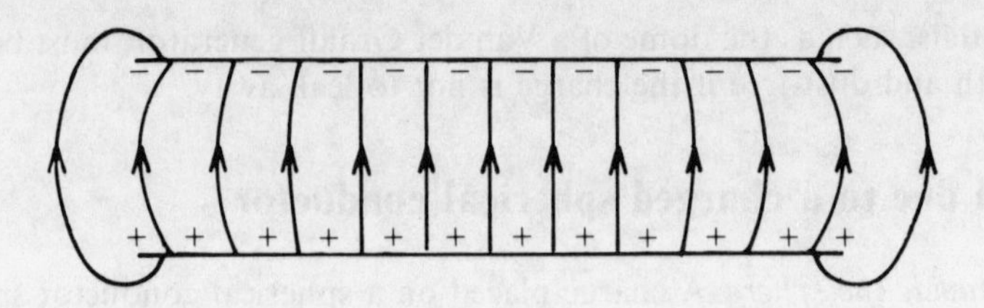

Fig. 115 The electric field between oppositely charged parallel plates

uniform and equal to V/d, where V is the potential difference between the plates and d is their separation.

The lines of force bulge outwards around the edges of the plates and the field there is no longer uniform.

EXERCISES

1 Calculate the size of two equal charges if they repel one another with a force of 0.1 N when situated 0.50 m apart in a vacuum.

2 What is the strength of the electric field due to a point charge of 10^{-6} C, at a point 0.1 m away in a medium of permittivity 5×10^{-11} F m^{-1}?

3 Calculate the greatest charge that can be placed on an isolated conducting sphere of radius 0.20 m if electrical breakdown occurs in the surrounding air when the electric intensity exceeds 3×10^{6} V m^{-1}.

Potential due to a point charge

To calculate the potential at some point P due to a point charge $+q$, suppose a charge of 1 coulomb has been brought up from infinity to a distance x from the point charge, as shown in Fig. 116.

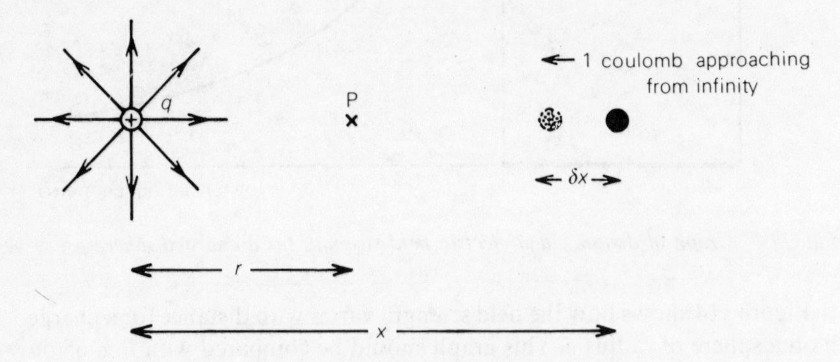

Fig. 116 Calculation of potential due to a point charge

A force of repulsion acts between the two charges. At a separation of x, the size of this force is given by equation (1) as

$$F = \frac{1}{4\pi\varepsilon}\frac{q}{x^2},$$

since the second charge is 1 unit.

The work which must be done to bring the approaching charge nearer by a distance δx is given by

$$-F\delta x = -\frac{1}{4\pi\varepsilon}\frac{q}{x^2}\,\delta x.$$

(The minus sign indicates that the work is being done *against* the force F.)

Hence the total work required to bring the charge up from infinity to a distance r is

$$\int_{\infty}^{r} -F\,\mathrm{d}x = \frac{1}{4\pi\varepsilon}\int_{\infty}^{r} -\frac{q}{x^2}\mathrm{d}x$$

$$= \frac{1}{4\pi\varepsilon}\left[\frac{q}{x}\right]_{\infty}^{r}$$

$$= \frac{1}{4\pi\varepsilon}\frac{q}{r}.$$

This expression gives the potential at a distance r from the point charge q.

In general, the potential V at any point in an electric field is related to the field strength E by

$$V = \int -E\,\mathrm{d}x$$

or

$$E = -\frac{\mathrm{d}V}{\mathrm{d}x}.$$

Figure 117 shows the variation of potential with distance from a charge q on a sphere of radius a.

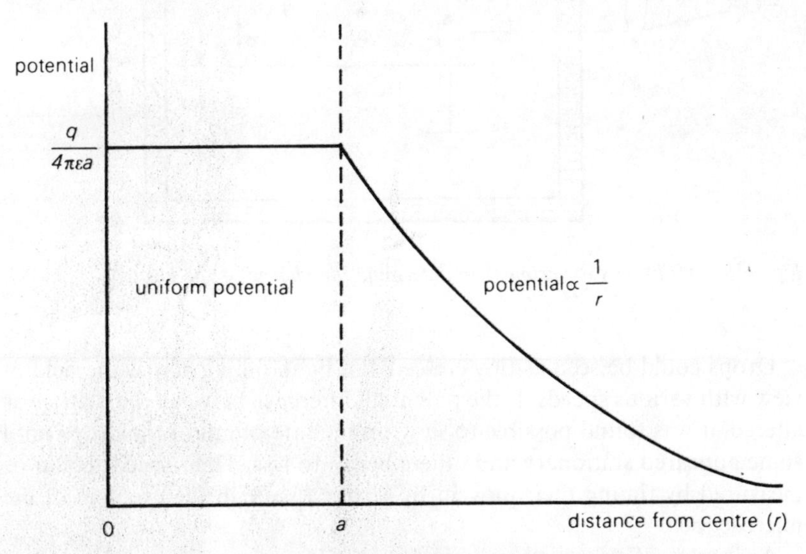

Fig. 117 Graph of distance and potential for a charged sphere

It is often easier to solve problems by considering potential rather than field strength, since potential is a scalar while fields must be treated by vector arithmetic. In the same way, it is sometimes convenient to consider a gravitational field in terms of gravitational potential, which would have units of $\mathrm{J\,kg^{-1}}$.

EXERCISE

4 What is the potential at a point 20 mm away from a point charge of 2×10^{-7} C in a vacuum?

Millikan's oil-drop experiment

This experiment, designed in 1909, was probably the first to obtain a value for the charge of a single electron.

A cloud of tiny drops of oil was formed in an electric field set up between two metal plates about 200 mm in diameter and 15 mm apart. The space was brightly illuminated from the side, and viewed through a low-power microscope (Fig. 118).

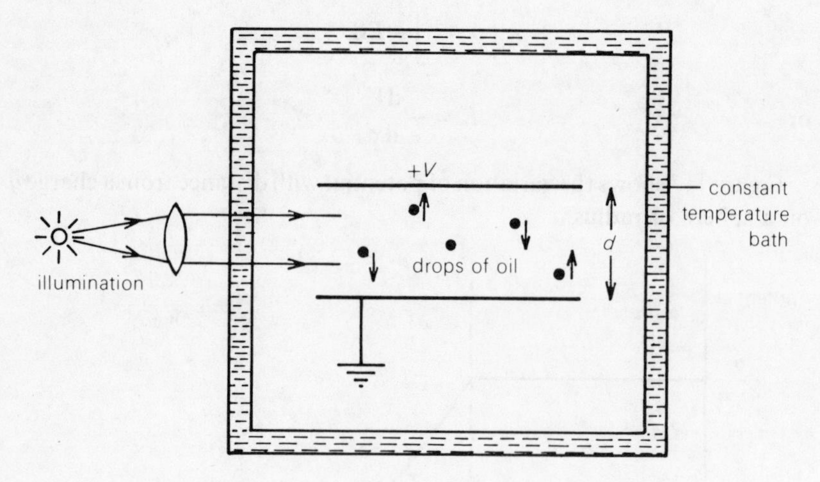

Fig. 118 Millikan's experiment to determine the charge of the electron

Drops could be seen as tiny crescents of light, falling across the field of view with various speeds. If the potential difference between the plates was altered, it was found possible to slow up the rate of fall of the drops until some appeared stationary and others began to rise. Their speeds could be measured by timing their movement across a scale in the eyepiece of the microscope.

A droplet of oil which remains *at rest* must have its weight exactly balanced by an upward force due to electrostatic attraction towards the top (positively charged) metal plate. This can only occur if the drop has picked up a negative charge from friction with the air.

The strength of the electric field between the plates is V/d V m^{-1} as above. So if there is a charge $-q$ on the drop, the upward force on it will be Vq/d newtons, which must equal its weight.

Millikan's experiment involved two separate measurements for a single droplet. After recording the value of V for which the droplet was held stationary, the voltage was then reduced so that it fell slowly with uniform speed v. Under these conditions, part of the weight of the droplet was balanced by the resisting force due to the viscosity of the air (p. 96).

$$\text{weight} = (\text{upward electrostatic force}) + (\text{viscous resistance}).$$

By considering the difference between the two sets of results, the radius of the particular droplet may be calculated and hence its mass, knowing the density of the oil used. Using this together with the value of V applying in the first part of the experiment then yields a value for the charge q on the drop. A small correction can be made if desired for the buoyancy of the air.

Millikan's results showed that the charge was always a simple multiple of the value 1.602×10^{-19} C, and he deduced that this corresponded to the charge of a single electron. Drops with higher charges must have picked up two, three, or more electrons. X-ray apparatus was used to increase the charges available, by ionising the air inside the chamber.

The observation chamber was kept at constant temperature during the experiment to prevent variation in the density of the oil or the viscosity of the air.

Electron motion in an electric field

In Figure 119, a uniform electric field E is set up between two parallel plates and a fine beam of electrons is injected midway between them.

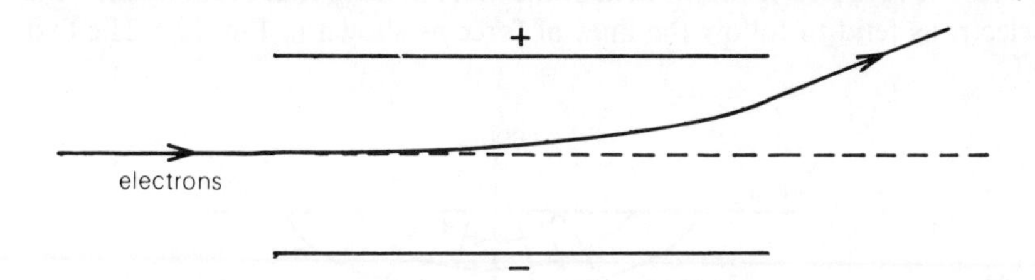

Fig. 119 Displacement of an electron beam in a uniform electric field

If the charge on an electron is e, it will be attracted towards the positive plate with a force Ee newtons. Its motion will then be similar to that of a projectile moving under the force of gravity, i.e. **its path will be a parabola**.

Calculations of the sideways displacement of the electrons as they move through the field is done by considering their forwards and sideways motion independently (as for a projectile):

Motion parallel to the plates: no acceleration, constant velocity.

Motion towards positive plate: initial velocity zero, accelerating force Ee. The mass of the electron must be brought into the equation, but it is actually sufficient to know only the ratio charge/mass to be able to solve the problem.

EXERCISES

5 Taking the electronic charge to be -1.60×10^{-19} C, find the potential difference needed between two horizontal metal plates, one 5 mm above the other, so that a small oil drop of mass 1.31×10^{-11} g, with two electrons attached to it, remains in equilibrium between them.

6 An electron of charge -1.6×10^{-19} C is situated in a uniform electric field of intensity 120 V mm^{-1}. Find the force on the electron, its acceleration, and the time it takes to travel 20 mm from rest. (Electronic mass $= 9.10 \times 10^{-31}$ kg).

7 Two plane metal plates 40 mm long are held horizontally 30 mm apart in a vacuum, one being vertically above the other. The upper plate is at a potential of 300 V and the lower is earthed. Electrons having a velocity of 1.0×10^7 m s^{-1} are injected horizontally midway between the plates and in a direction parallel to the 40 mm edge. Calculate the vertical deflection of the electron beam as it emerges from the plates (e/m for electron $= 1.8 \times 10^{11}$ C kg^{-1}).

Cathode-ray oscilloscope

Electric fields are used in the cathode-ray oscilloscope to deflect the electron beam in the X and Y directions, and also to focus the beam by an arrangement of two cylindrical anodes forming an 'electron-lens'.

As the rays pass from the field of the first anode to that of the second, the electrons tend to follow the lines of force as shown in Fig. 120. The first

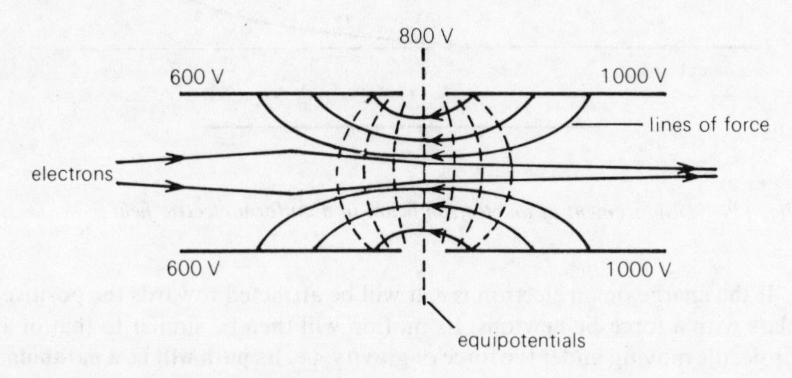

Fig. 120 The action of an 'electron-lens'

cylinder converges the beam and the second tends to diverge it, but as the electrons are moving with high speeds and continuing to accelerate, they are less affected by the second field than the first, and the beam emerges slightly convergent. The effect can be controlled by adjusting the potential of the first anode so that the size of the spot on the fluorescent screen can be altered. This is the function of the control knob marked 'Focus'.

Examination questions 25

1 (a) Explain what is meant by *electric field intensity* and *electric potential* at a point in space. State how they are related.
(b) Two point charges of $+q$ and $-q$ are separated by a distance $2a$. Calculate (i) the electric potential and (ii) the electric field intensity at the point midway between the charges. [AEB, Nov. 1975]

2 Describe, giving all relevant equations, how the electron charge may be found from observations of the motion of a charged oil drop moving vertically in a vertical electric field.

An oil drop of mass 2.0×10^{-15} kg falls at its terminal velocity between a pair of *vertical* parallel plates. When a potential gradient of 5.0×10^{4} V m^{-1} is maintained between the plates, the direction of fall becomes inclined at an angle of $21^\circ 48'$ to the vertical. Draw vector diagrams to illustrate the forces acting on the drop (a) before, and (b) after, the field is applied. Give formulae for the magnitude of the vectors involved. (Stokes' law may be assumed and the Archimedes' upthrust ignored.)

Calculate the charge on the drop.

$(g = 10 \text{ m s}^{-2})$ [C]

3 A charge of 3 C is moved from infinity to a point X in an electric field. The work done in this process is 15 J. The electric potential at X is
A 45 V B 22.5 V C 15 V D 5 V E 0.2 V [C]

4 In Millikan's experiment an oil drop of mass 1.92×10^{-14} kg is stationary in the space between the two horizontal plates which are 2.00×10^{-2} m apart, the upper plate being earthed and the lower one at a potential of -6000 V. State, with the reason, the sign of the electric charge on the drop. Neglecting the buoyancy of the air, calculate the magnitude of the charge.

With no change in the potentials of the plates, the drop suddenly moves upwards and attains a uniform velocity. Explain why (a) the drop moves, (b) the velocity becomes uniform. [JMB]

5 Draw a labelled diagram of the structure of a cathode ray tube as used in an oscilloscope. Explain how the electron beam is produced, focused, deflected and detected.

Draw a block diagram showing the essential units of a cathode ray oscilloscope and briefly explain their functions. (Details of the circuitry are *not* required.)

Explain the effect on the sensitivity of the oscilloscope of varying the accelerating potential. [L]

6 When a homogeneous electron beam enters a uniform electrostatic field which is at right angles to the original direction of the beam, its path in the field is

A a straight line.
B an arc of a circle.
C part of a helix.
D part of a parabola.
E none of the above. [O]

26 Capacitance

If an electric charge is given to a conducting body, it spreads out over the surface (because of the repulsion between every part of the charge) until the conductor is at the same potential everywhere.

The ratio charge Q/potential V for the conductor is called its *capacitance*, and if the charge is in coulombs and the potential in volts, then the capacitance will be in the units called *farads* (F).

Smaller units, more frequently used in practice, are
the microfarad $\mu F = 10^{-6}$ F
the nanofarad $nF = 10^{-9}$ F
and the picofarad $pF = 10^{-12}$ F.

Combination of capacitors

This section should be compared with the similar arguments relating to the combination of resistors (p. 37)

In parallel

Suppose a total charge $+Q$ given to the insulated side of the system shown in Fig. 121 produces a potential difference V across it. This will be the same for each capacitor, but the charge held on each will be different.

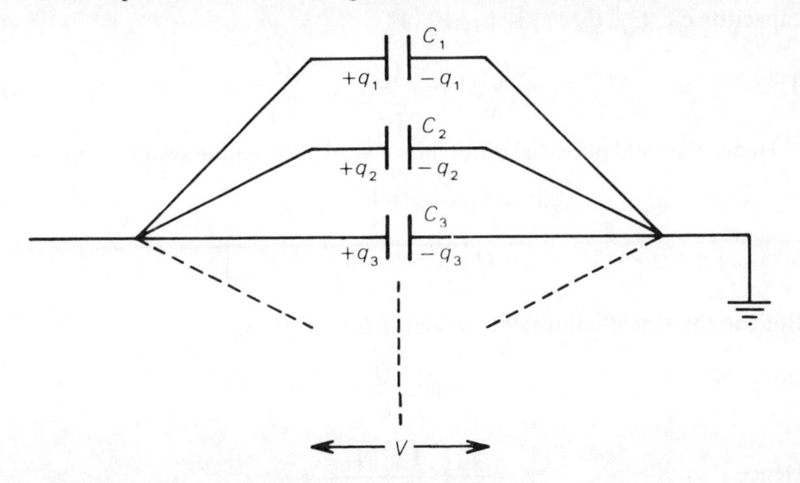

Fig. 121 Combination of capacitors in parallel

Suppose the charge held on each capacitor $C_1, C_2, C_3 \ldots$ is $q_1, q_2, q_3 \ldots$.

Then $$q_1 = C_1V, q_2 = C_2V, q_3 = C_3V \ldots$$
But $$q_1 + q_2 + q_3 + \ldots = Q.$$
Hence $$Q = (C_1 + C_2 + C_3 + \ldots)V.$$

For the single equivalent capacitor C,

$$Q = CV.$$

Hence $$C = C_1 + C_2 + C_3 + \ldots.$$

In series

Consider the system of capacitors shown in Fig. 122. If a charge $+Q$ is given to the insulated plate of C_1, then it will induce a negative charge $-Q$ on its opposite plate, and the equal and opposite charge $+Q$ will be repelled on to the connecting plate of C_2. This process will be continued throughout the system until the last positive charge flows to earth, leaving the system charged as shown in the figure.

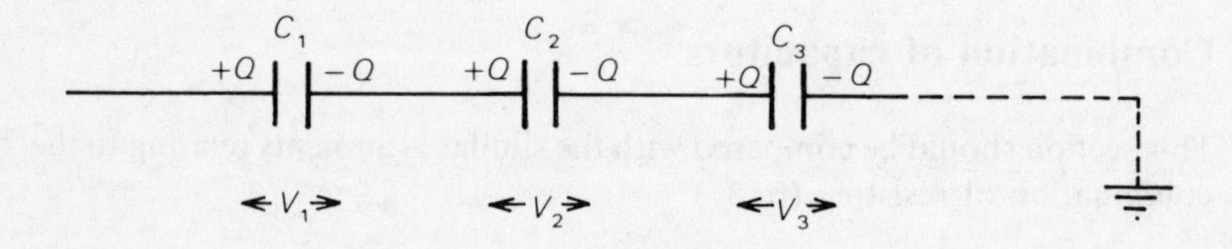

Fig. 122 Combination of capacitors in series

Suppose the potential difference created by these charges across each capacitor $C_1, C_2, C_3 \ldots$ is $V_1, V_2, V_3 \ldots$.

Then $$V_1 = \frac{Q}{C_1}, V_2 = \frac{Q}{C_2}, V_3 = \frac{Q}{C_3} \ldots.$$

Hence the total potential difference V across the whole system is given by

$$V = V_1 + V_2 + V_3 + \ldots$$
$$= Q\left[\frac{1}{C_1} + \frac{1}{C_2} + \frac{1}{C_3} + \ldots\right]$$

But for the single equivalent capacitor C,

$$V = \frac{Q}{C}.$$

Hence $$\frac{1}{C} = \frac{1}{C_1} + \frac{1}{C_2} + \frac{1}{C_3} + \ldots.$$

Factors affecting capacitance

The capacitance of a conductor depends on three factors.
(a) *Surface area.* A large conductor has a high capacitance.
(b) *Neighbouring conductors.* The proximity of an earthed conductor increases the capacitance of the original one, as an opposite charge will be induced on the earthed conductor which tends to neutralise the effect of the first charge.
(c) *The medium surrounding the conductor.* The capacitance increases if an insulating medium other than air surrounds the charged conductor. In this situation, such an insulator is called a *dielectric*, and the ratio of its permittivity to that of a vacuum is called the *dielectric constant* (or *relative permittivity*) of the medium. The practical values of dielectric constants range between 1 and 10.

A simple piece of apparatus known as the Aepinus air condenser can be used to study the effects of these three factors. It consists of a pair of parallel metal plates whose separation can be altered, supplied with several different insulating sheets to place between them. A fixed charge is given to one plate, and the resulting potential difference set up between the plates can be investigated using an electrostatic voltmeter or an oscilloscope.

The parallel-plate capacitor

A capacitor comprising two similar parallel metal plates separated by a thickness of dielectric has capacitance given by the expression

$$\frac{(\text{permittivity of dielectric}) \times (\text{area of plate})}{\text{distance between plates}} = \frac{\varepsilon A}{d}. \qquad (1)$$

This is only true if the lines of force between the plates are strictly parallel. Generally, as shown in Fig. 115 (p.202) in the previous chapter, the field is not uniform near the edges of the plates. A guard-ring arrangement must be added if such a capacitor is to be used in work of high accuracy.

The expression $\varepsilon A/d$ can be deduced theoretically, but the mathematics involved (Gauss' theorem) will not be considered in this book.

Measurement of permittivity

The permittivity ε of a medium was defined in the previous chapter (p. 200). By measuring the capacitance of a parallel-plate system and then applying equation (1), it is possible to determine the permittivity of the dielectric in use. This is the way in which the value of ε_0 was obtained (the permittivity of a vacuum).

A suitable circuit for this experiment is shown in Fig. 123. The capacitor is to be alternately charged and discharged so rapidly that the pulses of charge passing through the milliammeter give a steady reading. A special vibrating

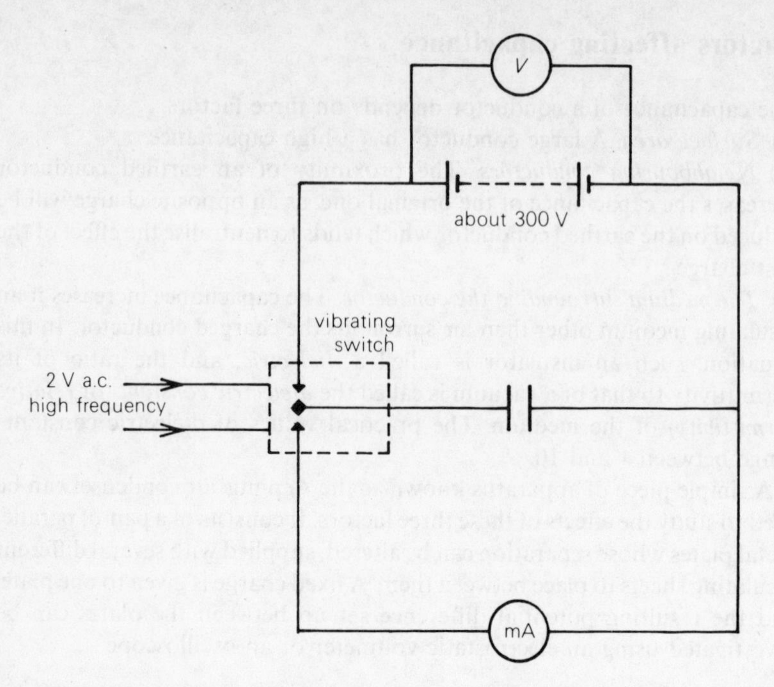

Fig. 123 Circuit diagram for the measurement of permittivity

switch can be obtained for this purpose, which works from a 2 V a.c. supply; this should be controlled by an oscillator so that its frequency is known.

The voltage supplied to the capacitor should be about 200–300 V, and its value is read from the voltmeter connected across it.

The parallel-plate system consists of two metal plates each about 0.15 m square, separated by a sheet of dielectric about 2 mm thick. The dimensions of the plates may be found by using a millimetre rule, but measurement of the thickness of the dielectric requires the extra accuracy obtained by using a travelling microscope.

The capacitance of a system of this size will be only about 100 pF if the dielectric is air, so that the charge given to it by a 300 V supply will be very small.

The average current observed when the capacitor is discharged through the vibrating switch will be (charge on capacitor) × (frequency of operation of switch), since this gives coulombs per second, or amps. A sensitive milliammeter of the reflecting type (p. 50) will be needed to measure it accurately.

EXERCISES

1 Capacitors of 2 μF and 3 μF are connected in parallel, and a charge of 1×10^{-4} C is given to the system. What p.d. will be produced across it?

2 Capacitors of 10^{-10} F and 2×10^{-10} F are connected in series, and a p.d. of 300 V is applied across them. Calculate the resulting charge on each capacitor.

3 Calculate the capacitance of a parallel-plate system consisting of two square plates of side 300 mm separated by a glass sheet 0.5 mm thick, if the dielectric constant of the glass is 6.0 and the permittivity of free space is $8.85 \times 10^{-12}\ \mathrm{F\,m^{-1}}$.

4 In an experiment to determine the permittivity of air as described above, the following results are obtained.

Dimensions of metal plates = 148 mm × 151 mm
Separation of plates = 3.25 mm
Frequency of oscillator = 300 Hz
Voltage of supply = 250 V
Current observed = 4.6 μA.

Deduce a value for the permittivity of air.

Behaviour of a dielectric

The atoms of an insulator normally appear to be uncharged since the number of protons in the nucleus is balanced by the number of circulating electrons. However, it is thought that the presence of an external electric field distorts the atoms slightly, the electron cloud being attracted away from the nucleus as illustrated in Fig. 124.

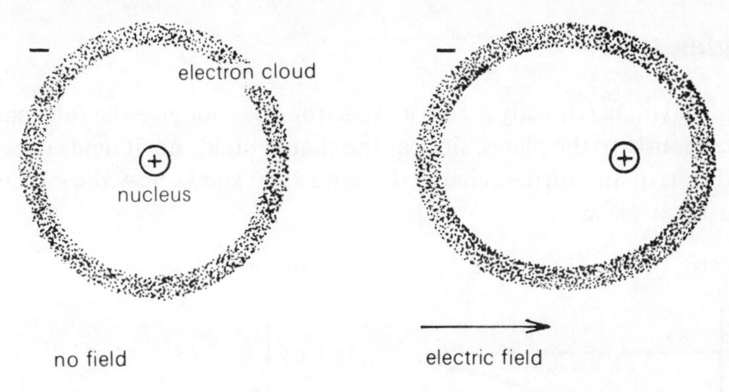

Fig. 124 Distortion of an atomic electron cloud by an external field

This separation of the charges throughout the dielectric medium virtually decreases the distance between the plates of the capacitor, and so increases its capacitance.

Energy must be supplied to the dielectric to produce this effect (called *polarisation*) and it is stored as electric potential energy.

Energy of a charged capacitor

Consider a system of capacitance C, and suppose that at some moment it has a charge $+q$ on the insulated plate. This will give the plate a potential $V = q/C$.

To add a further small charge dq, the amount of work necessary is Vdq (from the definition of potential, p. 199). Hence, for the whole process of charging the capacitor from zero to a final charge $+Q$,

$$\begin{aligned} \text{total work necessary} &= \int_0^Q V\,dq \\ &= \int_0^Q \frac{q}{C}\,dq \\ &= \frac{1}{C}\left[\frac{1}{2}q^2\right]_0^Q \\ &= \frac{1}{2}\frac{Q^2}{C}. \end{aligned} \qquad (2)$$

This amount of energy is stored in the capacitor and may be recovered if the system is allowed to discharge.

Equation (2) may alternatively be expressed in terms of the final potential V of the capacitor:

$$\text{energy of charged capacitor} = \frac{1}{2}\frac{Q^2}{C} = \frac{1}{2}QV = \frac{1}{2}CV^2.$$

Charging time

Applying a voltage directly across a capacitor does not give the full charge instantaneously to the plates, since as the charge builds up, it tends to repel the addition of any further charge. Figure 125(a) shows how the charging process takes place.

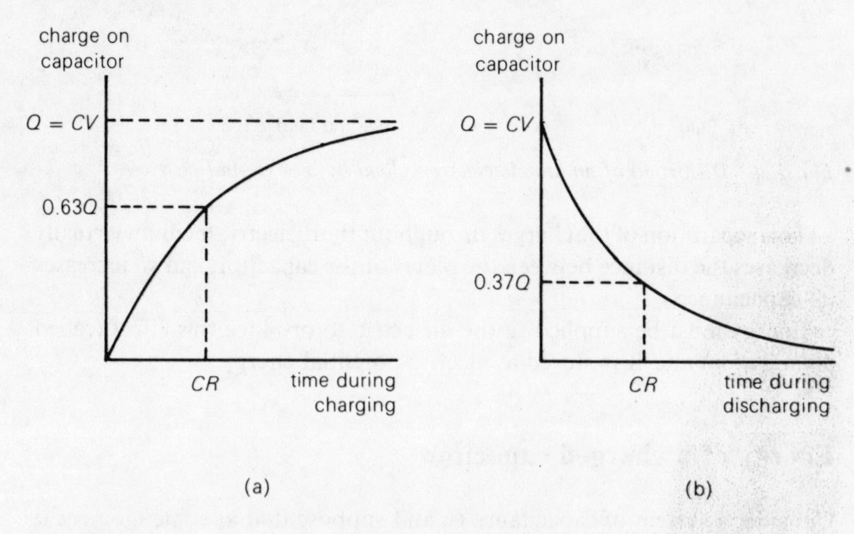

Fig. 125 Graphs of time and charge on a capacitor during (a) charging, and (b) discharging

The graph is an exponential curve, beginning at the moment that the circuit is closed, and rising asymptotically to the value given for the charge = capacitance × voltage.

The equation of the curve may be calculated by assuming that the total resistance of the circuit is R, the applied voltage is V, and the capacitance C. Then the charge on the capacitor at time t is given by

$$\text{charge } Q = CV(1 - e^{-t/CR}).$$

The term CR is called the *time-constant* of the circuit. After a time $t = CR$, the charge on the capacitor will have risen to a value given by

$$Q = CV(1 - e^{-1}).$$

Since $e = 2.718$, $e^{-1} = 0.37$ approximately and $(1 - e^{-1}) = 0.63$, so that **after a time equal to the time-constant of the circuit, the charge on the capacitor** (and hence the potential difference across it) **will have reached 0.63 of its final value.**

The discharging process follows the inverse curve (Fig. 125(b)), and the charge remaining on the capacitor at time t is given by

$$Q = CVe^{-t/CR}$$

where V was the potential difference across the capacitor when it was fully charged.

If a large capacitor, about 100 μF, is used with a radio-type resistor of about 20 000 Ω, then $CR = 2$ s. A long time-constant such as this may be studied experimentally.

The charge on a capacitor may be measured by discharging it through a ballistic galvanometer (p. 53). The result will only be correct if the time-constant of the capacitor circuit is very small compared with the time of swing of the galvanometer needle, but this will naturally be true if the galvanometer leads are connected directly to the plates of the capacitor.

Circuits with long time-constants are used in many practical applications, e.g. to activate the flashing lights set up near roadworks, and the regular sound pulses emitted by sonar.

EXERCISES

5 A 0.5 μF capacitor is charged to 100 V. Find the energy stored in it.

6 A 0.1 μF capacitor is given a charge of 80 μC. Find the energy stored in it.

7 A 50 μF capacitor is being charged from a 20 V supply through a resistance of 1000 Ω. Calculate how great a charge it will have after 0.01 s.

Examination questions 26

1 (a) Define *capacitance*.

Describe an experiment to show that the capacitance of an air-spaced parallel-plate capacitor depends on the separation of the plates.

State three desirable properties of a dielectric material for use in a high-quality capacitor.

(b) Two plane parallel-plate capacitors, one with air as dielectric and the other with mica as dielectric, are identical in all other respects. The air capacitor is charged from a 400 V d.c. supply, isolated, and then connected across the mica capacitor which is initially uncharged. The potential difference across this parallel combination becomes 50 V. Assuming the relative permittivity of air to be 1.00, calculate the relative permittivity of mica.

(c) Compare the energy stored by the single charged capacitor in part (b) with the energy stored in the parallel combination.

Comment on these energy values. [AEB, Specimen paper 1977]

2 A 2 μF capacitor is charged to a potential of 200 V and then isolated. When it is connected in parallel with a second capacitor which is initially uncharged, the common potential becomes 40 V. The capacitance of the second capacitor is

A 2 μF B 4 μF C 6 μF D 8 μF E 16 μF [C]

3 By means of a two-way switch, a 10 μF capacitor is connected first across a battery of e.m.f. 2.0 V and negligible internal resistance and then quickly across a 5.0 MΩ resistor R.

(a) Calculate the charge on the capacitor and the energy stored in it when connected to the battery.

(b) Calculate the charge on the capacitor when it has been connected across R for 50 s.

(c) Using the same axes, sketch graphs showing how the charge on the capacitor varies with time when the resistance of R is (i) 5.0 MΩ, (ii) 10.0 MΩ. Label your graphs (i) and (ii) respectively. e = 2.718. [JMB]

4 Explain what is meant by the *relative permittivity* of a material. How may its value be determined experimentally?

A capacitor of capacitance 9.0 μF is charged from a source of e.m.f. 200 V. The capacitor is now disconnected from the source and connected in parallel with a second capacitor of capacitance 3.0 μF. The second capacitor is now removed and discharged. What charge remains on the 9.0 μF capacitor? How many times would the process have to be performed in order to reduce the charge on the 9.0 μF capacitor to below 50 % of its initial value? What would the p.d. between the plates of the capacitor now be? [L]

5 (a) The two plates X and Y of a parallel-plate air capacitor, each 1.5×10^{-2} m^2 in area, are 4 mm apart. One plate is earthed and the other is raised to a potential of +4 kV and then insulated. What is the electric field strength between the plates?

(b) The charge on the capacitor, measured with a ballistic galvanometer, is found to be 0.13 μC. Calculate its capacitance, the energy stored in the charged capacitor and the value of ε_0, the permittivity of free space.

(c) With the capacitor charged afresh from the 4 kV supply, and insulated, an insulated aluminium sheet 2 mm thick and of the same shape and size as the capacitor plates is inserted symmetrically between these plates so that it is parallel to them and does not project beyond the region between them. How does the insertion of this sheet affect the electric field in the space between X and Y? Explain as fully as you can. [O]

27 Magnetic fields

The concept of a magnetic field is a familiar one, being originally described as the region around a bar magnet in which its influence can be felt. Since the basis of magnetism is now considered to be the movement of electrons in their orbits about the nucleus (Chap. 7), the modern approach to the subject is via electromagnetism, and a magnetic field is now defined in terms of the force which it exerts on a current-carrying conductor.

Field strength

The unit of magnetic field strength is the *tesla* (T).

A force of 1 newton is exerted by a magnetic field of 1 tesla acting on a conductor 1 metre long carrying a current of 1 ampere, placed at right angles to the field (Fig. 126(a)).

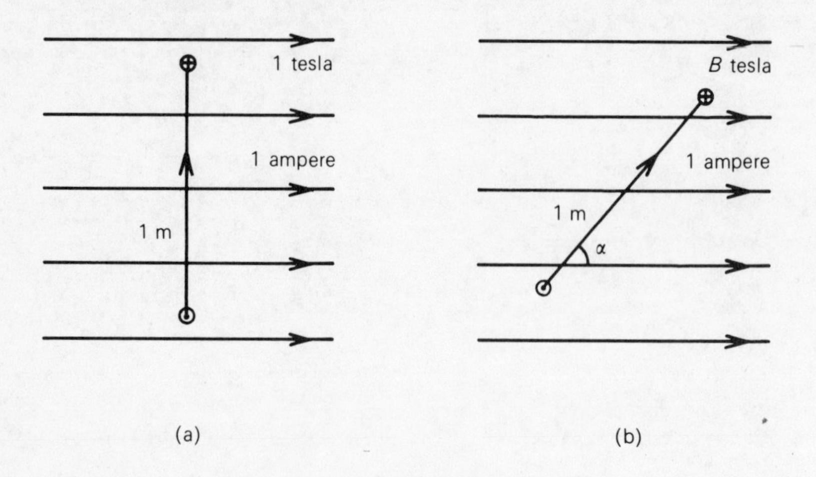

Fig. 126 Current-carrying conductors in a magnetic field

Magnetic field strength is sometimes called *magnetic induction* or *flux density*, and is given the symbol B.

The direction of the force is given by Fleming's left-hand rule. The conductor shown in Fig. 126(a) would experience a force downwards into the plane of the diagram.

If the conductor cuts the magnetic field B at an angle α as shown in Fig. 126(b), the force is reduced to $\mathbf{B} \sin\alpha$ newton per amp per metre. This relation was discovered by Ampère. Hence there is no force exerted on a conductor which lies parallel to the lines of force.

Definition of the ampere

The preceding theory is used to define the unit of current, as follows:

Two currents of 1 ampere, flowing in infinitely long parallel straight wires 1 metre apart in a vacuum, produce a force between the wires of 2×10^{-7} newton per metre.

Force on a rectangular coil in a magnetic field

Consider a current I flowing around a rectangular coil of wire PQRS. A magnetic field of intensity B acts as shown in Fig. 127, making an angle α with the plane of the coil.

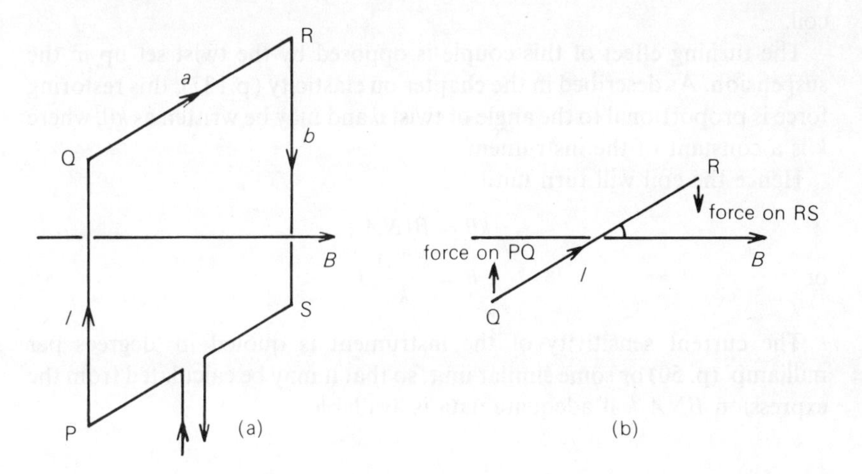

Fig. 127 Couple acting on a rectangular coil carrying a current in a magnetic field

This field will exert forces on the four sides of the coil. On QR and SP, the forces will be equal to $BIa \sin\alpha$, where QR = SP = a, and will act inwards towards the centre of the coil. If the coil is rigid, the two forces will balance each other.

On PQ and RS, the forces will be equal to BIb, where PQ = RS = b, and will act in the directions shown in Fig. 127(b), at right angles to the sides of the coil and also to the field. This will have a turning effect on the coil, tending to rotate it clockwise. Such a pair of forces forms a couple.

The torque of the couple is measured by the product (p. 107) force × (perpendicular distance between lines of action of forces)

$$= BIb.\ a \cos \alpha$$
$$= \mathbf{BIA \cos \alpha}$$

where $A = ab$, the area of the plane of the coil.

If the coil is free to turn, it will do so until $\cos\alpha = 0$, i.e. $\alpha = 90°$, and will set in equilibrium with its plane at right angles to the field.

The moving-coil milliammeter

Referring back to Fig. 25 of Chap. 6, it will be recalled that the magnetic field used in a moving-coil meter is shaped so that the lines of force lie along the radii of the cylindrical core of the coil. The plane of the coil is thus always in the direction in which the field is acting, and the couple exerted on the coil will have the constant value

$$BINA,$$

where B is the magnetic field strength, I is the current flowing through the coil, N is the number of turns in the coil, and A is the area of the plane of the coil.

The turning effect of this couple is opposed by the twist set up in the suspension. As described in the chapter on elasticity (p.137), this restoring force is proportional to the angle of twist θ and may be written as $k\theta$, where k is a constant of the instrument.

Hence the coil will turn until

$$k\theta = BINA$$

or

$$\theta = \frac{BNA}{k} I.$$

The current sensitivity of the instrument is quoted in 'degrees per milliamp' (p. 50) or some similar unit, so that it may be calculated from the expression BNA/k if adequate data is available.

Magnetic moment

The current-carrying coil considered above may be seen simply as being an electromagnet. In an external field, the coil will tend to turn until its own magnetic axis lines up with this field, which happens when the plane of the coil is at right angles to the field.

The expression IA (current × area of coil) represents the strength of this electromagnet, and it is called the *magnetic moment* of the coil. If there are N turns of wire making up the coil, its **total magnetic moment will be NIA** ampere/m^2 or joule/tesla.

A permanent magnet also has a magnetic moment and, if free to turn, will line up with its axis in the direction of any external field. Its magnetic moment m is defined in the same way as that for a coil carrying a current, from the strength of the couple acting on the magnet when in an external field B.

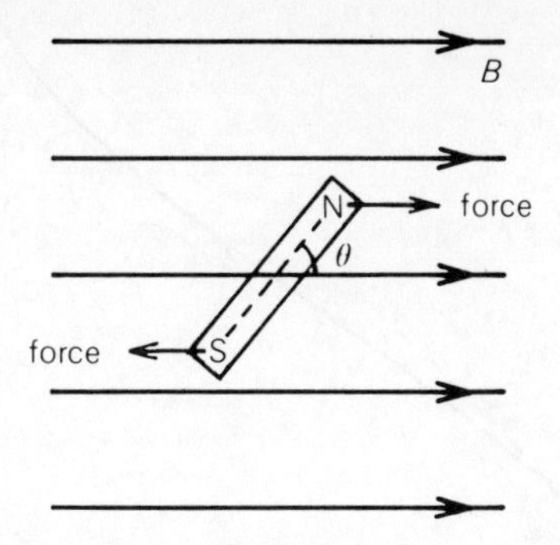

Fig. 128 Couple acting on a permanent magnet in a magnetic field

If the axis of the magnet makes an angle θ with the direction of the field (Fig. 128) then

$$\begin{aligned}\text{couple acting on magnet} &= (\text{field strength}) \times (\text{magnetic moment}) \sin\theta \\ &= Bm \sin\theta.\end{aligned}$$

EXERCISES

1 How great a force would be exerted on a straight conductor 2 m long, carrying a current of 0.5 A, by a magnetic field of 2×10^{-5} T acting at right angles to the conductor?

2 A rectangular loop of wire which encloses an area of 0.1 m^2 carries a current of 2 A. If its plane is at an angle of 45° to a magnetic field of 2 T, calculate the torque of the resultant couple acting on the loop.

3 An electromagnet is held in equilibrium with its axis at right angles to an external magnetic field of 0.4 T. If the couple needed to do this is 2×10^{-2} N m, find the magnetic moment of the electromagnet.

Biot–Savart law

Consider the magnetic field strength B produced at a point P by a current I flowing in a small straight section x of a wire, distant r from P and set at an angle α as shown in Fig. 129. The field strength at P will depend on several factors, and probably vary:

(a) proportionally to the current, I,
(b) proportionally to the length of the section of wire, x,
(c) according to the inverse square law of distance, $1/r^2$,
(d) with the sine of the angle α (Ampère's law, p. 219).

The Biot–Savart law,

$$B \propto \frac{Ix \sin\alpha}{r^2}$$

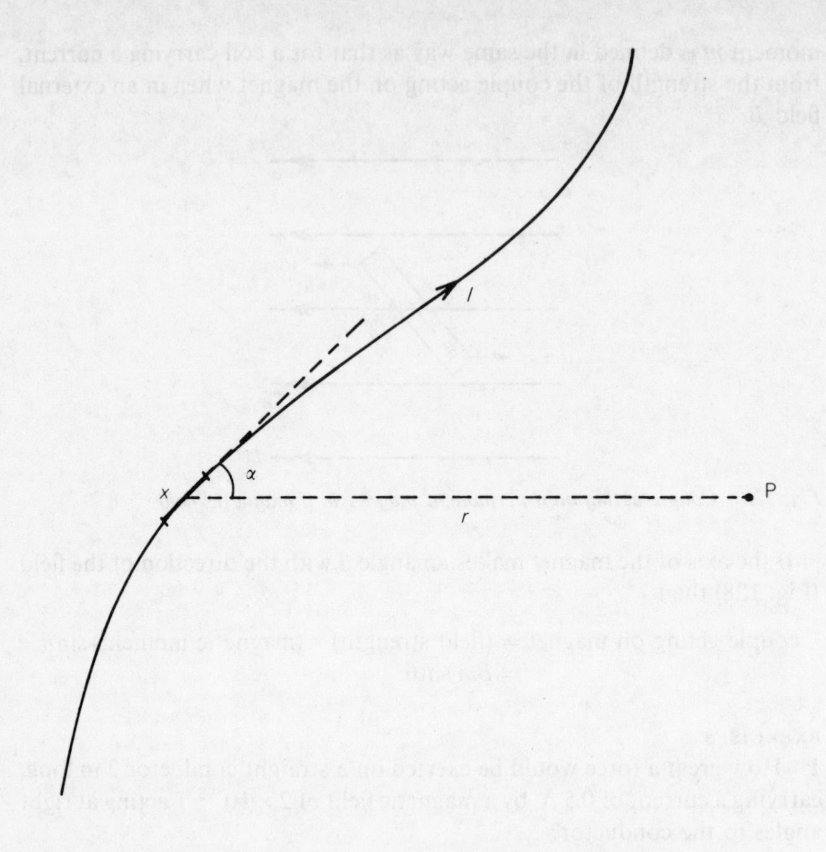

Fig. 129 Calculation of magnetic field strength due to a current

cannot be proved directly, but its application to particular shapes of current-carrying conductor gives results which agree with experiment.

As in the case of electric fields (p. 200), the relation is made into an equation by introducing a numerical constant $\mu/4\pi$, where μ depends on the medium surrounding the conductor, and is called its *permeability*. The equation is then

$$B = \frac{\mu}{4\pi}\frac{Ix\sin\alpha}{r^2}.$$

For a vacuum (and so approximately for air) μ has the special symbol $\mu_0 = 4\pi \times 10^{-7}$ henry metre^{-1}. This value follows from the definition of the ampere.

For other substances, their *relative permeability* is often given; this is the ratio μ/μ_0.

It is possible, using the Biot–Savart equation, to calculate the magnetic field produced when a current flows around circuits of particular shapes.

Field at the centre of a circular coil

For each element of the circle of radius a, I and r are constant, and the angle α is always 90°. If the length of the small section considered is written as dx, using calculus notation,

$$\text{field at P} = \frac{\mu}{4\pi}\frac{1}{a^2}\int_0^{2\pi a} dx$$

$$= \frac{\mu}{4\pi}\frac{I}{a^2}2\pi a$$

$$= \frac{\mu I}{2a}.$$

If there are N turns in the coil, but it is still thin enough to be considered of uniform radius, then

$$\text{field at P} = \frac{\mu NI}{2a}$$

With the direction of the current as shown in Fig. 130, the direction of the field will be *into* the plane of the diagram.

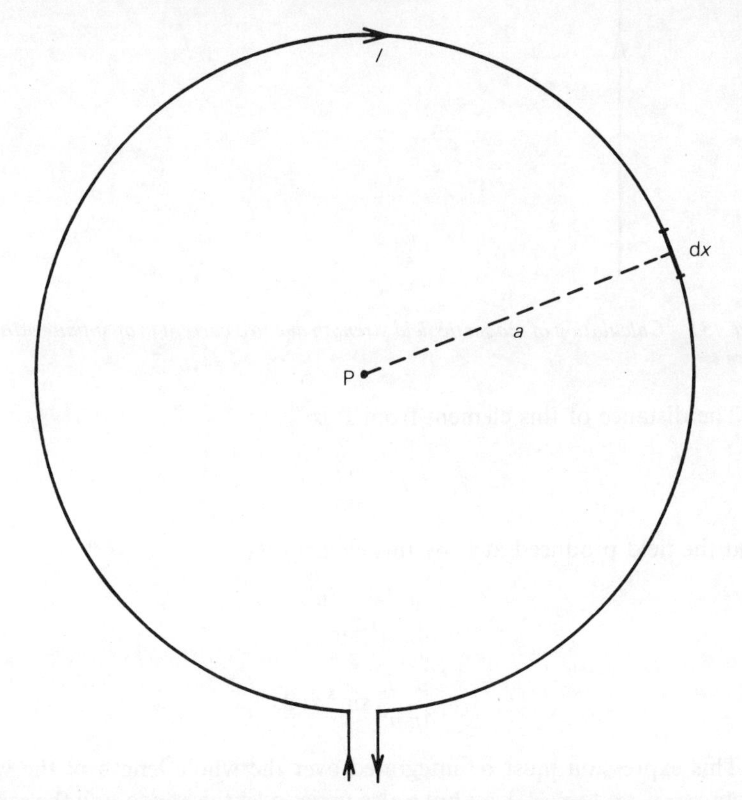

Fig. 130 Calculation of magnetic field strength due to a current in a circular coil

Field due to a long straight wire

To calculate the field strength at a point P, distance a from the wire, consider a small element of the wire, dx, at a distance x from the foot of the perpendicular from P to the wire.

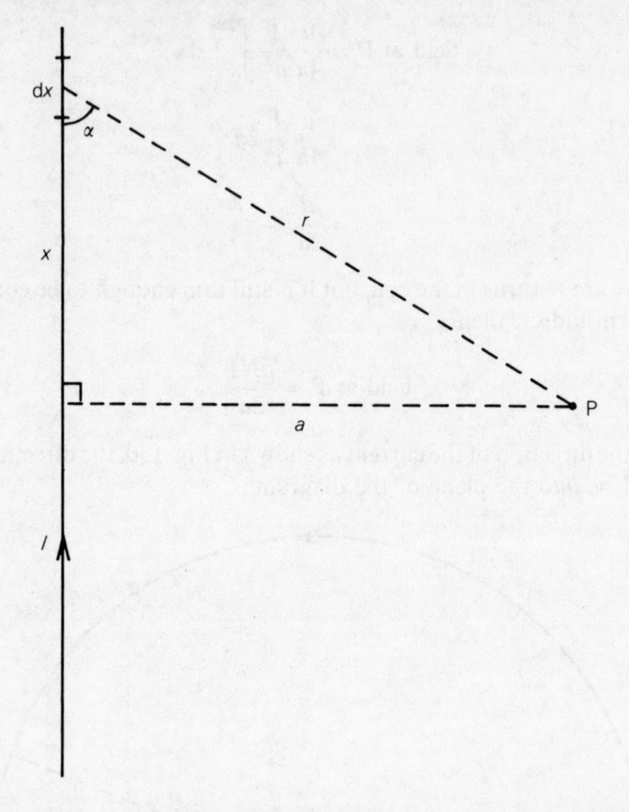

Fig. 131 Calculation of magnetic field strength due to a current in an infinite straight wire

The distance of this element from P is

$$r = \frac{a}{\sin\alpha}$$

and the field produced at P by this element is

$$\frac{\mu}{4\pi}\frac{I\,dx\sin\alpha}{a^2/\sin^2\alpha}$$

$$= \frac{\mu}{4\pi}\frac{I}{a^2}\sin^3\alpha\,.\,dx. \tag{1}$$

This expression must be integrated over the whole length of the wire, from $x = -\infty$ to $x = +\infty$, but α also varies over this range, and the easiest procedure is to change the variable from dx to $d\alpha$.

Since $a/x = \tan\alpha$, differentiating both sides of the equation gives

$$-\frac{a}{x^2}\mathrm{d}x = \sec^2\alpha\,\mathrm{d}\alpha.$$

Substituting $\tan\alpha$ for a/x,

$$-\frac{1}{a}\tan^2\alpha\,\mathrm{d}x = \sec^2\alpha\,\mathrm{d}\alpha$$

or

$$-\frac{1}{a}\frac{\sin^2\alpha}{\cos^2\alpha}\mathrm{d}x = \frac{1}{\cos^2\alpha}\mathrm{d}\alpha$$

$$\Rightarrow \quad \mathrm{d}x = -\frac{a}{\sin^2\alpha}\mathrm{d}\alpha.$$

The minus sign occurs because α decreases as x increases. Expression (1) can now be rewritten entirely in terms of α:

$$\text{field at P due to element of wire} = -\frac{\mu}{4\pi}\frac{I}{a^2}\sin^3\alpha\frac{a}{\sin^2\alpha}\mathrm{d}\alpha$$

$$= -\frac{\mu}{4\pi}\frac{I}{a}\sin\alpha\,\mathrm{d}\alpha.$$

For the whole of the wire, this must be integrated for α from 0° to 180°, giving

$$\text{field at P} = \frac{\mu}{4\pi}\frac{I}{a}\int_0^{180^\circ}\sin\alpha\,\mathrm{d}\alpha$$

$$= \frac{\mu}{4\pi}\frac{I}{a}\Big[-\cos\alpha\Big]_0^{180^\circ}$$

$$= \frac{\mu\mathbf{I}}{2\pi\mathbf{a}}. \qquad (2)$$

With the direction of the current as shown in Fig. 131, the direction of the field will be *into* the plane of the diagram.

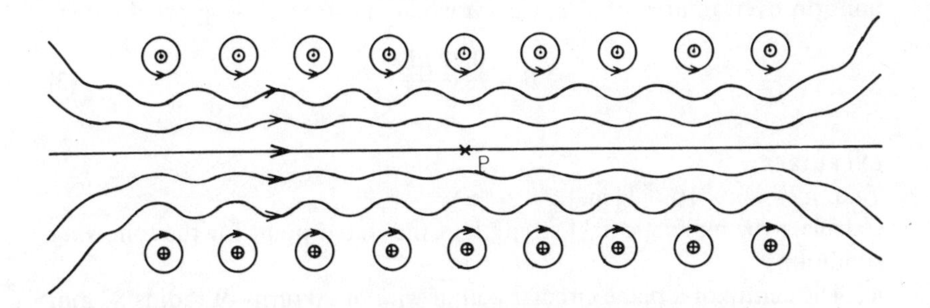

Fig. 132 Magnetic field due to a solenoid

Field inside a long solenoid

A solenoid consists of a large number of loops of wire side by side. If there are N turns per metre and the current flowing through the coils is I, then the field at the centre of the solenoid is

$$B = \mu NI.$$

This expression is strictly true only if the coils are thin and closely packed, and the solenoid is infinitely long. The proof of the formula involves calculus methods (Ampère's theorem).

Field due to Helmholtz coils

These are two coils of equal radius a carrying the same current I, flowing in the same direction. They are spaced apart a distance equal to their radius.

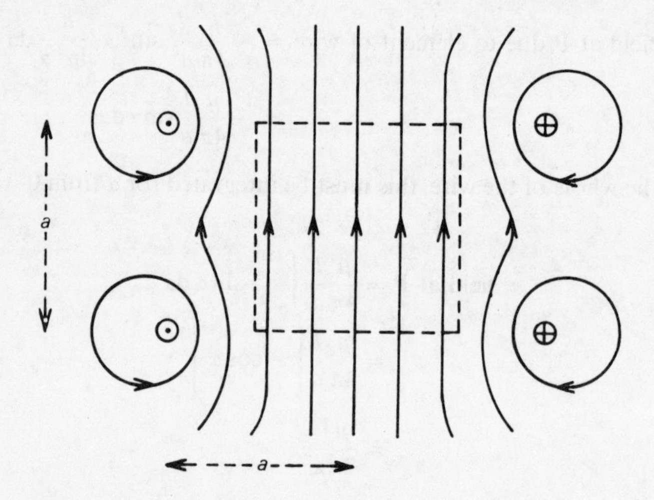

Fig. 133 Magnetic field produced by Helmholtz coils

The field between them is as shown in Fig. 133, and may be considered uniform over an area of about $a \times a$ where its strength is approximately

$$B = \frac{0.72\,\mu I}{a^2}. \tag{3}$$

EXERCISES

Take $\mu_0 = 4\pi \times 10^{-7}\ \mathrm{H\,m^{-1}}$.

Calculate the magnetic field strength at the stated point for the following conditions:

4 The centre of a plane circular coil of wire of 20 turns of radius 80 mm carrying a current of 0.25 A.

5 0.1 m away from a single straight wire carrying a current of 1 A.

6 At the centre of a long solenoid of 1 turn per mm wound on a soft-iron core of relative permeability 60, carrying a current of 0.05 A.
7 In the uniform region of the field due to a pair of Helmholtz coils each of 100 turns of average radius 60 mm, carrying a current of 0.1 A.

The current balance

This apparatus is designed to measure a current in absolute terms, by balancing the repulsion between two equal but opposite currents against a small weight on the other side of the balance.

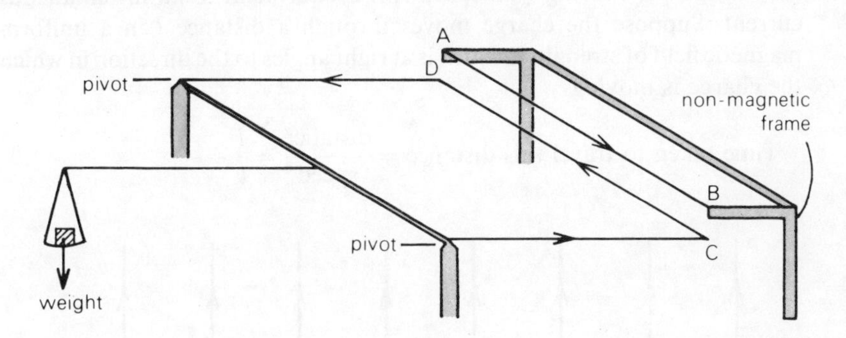

Fig. 134 A simple form of current-balance

A simple form of the apparatus is shown in Fig. 134. AB and CD are the two conducting wires. AB is fixed to a rigid non-magnetic frame, and CD is supported parallel and close to AB so that it is free to turn about a horizontal axis. Connections are made to the wires so that the same current flows through them in series.

The apparatus is first balanced when no current is flowing, by placing a suitable load in a scale pan hanging from an arm on the far side of the axis. When the current is switched on, CD will be repelled downwards by the magnetic field set up around AB. The balance is restored to its original position by adding an additional small mass m to the scale pan, only a few milligrams in practice.

The equilibrium distance between AB and CD should be measured using vernier callipers, and then equation (2) may be applied to give an expression for the field strength at CD due to the current I in AB (this is only true if the wires are set close enough together to appear infinitely long).

The force exerted on CD by this field is given by [$I \times$ (field strength)] newtons per metre length of CD. The length of the wire may be measured with a millimetre rule.

For simplicity, the balance is constructed with arms of equal length, so that in the equilibrium position the repelling force between the two parts of the current will equal the additional weight mg which was added to the scale pan.

In this form, the experiment cannot be highly accurate. A different form of apparatus is used at the National Physical Laboratory to standardise the ampere. The magnetic field in this is set up between two cylindrical coils, and a small coil is suspended inside them; the geometry of this arrangement makes the calculation of the field simpler. The apparatus is enclosed in a glass case and its temperature is carefully controlled. An accuracy of about 0.001 % is claimed for results recently obtained.

Force on a moving charge

If a charge q is moving with speed v, it is equivalent to an instantaneous current. Suppose the charge moves through a distance l in a uniform magnetic field of strength B which is at right angles to the direction in which the charge is moving.

$$\text{Time taken to travel this distance} = \frac{\text{distance}}{\text{speed}} = \frac{l}{v},$$

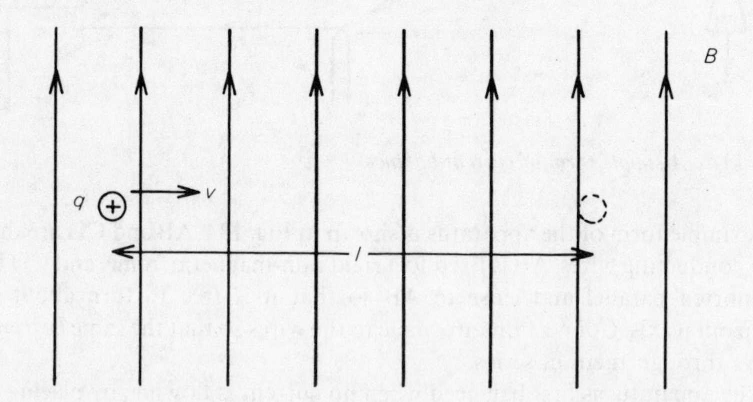

Fig. 135 Calculation of the force on a charge moving in a magnetic field

and during this time it is equivalent to a current

$$I = \frac{\text{charge}}{\text{time}} = \frac{q}{l/v} = \frac{qv}{l}.$$

The force exerted by the field B on this current is
(field strength) × (current) × (length of conductor)

$$= B\frac{qv}{l}l$$

$$= \boldsymbol{Bqv}. \qquad (4)$$

The direction of this force is at right angles to both the field and the direction of motion of the charged particle, and its result will be to deflect the particle to move in **an arc of a circle**, without any change of speed.

This is the principle used in the mass spectrograph (Chap. 28) which sorts a beam of charged particles according to their different masses. It is also applied in J. J. Thomson's classic experiment by which the ratio charge/mass for the electron was first determined, using a method similar to the following.

Determination of *e/m* for a charged particle

Figure 136 shows the Teltron vacuum tube available for the measurement of the ratio (electron charge/electron mass) in the school laboratory.

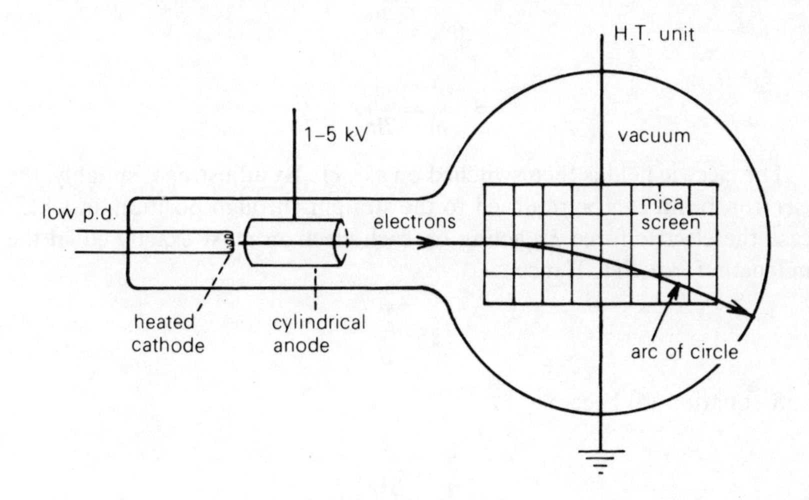

Fig. 136 Teltron vacuum tube used for measurement of e/m for electrons

The beam of electrons is obtained by thermionic emission from a heated cathode, and accelerated down the tube by an anode potential of a few kilovolts. The anode is cylindrical, so that the electrons pass through it and enter the glass bulb which is about 0.2 m across. A mica screen is mounted in the centre of the bulb. The face of the screen is marked off in centimetres and its back is coated with a phosphor so that it will fluoresce when struck by the electron beam.

Horizontal plates at the top and bottom of the screen can be connected to a high p.d. so that an electric field E can be set up between them. A pair of Helmholtz coils are also available to be used with this tube. They can be arranged one on each side of the bulb to produce a uniform horizontal magnetic field B throughout the central region.

When the apparatus has been checked, the current in the Helmholtz coils is switched on. The magnetic field causes the beam of electrons to be deflected into the arc of a circle, and readings are taken from the luminous trace showing up against the mica screen, so that the radius r of this circle may be calculated.

From equation (4),

$$\text{force on each electron} = Bev,$$

where e is the electronic charge and v is the speed with which it is moving.

From equation (2) of Chap. 13 (p. 111), the relation between the force-towards-the-centre and the speed of an electron moving in a circle is

$$\text{force} = \frac{mv^2}{r}$$

where m is the electronic mass. Hence,

$$Bev = \frac{mv^2}{r}$$

$$\Rightarrow \quad \frac{e}{m} = \frac{v}{Br}. \tag{5}$$

The electric field is then switched on as well. By adjusting it suitably, the electron beam can be returned to the straight-through position, in which case the electric force Ee acting on each electron must exactly equal the magnetic force Bev. Hence

$$v = \frac{E}{B}$$

and equation (5) becomes

$$\frac{e}{m} = \frac{E}{B^2 r}.$$

E is calculated from the p.d. between the two deflector plates, read from a voltmeter, and their separation, indicated from the grid on the mica screen. B is calculated from the dimensions of the Helmholtz coils, and the current through them, read from an ammeter, applying equation (3) of this chapter.

In practice, the beam of electrons produces a rather poorly defined trace as some of the electrons lose energy and speed through causing fluorescence.

EXERCISES

8 A fine beam of electrons is moving at $3 \times 10^7\ \text{m s}^{-1}$ in a vacuum in a magnetic field of 10^{-3} T. If e/m for the electron is $1.76 \times 10^{11}\ \text{C kg}^{-1}$, find the radius of the circle in which the electrons are travelling.

9 In an experiment of the Thomson type, a beam of electrons is observed to be undeflected when travelling through combined magnetic and electric fields of strengths 2×10^{-4} T and $10\ \text{kV m}^{-1}$ respectively. Calculate the average speed of the electrons.

Examination questions 27

1 Electrons, accelerated from rest through a potential difference of 3000 V, enter a region of uniform magnetic field, the direction of the field being at right angles to the motion of the electrons. If the flux density is 0.010 T, calculate the radius of the electron orbit.
(Assume that the specific charge, e/m for electrons $= 1.8 \times 10^{11}\,\mathrm{C\,kg^{-1}}$.)
[AEB, Specimen paper 1977]

2 An electron moves in a circular path in a vacuum, under the influence of a magnetic field. The radius of the path is 10^{-2} m and the flux density is 10^{-2} T. Given that the specific charge of the electron is $-1.76 \times 10^{11}\,\mathrm{C\,kg^{-1}}$, calculate (a) the period of its orbit, (b) the period if the electron had only half as much energy. [C]

3 (a) A long straight wire of radius a carries a steady current. Sketch a diagram showing the lines of magnetic flux density (B) near the wire and the relative directions of the current and B. Describe, with the aid of a sketch graph, how B varies along a line from the surface of the wire at right-angles to the wire.
(b) Two such identical wires R and S lie parallel in a horizontal plane, their axes being 0.10 m apart. A current of 10 A flows in R in the opposite direction to a current of 30 A in S. Neglecting the effect of the earth's magnetic flux density, calculate the magnitude and state the direction of the magnetic flux density at a point P in the plane of the wires if P is (i) midway between R and S, (ii) 0.05 m from R and 0.15 m from S.

The permeability of free space, $\mu_0 = 4\pi \times 10^{-7}\,\mathrm{H\,m^{-1}}$. [JMB]

4 Two long parallel straight wires P and Q form part of a closed circuit carrying an alternating current. The current flows out through P and returns to the source through Q. The net mutual effect between wires P and Q is

A attraction at all times.
B repulsion at all times.
C alternate repulsion and attraction.
D alternate contraction and elongation.
E zero. [L]

5 A straight wire 2 m long lies at 30° to a uniform magnetic field of flux density 2×10^{-5} T and carries a current of 0.02 A. The magnitude of the force experienced by the wire is

A 10^{-7} N
B 2×10^{-7} N
C 3.36×10^{-7} N
D 4×10^{-7} N
E 6.72×10^{-7} N
[O]

6 Write down a formula for the magnitude of the force on a straight current-carrying wire in a magnetic field, explaining clearly the meaning of each symbol in your formula.

Derive an expression for the couple on a rectangular coil of N turns and dimensions $a \times b$ carrying a current I when placed in a uniform magnetic field of flux density B at right angles to the sides of the coil of length a and at an angle θ to the sides of length b.

Describe briefly how you would demonstrate experimentally that the couple on a plane coil in a uniform field depends only on its area and not on its shape.

A circular coil of 50 turns and area $1.25 \times 10^{-3}\ \text{m}^2$ is pivoted about a vertical diameter in a uniform horizontal magnetic field and carries a current of 2 A. When the coil is held with its plane in a north–south direction, it experiences a couple of 0.04 N m. When its plane is east–west, the corresponding couple is 0.03 N m. Calculate the magnetic flux density. (Ignore the earth's magnetic field.) [OC]

28 Radioactivity and the nucleus

Natural radioactivity

The naturally radioactive elements occupy positions at the end of the periodic table. They have large nuclei which are made up of about 90 protons together with about 140 neutrons. Little is yet known for certain about the force which holds protons and neutrons together in the nucleus (much present-day research is devoted to this subject), but it seems to have reached its limit in uniting this number of nucleons, and from time to time an atom disintegrates.

Natural radioactive emissions are α-particles, β-particles, and γ-rays.

Artificial radioactivity

Through high-energy collisions, the atoms of most other elements have now been 'split'. Sometimes the result of a collision is the creation of a temporarily unstable nucleus, which may exist for some time before disintegrating. The *atomic piles* at Harwell and other research establishments are used to produce radioactive carbon, iodine, gold, etc., for use in industry and medicine.

Such radioactive substances emit α, β and γ-rays, and also may decay by *fission*. This word describes a splitting of the nucleus into two unequal parts, often accompanied by the emission of neutrons.

Symbols

In this area of study, the atom of a particular element E is described in the form

$$^{\text{mass number}}_{\text{atomic number}}\text{E}$$

e.g. $^{232}_{90}$Th refers to the element thorium, which occupies position 90 in the periodic table, having 90 protons in its nucleus, and correspondingly 90 orbiting electrons, in its neutral state. The atomic number, denoted by Z, determines the chemical behaviour of the element, its valency, etc.

The mass number, 232 in this example, means that the nucleus is made up of 232 *nucleons* (protons and neutrons together). The mass number is given the symbol A.

Isotopes

A particular element can have only one atomic number Z, but may exist in several forms known as *isotopes*, having different masses. An additional neutron or two in the nucleus may still give a stable structure, and most elements have been found to have more than one natural isotope. Artificially radioactive elements are unstable isotopes.

A particular isotope is referred to by adding its mass number after the name of the element, e.g. carbon-14.

α-emission

α-particles are positively charged and have mass number 4. Early experiments showed that these particles are the same as helium nuclei.

When an α-particle is emitted from a radioactive nucleus $^{A}_{Z}\mathrm{E}$, this reduces the mass number by 4 and the atomic number by 2. The atom no longer belongs to its original element E, but to the element E′ two places earlier in the periodic table; it has been transmuted. The new nucleus is not necessarily a stable form, and may disintegrate again.

The change is represented by

$$^{A}_{Z}\mathrm{E} \rightarrow {}^{(A-4)}_{(Z-2)}\mathrm{E}' + {}^{4}_{2}\mathrm{He}. \qquad (1)$$

β-emission

β-particles are negatively charged and of negligible mass in this connection. Experiments have proved that they are electrons.

When a β-particle is emitted from a radioactive nucleus $^{A}_{Z}\mathrm{E}$, this leaves the mass number unchanged but increases the atomic number by 1. This has been interpreted as a change inside the nucleus, one neutron becoming a proton. The atom is now an isotope of the element E″, one place further on in the periodic table.

This change is written as

$$^{A}_{Z}\mathrm{E} \rightarrow {}^{A}_{(Z+1)}\mathrm{E}'' + {}^{0}_{-1}\mathrm{e}. \qquad (2)$$

γ-emission

γ-rays are short-wave electromagnetic radiation, belonging to the extreme end of the general spectrum considered in Chap. 10.

Many radioactive distintegrations involve emission of a γ-ray at the same time as an α- or β-particle; γ-ray emission does not occur on its own. It appears to be necessary to balance the energy equation of the disintegration, but is not included in reaction equations such as (1) and (2).

Half-life

Radioactivity is a continuous but random process. There is no way of telling when a particular nucleus will disintegrate, but measurements in bulk can provide a value for the *half-life*, i.e. **the period of time during which one-half of the total number of nuclei will disintegrate.**

This varies widely from element to element. Some experimental results are given in Table 7.

Table 7

Element	Half-life
Uranium-238	4.5×10^9 years
Radium-226	1620 years
Radon-222	3.8 days
Gold-198	2.7 days
Iodine-131	8.0 days
Strontium-90	28 years
Cobalt-60	5.25 years
Phosphorus-30	2.5 mins
Carbon-14	5730 years

It is essential to realise that, although half of the radioactive element disappears during the half-life, it is only a transmutation, and the actual *mass* of the sample remains constant. Experiments on radioactive decay must involve either chemical analysis or, more easily, radioactivity measurements.

The number of disintegrations to be expected in any period of time is proportional to the number of radioactive nuclei present at the beginning of that period. This is a reasonable hypothesis, and is borne out by experiment. It gives an exponential relationship between time t and the number of nuclei, N, present at that time (Fig. 137. A similar graph applied to the discharging of a capacitor, p. 214).

The mathematical expression of this is

$$\frac{dN}{dt} \propto -N$$

which can be written

$$\frac{\mathbf{d}N}{\mathbf{d}t} = -\lambda N \tag{3}$$

where λ is a constant characteristic of the nucleus concerned, called the *radioactive decay constant.*

Equation (3) may be interpreted as

number of disintegrations observed per second (*activity* of specimen) $= \lambda \times$ (number of nuclei present).

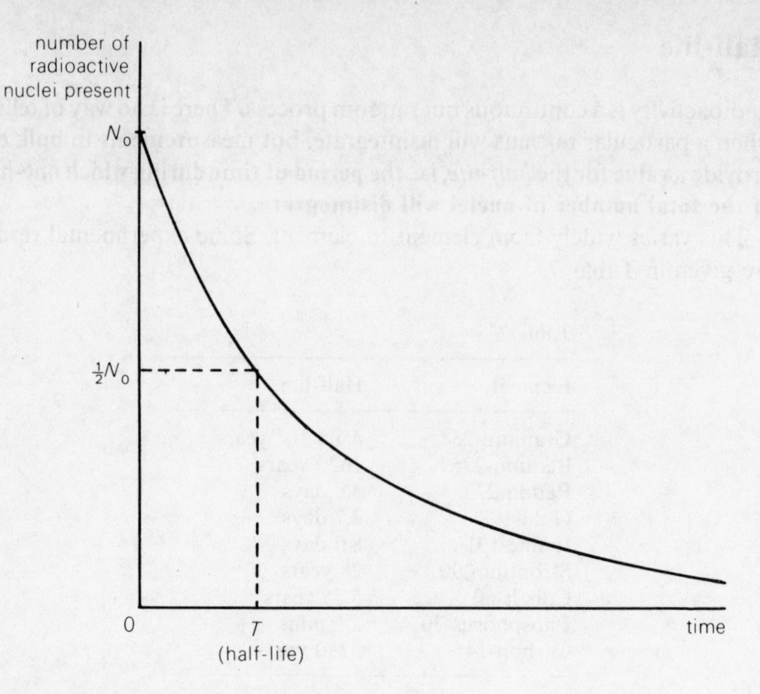

Fig. 137 Decay curve for a radioactive element

The solution to this equation is

$$N = N_0 e^{-\lambda t}$$

where N_0 was the number of nuclei present at the time chosen as $t = 0$.

The constant λ is related to the half-life T, which is the value of t when $N = \frac{1}{2}N_0$,

$$\text{i.e. } \tfrac{1}{2}N_0 = N_0 e^{-\lambda T}.$$

Taking logs to base e on both sides of this equation,

$$\log_e \tfrac{1}{2} + \log_e N_0 = \log_e N_0 - \lambda T$$

$$\Rightarrow \quad \log_e 1 - \log_e 2 = -\lambda T$$

$$\Rightarrow \quad \log_e 2 = \lambda T$$

$$\Rightarrow \quad \lambda = \frac{\log_e 2}{T} = \frac{0.693}{T}$$

or, in words, the decay constant $= \dfrac{0.693}{\text{half-life}}$.

Radioactive series

When uranium-238 decays, the new element formed is also radioactive and decays in its turn, and so on through a whole series of changes until it ends

Table 8

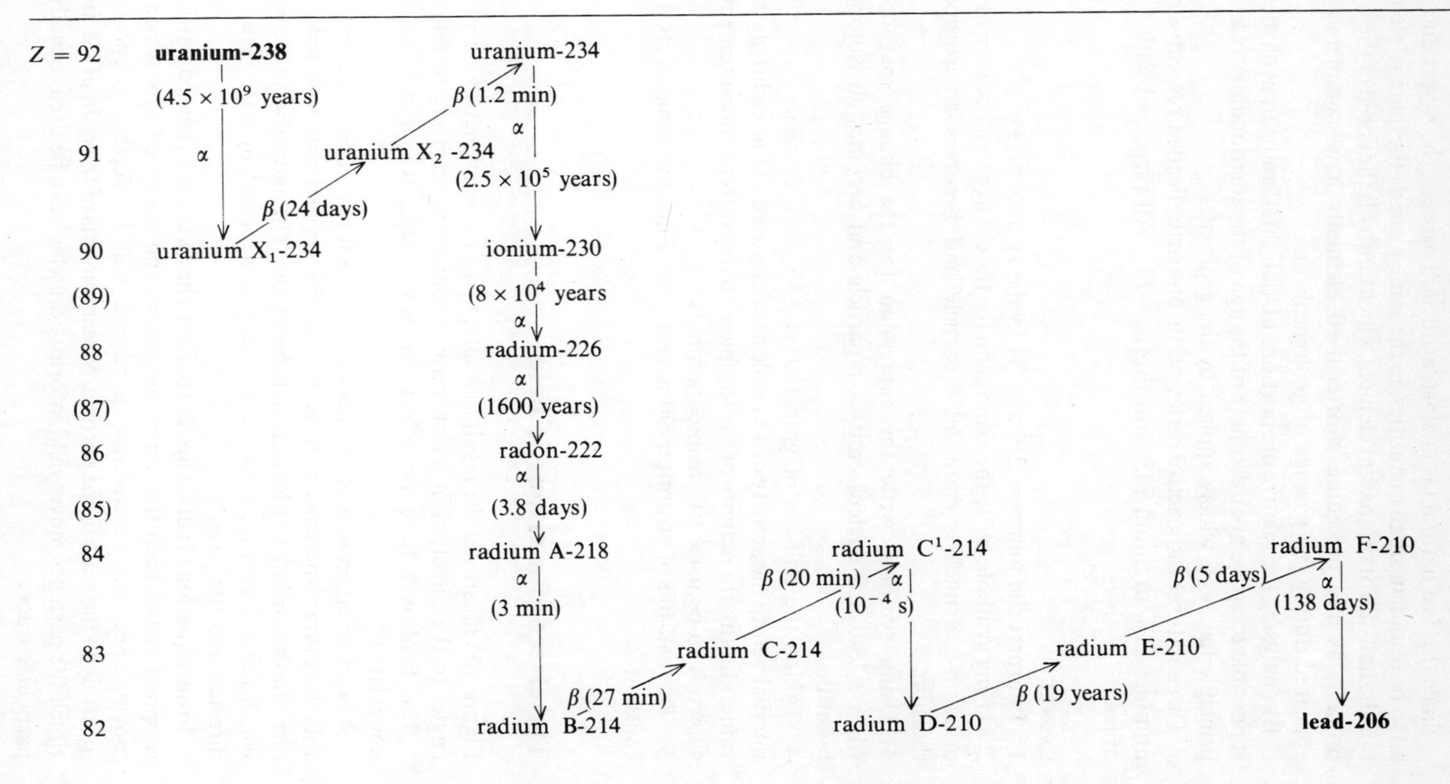

Z = 92
91
90
(89)
88
(87)
86
(85)
84
83
82
uranium-238
(4.5 × 10⁹ years)
α
uranium X₁-234
β (24 days)
uranium X₂ -234
β (1.2 min)
uranium-234
α
(2.5 × 10⁵ years)
ionium-230
(8 × 10⁴ years
α
radium-226
α
(1600 years)
radon-222
α
(3.8 days)
radium A-218
α
(3 min)
radium B-214
β (27 min)
radium C-214
β (20 min)
radium C¹-214
α
(10⁻⁴ s)
radium D-210
β (19 years)
radium E-210
β (5 days)
radium F-210
α
(138 days)
lead-206

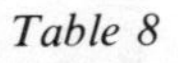

with a stable isotope of lead (Table 8. The isotopes in this table are shown with their original names, some of which are still used.)

Since this is a natural (as against 'artificial') process, the by-products are all to be found, in small quantities, in the same place as the 'parent' element. It is the half-life of this parent element which controls the speed of the whole process, as any preceding shorter-lived elements have completely disappeared during the passage of geological time.

By analysing the relative proportions of 'daughter' and parent element, it is possible to arrive at an estimate of the age of the surrounding rocks; this principle has been widely applied by the geologists.

Three natural radioactive series occur on earth, headed respectively by uranium-238, uranium-235, and thorium-232. All three end with forms of lead.

EXERCISES

1 Interpret the numbers 210 and 84 in the symbol $^{210}_{84}Po$.

2 Using symbols, write the expression for the change that occurs when an uranium-238 nucleus emits an α-particle and becomes an isotope of thorium.

3 Using symbols, write the expression for the change that occurs when a lead-210 nucleus emits a β-particle and becomes an isotope of bismuth.

4 Taking the half-life of gold-199 as 3.15 days, calculate (i) the time interval before the activity of a sample has decreased to one eighth its initial value, and (ii) the activity of the sample 12.6 days after a measurement has shown it to be 6.4×10^5 disintegrations/s.

5 The half-life of strontium-90 is 28.1 year. Find the value of its decay constant λ.

The effect of a magnetic field

Figure 138 illustrates the result of applying a strong magnetic field at right angles to the direction in which various radioactive rays are travelling.

The field will have no effect on any γ-rays, which will continue undeviated.

A beam of either α- or β-particles is equivalent to an electric current, and will therefore experience a force due to the applied magnetic field, in a direction according to Fleming's left-hand rule. The figure shows the way in which the different types of beam would be deflected by a magnetic field directed *into* the page.

From equation (4) of Chap. 27 (p. 228), the *force* on a charged particle is proportional to both the size of the charge and the speed with which it is moving, and the *acceleration* produced will be further proportional inversely to the mass of the particle. Measurements of the radii of the arcs in which the particles move, yield information which can be used to analyse a particular track.

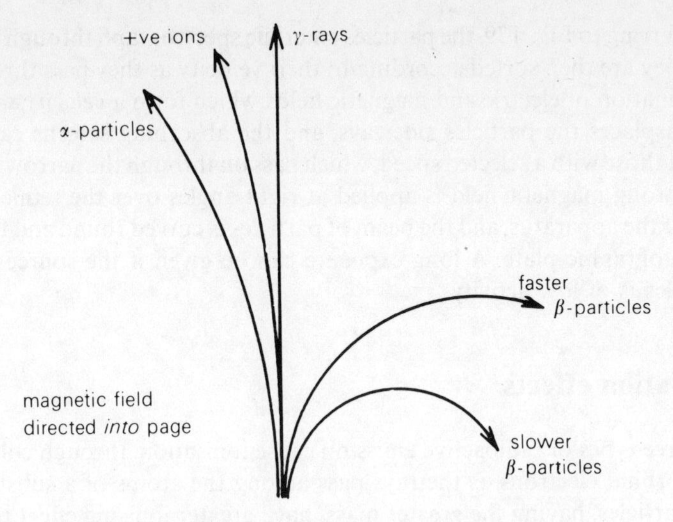

Fig. 138 Circular paths of charged particles in a magnetic field

The principle is the basis of the mass spectrograph, and is also used in conjunction with cloud-chambers and bubble-chambers.

The Bainbridge mass spectrograph

This apparatus is designed to sort a beam of charged particles according to their various masses, and is much used as an aid to analysis of fission products and radioactive waste.

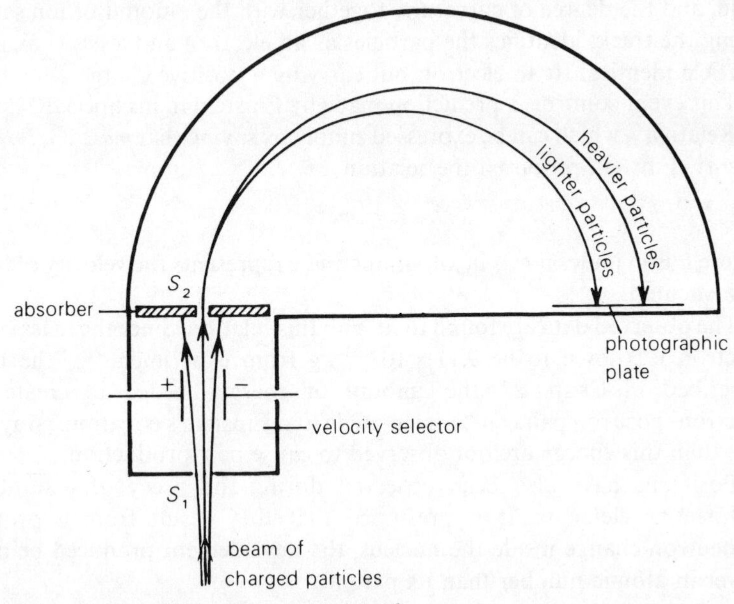

Fig. 139 The Bainbridge mass spectrograph

Referring to Fig. 139, the particles enter the spectrograph through the slit S_1. They are then sorted according to their velocity as they pass through a combination of electric and magnetic fields which form a *velocity selector*; this displaces the particles sideways, and the absorbing screens catch all except those with a selected speed, which pass on through the narrow slit S_2.

A strong magnetic field is applied at right angles over the semicircular area of the apparatus, and the beam of particles is curved round and falls on a photographic plate. A long exposure can be given if the source of the particles is of low activity.

Ionisation effects

All three types of radioactive emission cause ionisation, through collisions with orbital electrons as the rays pass among the atoms of a substance.

α-particles, having the greater mass, have greater ionising effect than β-particles, and these again have more effect than γ-rays. This helps to distinguish between particle tracks observed in bubble-chambers.

Conversely, γ-rays are more penetrating than either β- or α-particles.

Pair-production

During experimental work with bubble-chambers, it is sometimes found that a high-energy γ-ray track ends with the creation of two equal, oppositely charged particles, as shown in Fig. 140. The tracks of the particles have been curved in opposite directions by the applied magnetic field, and the degree of curvature, together with the amount of ionisation along the track, identifies the particles as an electron and a *positron*, i.e. a particle identical to an electron but carrying a positive charge.

This event confirms a prediction made by Einstein in his Special Theory of Relativity, which can be expressed simply by saying that *mass is a form of energy*. Einstein proposed the relation

$$E = mc^2 \tag{4}$$

where E is in joules if m is in kilograms and c represents the velocity of light in a vacuum.

The observed data are found to fit with this relation. Since the mass of an electron is known to be 9.11×10^{-31} kg from experiments of the type described in Chap. 27, the amount of energy needed to create an electron–positron pair can be calculated from Einstein's equation. γ-rays of less than this energy are not observed to cause pair-production.

Positrons have also been observed during the decay of man-made radioactive elements. It is presumed that they result from a proton $\rightarrow$ neutron change inside the nucleus, the new element produced being 1 lower in atomic number than its parent.

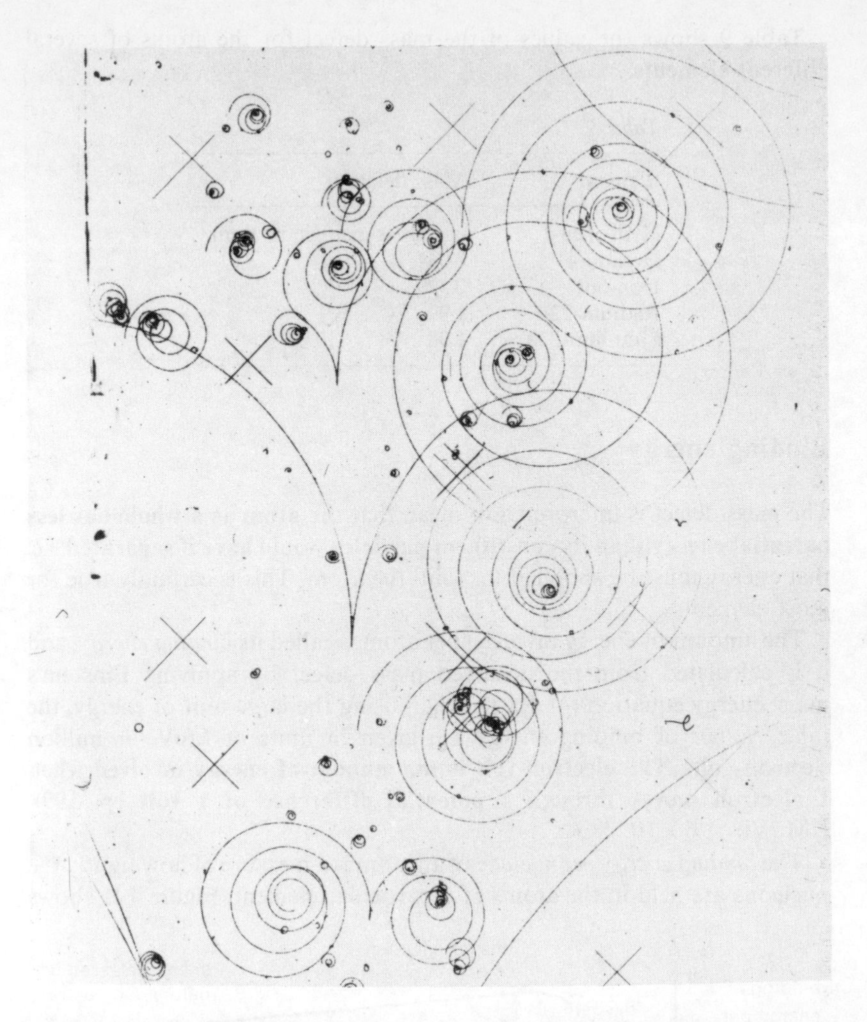

Fig. 140   Pair-production in a bubble chamber (The tracks of the incoming γ-rays do not show up in this photograph)

Mass defect

The masses of the atoms of nearly all elements have now been determined absolutely in kilograms, using the mass spectrometer. When these results are compared with the corresponding values calculated by adding together the masses of all the elementary particles (protons, neutrons and electrons) which make up an atom of a particular element, the two figures do not agree. The difference between them is called the *mass defect*.

This is of course a very small quantity, and instead of quoting it in kilogram units, the mass defect is generally thought of in relation to the unit of molecular mass (p. 260), i.e. 1/12 of the mass of the carbon-12 atom. This is given the symbol m_u or u (*unified atomic mass unit*). $1\ u = 1.66 \times 10^{-27}$ kg.

Table 9 shows the values of the mass defect for the atoms of several different elements.

Table 9

Element	Mass defect/u
Hydrogen-1	– (this is a single proton)
Helium-4	0.031
Iron-56	0.529
Radium-226	1.90
Uranium-238	1.93

Binding energy

The mass defect is interpreted to mean that **the atom as a whole has less potential energy than its constituent particles would have if separated**, i.e. that energy must be supplied to 'split' the atom. This is certainly true for most elements.

The amount of energy involved per atom is called its *binding energy*, and it is calculated from the observed mass defect by applying Einstein's mass–energy equation (4). Rather than using the large unit of energy, the joule, values of binding energy are given in units of MeV, or million electron-volts. **The electron-volt is the amount of energy involved when 1 electron moves through a potential difference of 1 volt** (p. 199). 1 MeV = 1.6×10^{-13} J.

The *binding energy per nucleon* of an atom is a measure of how tightly the nucleons are held in the atoms of a particular element. Figure 141 shows

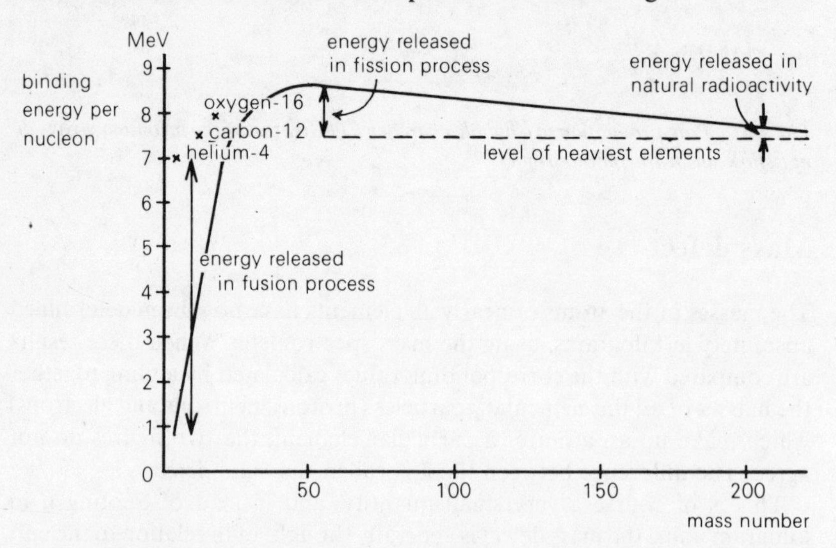

Fig. 141 Graph showing the variation of the binding energy per nucleon throughout the Periodic Table

how this quantity varies throughout the Periodic Table. The values lie on a smooth curve, with the exception of those for helium-4, carbon-12, and oxygen-16, which are slightly displaced and seem to be particularly stable nuclei. The maximum of the curve is around the mass number 50, though it falls only slowly towards higher numbers.

This result implies that, in general, the nuclei of elements near the maximum of the curve are the more stable, having less internal potential energy than those at either extreme. This is confirmed by the fact that all the naturally radioactive elements occur at the extreme high number end of the periodic table; their nuclei decay spontaneously, losing energy in order to become more stable structures.

It is possible to obtain greater energy from such a change if the nucleus can be made to divide into two more equal parts, both occurring in the middle range of the table. This is called *fission.* The present-day nuclear reactors release energy by fission.

At the other end of the table, it has been found possible to induce the nuclei of one of the hydrogen isotopes to combine into the more stable structure of helium. This *fusion* process releases immense amounts of energy, and has provided us with the hydrogen bomb. At present, the first steps are being taken towards controlling this reaction in a more useful way.

EXERCISES

6 If the mass of an electron is 9×10^{-31} kg, calculate the minimum energy necessary to produce an electron–positron pair. Give the answer (i) in joules, and (ii) in MeV. Take the velocity of light as 3×10^8 m s^{-1}.

7 Given the following data:

mass of 1 proton + 1 electron = 1.0078 u

mass of 1 neutron = 1.0087 u

mass of helium atom = 4.0026 u,

calculate the mass defect of the helium nucleus.

8 Using the values given in Table 9, p. 242, calculate the binding energy per nucleon (in MeV) of (i) iron-56 and (ii) radium-226.

Examination questions 28

1 (a) Outline briefly the properties of the particles and the radiation emitted from radioactive substances. Explain in general why most naturally occurring radioactive mineral ores emit a mixture of α, β and γ-radiation. (b) What is meant by the term *half-life* of a radioactive isotope?

Certain charcoal from an ancient camp site has an activity of 220 disintegrations kg^{-1} s^{-1} due to the presence of ^{14}C. Wood from a living tree has an average activity of 255 disintegrations kg^{-1} s^{-1}. If the decay constant for ^{14}C is 4.0×10^{-12} s^{-1}, estimate the age of the camp site, explaining carefully the basis of your calculation. [AEB, Nov. 1975]

2 In a nuclear reaction, energy equivalent to 10^{-11} kg of matter is released. The energy released is approximately

A 4.5×10^{-6} J B 9.0×10^{-6} J C 3.0×10^{-3} J
D 4.5×10^{5} J E 9.0×10^{5} J [C]

3 (a) In terms of the constituents of atomic nuclei, explain the meaning of (i) atomic number, (ii) mass number, (iii) isotope.
(b) Account for the fact that, although nuclei do not contain electrons, some radioactive nuclei emit beta particles.
(c) Cobalt has only one stable isotope, ^{59}Co. What form of radioactive decay would you expect the isotope ^{60}Co to undergo? Give a reason for your answer.
(d) The radioactive nuclei $^{210}_{84}$Po emit alpha particles of a single energy, the product nuclei being $^{206}_{82}$Pb.
(i) Using the data below, calculate the energy, in MeV, released in each disintegration.
(ii) Explain why this energy does not all appear as kinetic energy, E_α, of the alpha particle.
(iii) Calculate E_α, taking integer values of the nuclear masses.

Nucleus	*Mass (u)*
Po	209.936730
Pb	205.929421
α-particle	4.001504

(1 atomic mass unit, u = 931 MeV) [JMB]

4 What is meant by the term *binding energy* for an atomic nucleus and how can this be calculated?

A possible fusion process which has been suggested for the generation of energy is represented by the following equation:

$$^2_1\text{H} + {}^2_1\text{H} \rightarrow \text{X} + {}^1_0\text{n} + 3.27 \text{ MeV}.$$

Identify the nucleus X.

Using the data given below, (a) explain why $^2_1$H is stable, and (b) calculate the atomic mass of X.

mass of proton = 1.00783 u
mass of neutron = 1.00867 u
mass of $^2_1$H = 2.01410 u
1 u = 931 MeV. [L]

5 In a mass spectrograph, the beam contains singly charged neon ions of mass number 20(^{20}Ne$^+$) and doubly charged neon ions of mass number 22(^{22}Ne^{2+}) all moving with the same velocity. On entering the magnetic field the ^{20}Ne$^+$ ions describe a circular arc of radius 0.25 m. The radius of the arc described by the ^{22}Ne^{2+} ions is approximately
A 0.13 m B 0.24 m C 0.26 m D 0.44 m E 0.48 m [O]

6 A thoron nucleus ($A = 220$, $Z = 90$) emits an α-particle with energy 10^{-12} J. Write down the values of the mass number A and the atomic number Z for the resulting nucleus, calculate the momentum of the emitted α-particle, and find the velocity of recoil of the resulting nucleus.

(Take the mass of the thoron nucleus = 3.5×10^{-25} kg, and the mass of an α-particle = 6.7×10^{-27} kg.) [O]

29 Semiconductors

Semiconductors are a group of substances whose electrical conductivity puts them halfway between the metals and the insulators. They are made out of elements occurring in the middle columns of the periodic table, where the outer electron orbits are about half-full.

Intrinsic semiconductors

This name describes a particular small group of substances which are used in a very pure state. The two chief members of the group are silicon and germanium.

In their solid state, these two elements form tetrahedral crystals, held together by electron bonds between neighbouring atoms. At 0 K, there would be no movement of these atoms, but at normal room temperatures, they have a certain amount of thermal energy, vibrating about their mean position in the crystal lattice. Under these conditions, a few atoms lose hold of an outer electron, which wanders away through the crystal until it falls again, by chance, into the orbit of a different atom.

Semiconductor theory describes this state of the substance as containing equal numbers of free electrons and 'holes'. These holes, the spaces left empty by the wandering electrons, can be considered as though they were positive charge carriers.

Current flow through a semiconductor

If a small potential difference is applied between opposite faces of such a crystal, the free electrons will tend to drift towards the positive connection. This drift differs from the electron-gas movement in a metal (p. 32), as in a semiconductor electrons are continually falling into holes and disappearing, while other electrons appear free elsewhere. This explains why the conductivity of a semiconductor is only about 10^{-7} times that of a metal.

The external p.d. will be supplying electrons into the semiconductor from the negative connection, so that a small current will be maintained.

An increase in the temperature of the semiconductor will increase the number of electrons with sufficient thermal energy to break loose and become free. Experiments show that **the conductivity of semiconductors increases with temperature**, in contrast to the behaviour of pure metals.

A *thermistor* is a semiconductor specially made to show a marked drop in resistance as it gets warm. It is used in circuits to guard against overloading on first switching on, also to compensate for other elements of the circuit whose resistance increases as they warm up.

Extrinsic semiconductors

The conductivity of a semiconductor can be greatly increased by 'doping' the crystal with about one part in a hundred million of one of the elements in the next column of the periodic table.

This can be done in two different ways, according to whether the added element lies higher or lower in the table.

(a) *n-type.* Usual *donor* element: arsenic or bismuth added to germanium; phosphorus added to silicon.

A small amount of the foreign element is added to the host element during crystallization. The foreign atom takes a place in the crystal structure, but as it has five electrons in its outer orbit, and only four of these are needed to form the electron bonds, the extra electron becomes 'free'. The 'n' in the name of this type of semiconductor refers to the excess of *negative* charge.

Doping in this way in small proportions thus adds a large number of free electrons to the crystal without disturbing its lattice structure, and its conductivity can be made to increase by a factor of about 100.

(b) *p-type.* Usual *acceptor* elements: boron, aluminium, and indium.

This kind of doping works in the opposite manner. The foreign atom has only three electrons in its outer orbit, so that if this atom occupies a place in the crystal structure, a hole will be created. Such a semiconductor will have an excess of holes, or *positive* charge carriers, hence it is described as 'p' type.

Holes are not quite as mobile as free electrons, and the conductivity of a p-type semiconductor is not usually as good as that of an n-type.

The Hall effect

This is an example of the force produced on a current-carrying conductor, or on a stream of moving charges, by an applied magnetic field (p. 218). If a current is passing through a semiconductor, and a magnetic field is applied at right angles to the flow of the current, then a force is exerted on the moving charges in the direction given by Fleming's left-hand rule.

When applying this rule to the electron movement inside a semiconductor, it must be remembered that the conventional direction of the current is opposite to the direction of movement of the negative electrons.

Both the electrons and the holes will be affected by the magnetic field *in the same way.* In the example illustrated in Fig. 142, the direction of the resulting force would be upwards, towards the top surface of the semiconductor. In fact, charges are observed to drift under the influence of

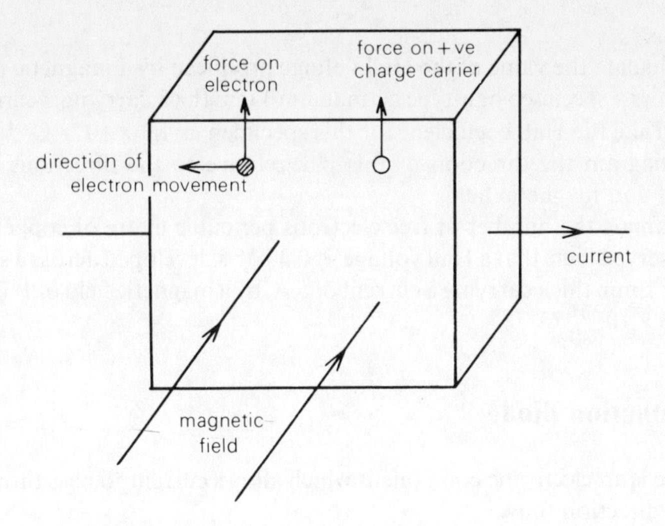

Fig. 142 The Hall effect in a semiconductor

this force until the number accumulated near the surface is sufficient to repel further drift.

The name of the *Hall effect* is given to **the existence of the e.m.f. developed at right angles to the current flow** in this way. Even in small specimens of semiconductor, the e.m.f. may be several millivolts; the Hall effect can also be observed in metals, but is much smaller, producing only microvolts of e.m.f.

The magnitude of the Hall voltage, V_H, developed across a thickness t of a conductor, depends on

(a) the current flowing, I,
(b) the magnetic field applied, B, and
(c) the number of charge-carriers per cubic metre, N.

The relation

$$V_H = \frac{BI}{Net}$$

where e is the charge of a single electron, may be derived by considering the current as made up of separate charges drifting through the lattice (p. 34), and equating the magnetic and electric forces acting on an individual charge as in the Thomson experiment on p. 230.

The term $1/Ne$, which is characteristic of a particular semiconductor, is called its *Hall coefficient*.

Since the density of charge-carriers, N, is much less in a semi-conductor than in a metal, the Hall voltage for a particular value of current is correspondingly greater.

The polarity of the observed Hall voltage in a specimen indicates whether it is n- or p-type, i.e. whether the majority of charge-carriers are negative or positive.

EXERCISES

1 Calculate the value of the Hall voltage produced by a magnetic field of 1 T across a specimen of n-type germanium 1 mm thick carrying a current of 2.0 A. Take the Hall coefficient for this specimen as $1.3 \times 10^{-5}\ C^{-1}$. Show on a diagram the direction of this p.d. relative to the directions of the current and magnetic field.

2 Estimate the number of free electrons per cubic metre of copper from the observed facts that a Hall voltage of 0.3 μV is developed across a strip of copper 1 mm thick carrying a current of 5 A, by a magnetic field of 1 T. Take $e = 1.6 \times 10^{-19}\ C$.

The junction diode

A diode is an electronic component which allows current to pass through it in one direction only.

Before the development of semiconductors, the only diodes were vacuum tubes using thermionic emission from a heated cathode. Many forms of semiconductor diode have now been produced; Fig. 143(a) shows one example of the elements that can be used to make a *junction* diode.

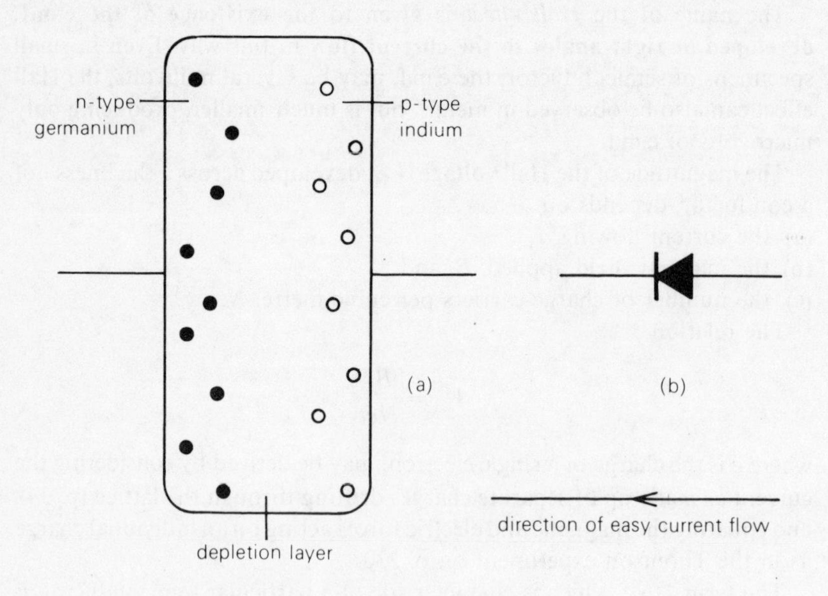

Fig. 143 A junction diode (a) structure, (b) symbol used

A small wafer of n-type germanium has a layer of indium melted on to one face. This produces a crystal which has an excess of free electrons at one end and an excess of holes at the other. At the junction, holes and electrons have mostly recombined during the manufacturing process, so that a thin neutral layer exists. This is called the *depletion layer*, and it holds apart the n- and p-regions of the crystal.

Only a small external p.d. need be applied in order to urge the free electrons across the junction in to the p-region, so that a current can be created. This flows easily from right to left in Fig. 143 (the opposite direction to the electron flow).

The diode is usually made as a thin slice of comparatively large cross-sectional area, so that its resistance (to current in the preferred direction) will be low. The symbol used for a diode in circuit diagrams is shown in Fig. 143(b).

Current can only be passed through the device in the reverse direction (left to right in Fig. 143) by the creation of new charge-carriers, so that in this direction the diode has the resistance of an undoped semiconductor.

In general for an n–p junction, current flows easily if an external voltage is applied

negative to the n-side
and positive to the p-side,

since this tends to drive the free electrons into the holes. Under these conditions, the diode is said to be *forward-biased* (Fig. 144(a)).

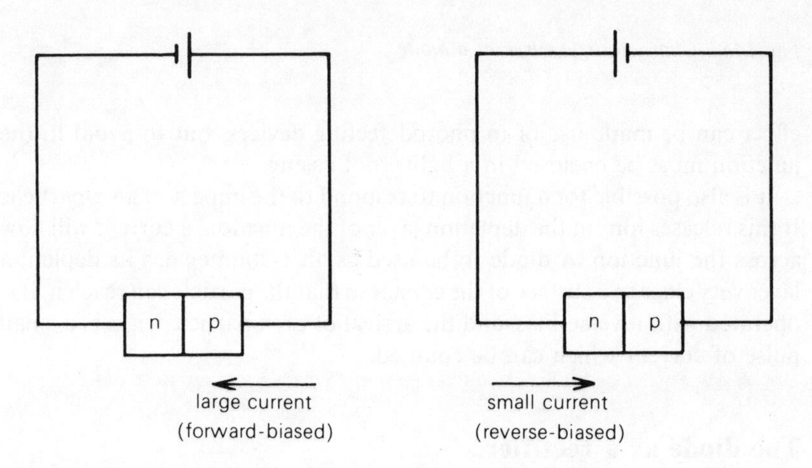

Fig. 144 (a) Forward-biased and (b) reverse-biased diodes

In the opposite conditions, shown in Fig. 144(b), only a very small current flows, and the junction is said to be *reverse-biased.*

Figure 145 shows how the current varies with the applied voltage. Such a graph is called the *characteristic* curve of the device.

Radiation detection

Junction diodes are generally sensitive to light and higher frequency radiation, since these release photoelectrons from the semiconductors. This

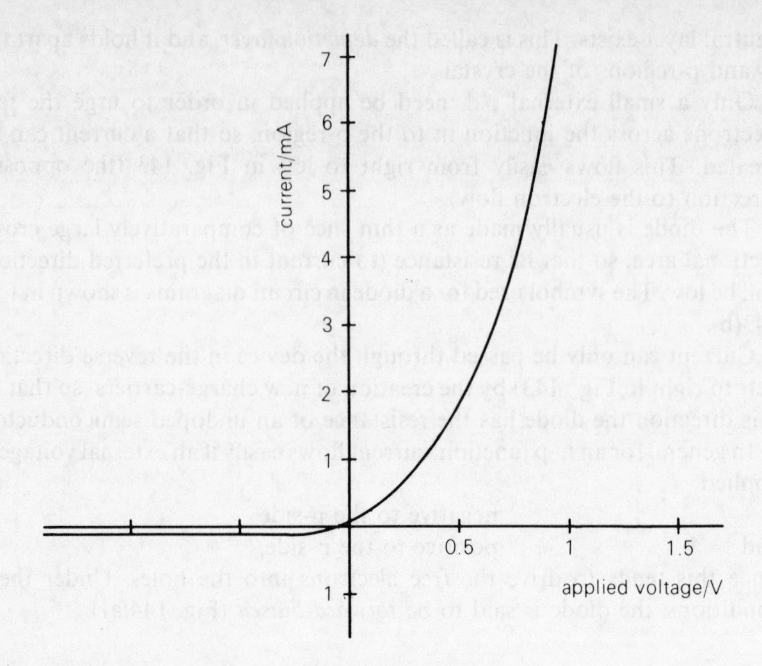

Fig. 145 Characteristic curve of a diode

effect can be made use of in photodetecting devices, but to avoid it, the junction must be enclosed in a lightproof casing.

It is also possible for a junction to respond to the impact of an α-particle. If this releases ions in the depletion layer of the junction, a current will flow across the junction. A diode to be used as an α-counter has its depletion layer very close to a surface of the crystal so that the particle can reach it. It is operated with reverse-bias, and the arrival of each particle triggers a small pulse of current which can be counted.

The diode as a rectifier

If a small alternating p.d. is applied across a diode, it will conduct current only during alternate half-cycles. This is called *half-wave rectification*, and the current transmitted is as shown in Fig. 146(b).

Full-wave rectification can be obtained by using a bridge circuit as shown in Fig. 147. The a.c. input is connected at A and D across a system of four diodes. During the half-cycles in which A is positive with respect to D, current can pass from A to B and similarly from C to D. If a connection is made between the output terminals, then half-wave rectified current will flow via the route ABCD.

The other half of the wave is obtained via the route DBCA, during the alternate half-cycles in which D is positive with respect to A, and the combined output is as shown in Fig. 146(c).

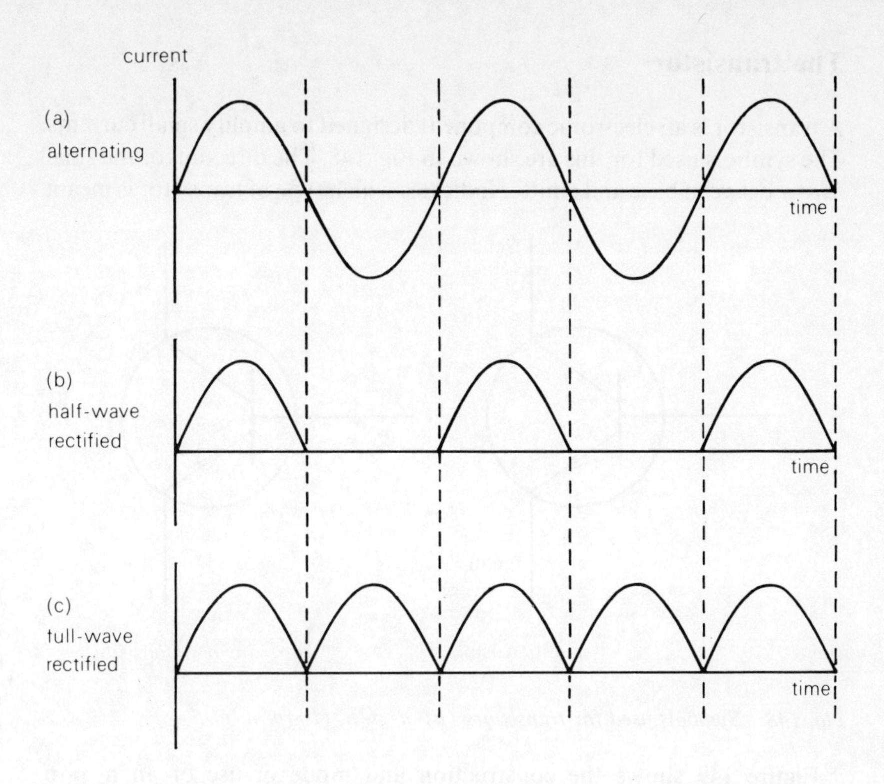

Fig. 146 Graphs of (a) alternating, (b) half-wave rectified and (c) full-wave rectified current

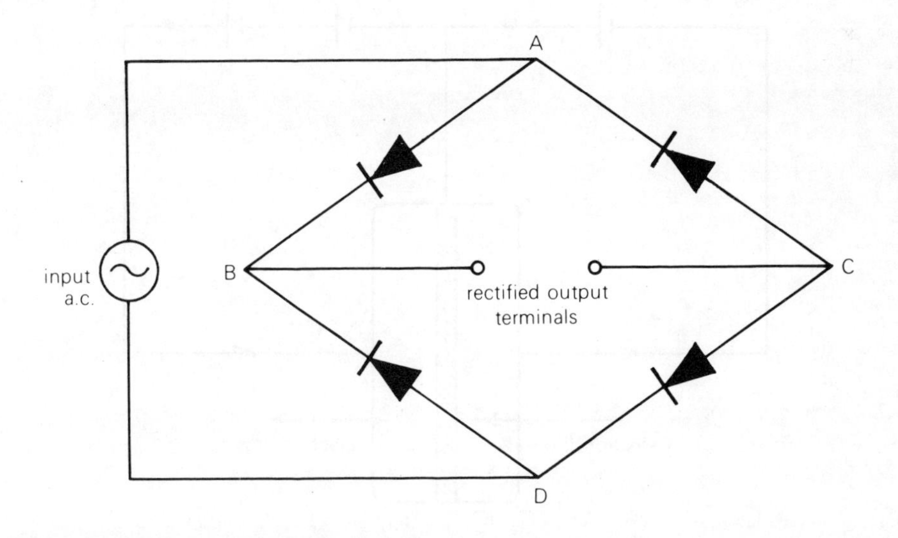

Fig. 147 A diode bridge rectifier circuit

The transistor

A transistor is an electronic component designed to amplify small currents. The symbols used for this are shown in Fig. 148. The direction of the small arrow between base and emitter indicates which type of transistor is meant.

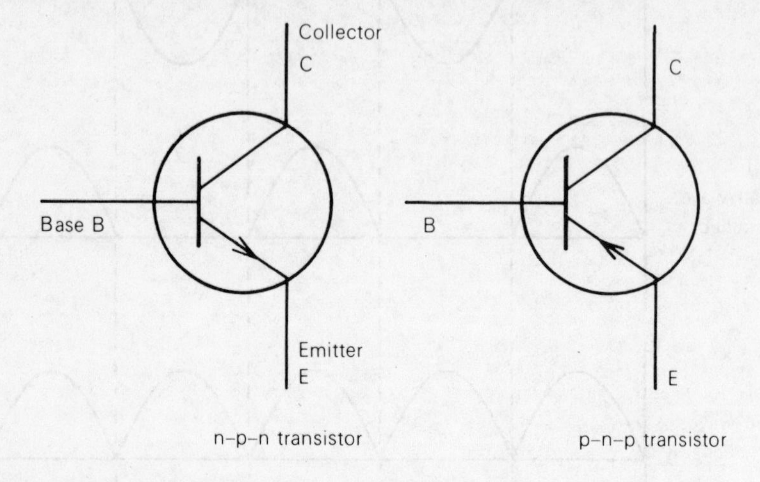

Fig. 148 Symbols used for transistors (a) n–p–n, (b) p–n–p

Figure 149 shows the construction and mode of use of an n–p–n transistor. It consists of two n–p junctions, one forward-biased and the other reverse-biased, linked back-to-back with a common p-layer.

Current flows easily in the forward-biased circuit, from base to emitter. This means that actually a good flow of electrons is passing into the p-layer,

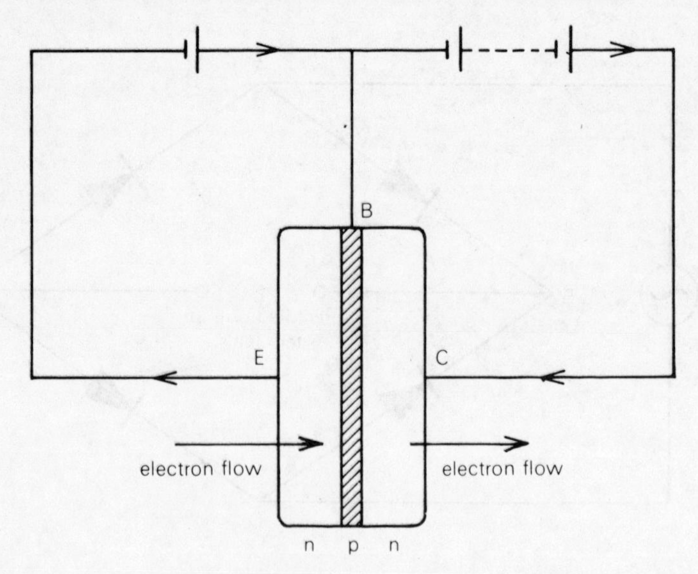

Fig. 149 Mode of use of a transistor

and as this is made very thin (less than 0.1 mm), many of the electrons cross over into the collector. These electrons partially cancel out the effect of the p-layer of the base–collector junction. The current in the second part of the circuit, drawn from a separate battery, is largely controlled by the current in the base–emitter circuit.

p–n–p transistors operate in a similar way, but as the holes are less mobile than free electrons they do not respond as well as n–p–n designs when used in high-frequency a.c. circuits.

Transistor amplification

The purpose of a current–amplifier is to produce a large change in output current from a small change in the input current. Any current passing out of the transistor through the base is wasted, in this connection, and in practice

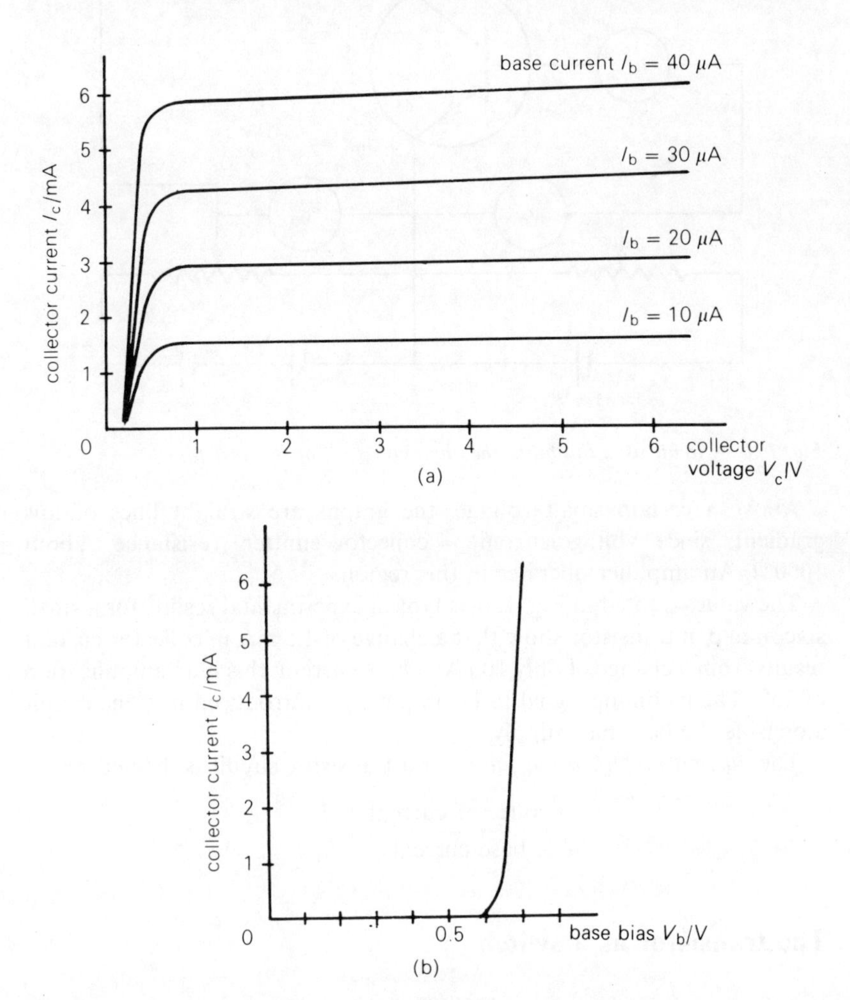

Fig. 150 Characteristic curves of a transistor

it is reduced by using, for the base layer, a semiconductor which is only very lightly doped. The bulk of the current then flows straight across the transistor.

The characteristics of a transistor are of the form shown in Fig. 150. The set of curves (a) are obtained by measuring the current I_c from the collector for various collector–emitter voltages V_c. Different curves result from varying the base current I_b by altering the bias voltage on the base. A suitable circuit for this experiment is shown in Fig. 151.

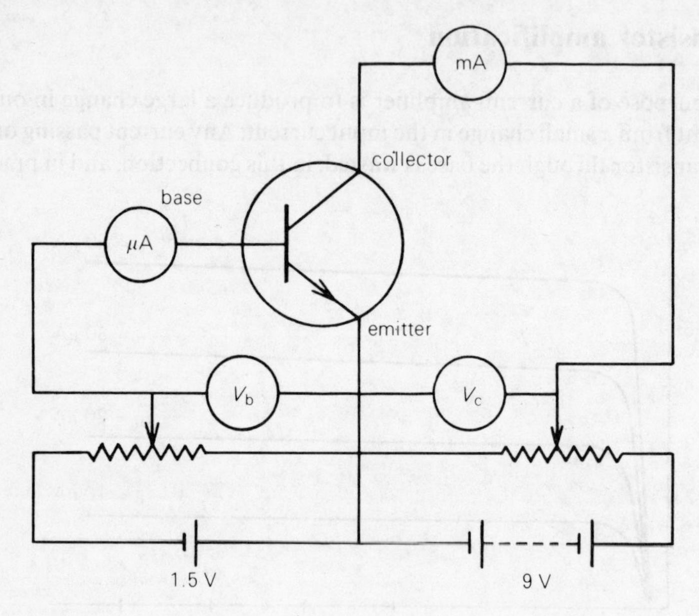

Fig. 151 Circuit used to obtain the characteristics of a transistor

Above a certain small voltage, the graphs are straight lines of low gradient, since voltage/current = collector–emitter resistance, about 1000 Ω. An amplifier operates in this region.

The values quoted in Fig. 150 as typical experimental results for a small silicon n–p–n transistor show that a change of 1.5 mA in collector current results from a change of only 10 μA in base current; this is an amplification of 150. The incoming signal to be amplified is introduced into the circuit alongside the base bias supply.

The *current amplification factor* of a transistor circuit is defined as

$$\frac{\text{collector current}}{\text{base current}} = \frac{I_c}{I_b}.$$

The transistor as a switch

In computers, transistors are used as switches, to divert a signal into one or other of two routes according to instructions given in binary code.

For this purpose, the transistor is operated near its cut-off point, at the knee of the characteristic. As can be seen from the scale of the graph (b) in Fig. 150, it only needs a change of about 0.1 V in the base–emitter voltage for the device to pass from a non-conducting to a conducting state.

When used in this way, a transistor is said to have 'bottomed' when it is conducting, since although it can be passing a fairly large current (perhaps 0.1 A), the collector–emitter voltage is nearly zero.

Examination questions 29

1 (i) Draw the input and output characteristics of a junction transistor in common-emitter connection.
(ii) Draw a circuit diagram for a common-emitter transistor amplifier suitable for amplifying small audio-frequency signals and explain the operation of the circuit.
(iii) Explain why a silicon transistor is normally used in preference to a germanium transistor for the above amplifier.
[AEB, Specimen paper 1977]

2 The electrical conductivity of an intrinsic semiconductor increases with increasing temperature. The reason for this is that
A the energy required to excite an electron is less at higher temperatures.
B the ratio of the number of electrons to the number of holes is greater at higher temperatures.
C the probability of thermal excitation of electrons is greater at higher temperatures.
D the drift velocity of the electrons and holes is greater at higher temperatures.
E the probability of collision of the electrons and holes with lattice ions is less at higher temperatures. [C]

3 Silicon has a valency of four (i.e. its electronic structure is 2:8:4). Explain the effect of doping it with an element of valency three (i.e. of electronic structure 2:8:3).

Explain the process by which a current is carried by the doped material.

Describe the structure of a solid-state diode.

Draw a circuit diagram showing a reverse-biased diode and explain why very little current will flow.

Suggest why a reverse-biased diode could be used to detect α-particles.
[L]

4 A rectangular block of p-type semiconductor material has n charge carriers per unit volume, each carrying charge q. A uniform current I is passed between conducting electrodes which cover opposite faces KLMN and K′L′M′N′ of the block.

KL = K′L′ = MN = M′N′ = a, and KN = K′N′ = LM = L′M′ = b.

Express the drift velocity of the charge carriers in terms of I, n, q, a and b.

A uniform magnetic field of flux density B is then applied parallel to the direction LK. Write down the magnitude and direction of the force the carriers experience from the magnetic field.

Hence show that, for the charge carriers to move in a direction perpendicular to the side KN, the face KK′L′L must be at a potential V higher than that face MM′N′N, where $V = BI/nqa$.

In general, Va/BI is known as the Hall coefficient R_H.

State and explain briefly

(a) why R_H is larger in semiconductors than in metals,

(b) what information concerning a material is given by the sign of R_H.

[OC]

Heat II

30 The experimental gas laws

The three variables of a gas, pressure, volume and temperature, are related by the following three experimentally derived laws.

(1) **The pressure law: pressure increases with temperature.** The apparatus used in the laboratory for this is shown in Fig. 1 (p. 5). The gas under test is contained in a glass bulb which is immersed in a water bath, and its pressure p is calculated from the difference in readings of a manometer.

If temperature T is measured on the Kelvin scale,

$$p \propto T \text{ if the volume of the gas remains constant}$$

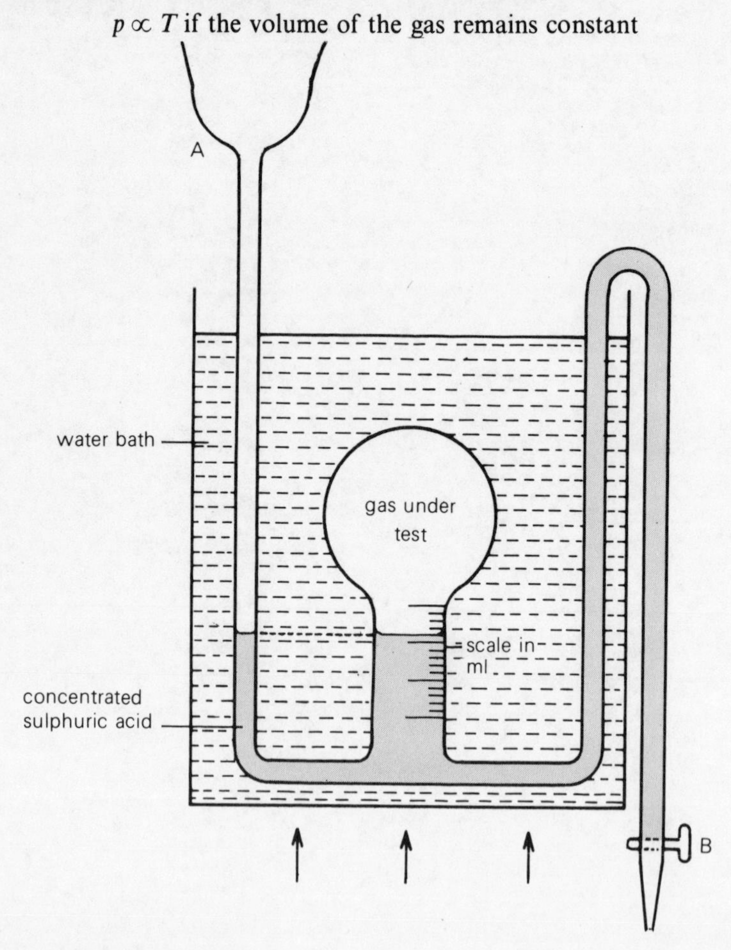

Fig. 152 Charles' law apparatus

(2) **Charles' law: volume increases with temperature.** Apparatus of the form shown in Fig. 152 is available to verify this law. The gas under test is trapped in the glass bulb by concentrated sulphuric acid which fills part of the tubing, and its pressure may be returned to atmospheric by adding more acid at A or withdrawing acid from the tap at B. The volume V of the gas can be read directly from the scale provided on the apparatus.

$V \propto T$ if the pressure of the gas remains constant.

(3) **Boyle's law: volume decreases as pressure increases.** The elementary experiment, using a J-tube containing mercury to enclose the gas and measure its pressure, gives results of sufficient accuracy to check this law.

pV = constant, if the temperature is constant.

These relations hold only if the gas is pure and perfectly dry, and are limited to fairly low pressures and large volumes (see below). A gas which obeys these laws is called an *ideal* gas.

The three separate relations are combined together into the *Equation of State*

$$\frac{pV}{T} = \textbf{constant } (R) \text{ for any given mass of gas.} \quad (1)$$

Van der Waals' law

When experiments were carried out at pressures of 100 atmospheres or above, some departure from Boyle's law was observed. Van der Waals, in 1879, suggested that this was because the mutual attraction between the molecules could not be ignored when they were forced close together at high pressure.

The force of attraction would depend on the density of the gas, and Van der Waals' calculations added a term a/V^2 to p, giving $(p \times a/V^2)$.

He also modified the Equation of State to include a term b to represent the combined volume actually occupied by the molecules themselves. This would become significant if the volume of the gas were small. The term V should be replaced by $(V - b)$.

Van der Waals' Equation of State is

$$\frac{(p + a/V^2)(V - b)}{T} = \text{constant.}$$

This agrees well, but not perfectly, with the observed experimental results.

Measurement of pressure

The pressure of a gas is defined as the force it exerts per unit area, and measured in *pascals*, where 1 pascal is a pressure of 1 newton/metre^2.

Atmospheric pressure is still often considered from the point of view of the reading of a mercury barometer, when standard pressure corresponds to a column of mercury 760 mm long, known as 760 torr. The equivalent value in pascals is 1.013×10^5 Pa.

The mole

The quantity of any gas known as 1 *mole* is a unit particularly used in chemistry, and so must be considered here since the topic of the gas laws is common to both physics and chemistry.

1 mole is the quantity of a gas which contains Avogadro's Number, N_A, of molecules and has a mass equal to the relative molecular mass of the gas, in grams.

It is derived from Avogadro's hypothesis, now confirmed by experiment, that equal volumes of all gases at the same temperature and pressure contain the same number of molecules; the *mass* of the gas will depend upon the mass of its molecule.

Molecular masses are defined in terms of a unit (u) equal to 1/12 of the mass of the carbon atom. The hydrogen atom has a relative atomic mass of 1.008, so its molecule (two atoms) has molecular mass 2×1.008 u $= 2.016$ u, and 1 mole of hydrogen will have a (molar) mass of 2.016 g. Similarly the mole of oxygen has a mass of 32.0 g (two atoms, each of relative atomic mass 16.0).

A larger unit, the kilomole, corrcsponds to the use of the molecular mass number of kilograms.

The gas constant

1 mole of any gas is found experimentally to occupy a volume of 22.4 litre at s.t.p. This leads to a value for the numerical constant figuring in the Equation of State (1), which will apply to the molar mass of any gas.

$$\text{At s.t.p.,}\ T = 273\ \text{K}$$
$$p = 1.013 \times 10^5\ \text{Pa}$$
$$V = 22.4\ \text{litre} = 22.4 \times 10^{-3}\ \text{m}^3$$

$$\frac{pV}{T} = \frac{1.013 \times 10^5 \times 22.4 \times 10^{-3}}{273} = 8.31.$$

The units of this expression are (newtons/metre^2) $\times$ (cubic metres) $\div$ (degrees K) which is equivalent to 'joules per degree K', and this calculated value, called the molar gas constant, is denoted by the letter $\mathcal{R}$.

$$\mathcal{R} = 8.31\ \text{J K}^{-1} \quad \text{for 1 mole of any gas.}$$

The value of the gas constant R for other quantities of gas can be found from this by relating the actual mass of the gas to its molar mass. If there are n moles of gas, $R = n\mathscr{R}$.

EXERCISES

Take $\mathscr{R} = 8.31 \text{ J K}^{-1}$ where appropriate.

1 A certain mass of gas occupies a volume of 0.1 m^3 at $0°\text{C}$, its pressure being 1×10^5 Pa. What will its volume be at $50°\text{C}$ if the pressure increases to 2×10^5 Pa?

2 What volume will be occupied by 1 mole of a gas at 400 K under a pressure of 1.3×10^5 Pa?

3 If the molecular mass of carbon is 12, and that of oxygen 32, how many moles of carbon dioxide are there in 1 kg?

Specific heat capacities of a gas

When a gas is heated, in general both its volume and its pressure will increase, depending on the conditions of the heating, and these conditions must be defined if the term 'specific heat capacity' is to have a precise meaning. The two simplest conditions are those where the gas is kept (a) at constant volume, and (b) at constant pressure, during the change of temperature.

(a) *Volume constant.* If the gas is prevented from expanding, the quantity of heat required *per mole* to increase its temperature by 1 K is called $\mathbf{C_v}$.

(b) *Pressure constant.* If the gas is allowed to expand freely so that its pressure remains constant, the quantity of heat required *per mole* to increase its temperature by 1 K is called $\mathbf{C_p}$.

C_p is always greater than C_v, because the gas does work against the external pressure while expanding, and heat energy must be supplied to do this work in addition to the heat taken in to speed up the molecules of the gas and show a rise in temperature. This statement is an application of the principle of the conservation of energy known as the *first law of thermodynamics*, which says generally:

$$\begin{pmatrix}\text{heat supplied}\\ \text{to system}\end{pmatrix} = \begin{pmatrix}\text{change in internal}\\ \text{energy}\end{pmatrix} + \begin{pmatrix}\text{external work}\\ \text{done}\end{pmatrix}.$$

Work done during expansion of a gas at constant pressure

Suppose the gas occupies a spherical volume of radius r. The total force acting on its surface due to its pressure p is $4\pi r^2 . p$.

If the expansion causes the sphere to increase in radius by δr, then the work done by this force is

$$\text{force} \times (\text{distance moved}) = 4\pi r^2 p\, \delta r$$

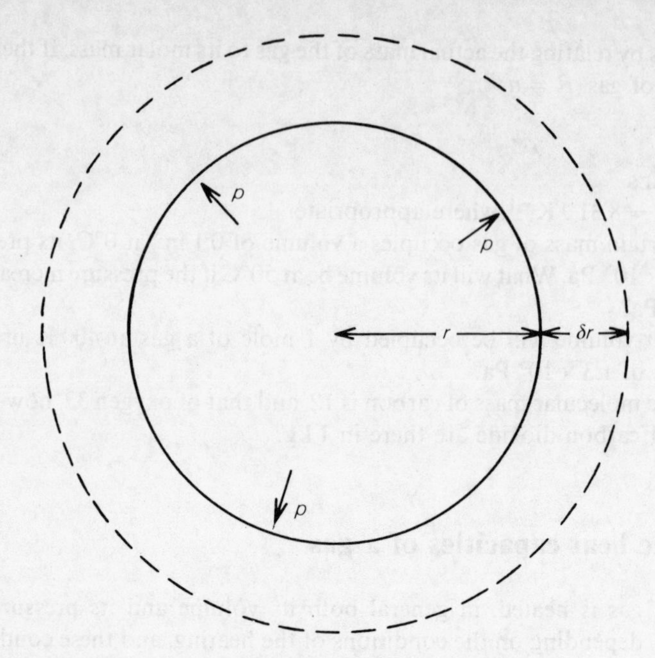

Fig. 153 Calculation of work done when a gas expands

which can be expressed as

$$p \times (\textbf{change in volume}) = p\delta V$$

and this result holds whatever shape the gas actually occupies.

Difference between C_p and C_v

Suppose 1 mole of a gas is heated through an increase of temperature of 1 K.

If the gas is not allowed to expand, then

$$\text{heat supplied} = C_v = \text{change in internal energy.}$$

If the gas expands through a volume of δV while its pressure remains constant at p, then

$$\text{heat supplied} = C_p = \text{change in internal energy} + p\delta V$$

$$\Rightarrow \quad C_p = C_v + p\delta V.$$

Now, from the Equation of State (1),

$$pV = \mathscr{R}T$$

and $\quad p(V+\delta V) = \mathscr{R}(T+1),$

Subtracting, $\quad p\delta V = \mathscr{R}.$

So $\quad C_p = C_v + \mathscr{R}$

or $\quad C_p - C_v = \mathscr{R} \qquad (2)$

Note that the conclusion in this form applies only to 1 mole of a gas, but it may be adapted to use with other masses provided that the appropriate data is available for R and the specific heat capacities.

Isothermal and adiabatic changes

'Isothermal' means equal temperature, so **in an *isothermal change* the temperature of the gas remains constant**, and the gas will obey Boyle's law:

$$pV = \text{constant.}$$

This type of change does not occur very often in ordinary life. If the temperature of the gas is not to alter, the change must be made quite slowly and the material enclosing the gas must be a good conductor so that heat can pass easily in or out.

An *adiabatic change* is one during which no heat passes into or out of the gas.

This condition holds for rapid changes of volume and/or pressure, and will be accompanied by some change of temperature. The changes in pressure produced by a sound wave are adiabatic, so are those occurring in explosions, and the pumping-up of a bicycle tyre may be considered almost adiabatic as the heat developed is very noticeable.

Changes in a *well-lagged* system are adiabatic.

The equation of a reversible adiabatic change is

$$pV^{\gamma} = \text{constant, where } \gamma = \frac{C_p}{C_v}. \qquad (3)$$

This equation can be derived mathematically by considering the slopes of the set of isothermal and adiabatic relations for a perfect gas. (Fig. 154)

The value of γ has been determined experimentally and is found to depend on the number of atoms comprising the molecule of the particular gas.

For monatomic gases, e.g. helium, argon $\gamma = 1.67$

For diatomic gases, e.g. hydrogen, oxygen, nitrogen $\gamma = 1.40$

For most triatomic gases, e.g. carbon dioxide γ lies between 1.29 and 1.33.

These values can be justified by the arguments of the kinetic theory (Chap. 31).

Combining equation (3) with the Equation of State (1) gives

$$TV^{\gamma-1} = \text{constant}$$

which is sometimes a more convenient form to use.

It has been mentioned previously that the rapid pressure changes of a sound wave are adiabatic changes. The quantity γ can therefore be expected to appear in the theory of sound waves, and it does in fact occur in the equation governing the speed of sound in a gas (Chap. 38). The values just quoted for γ result from experiments in this field.

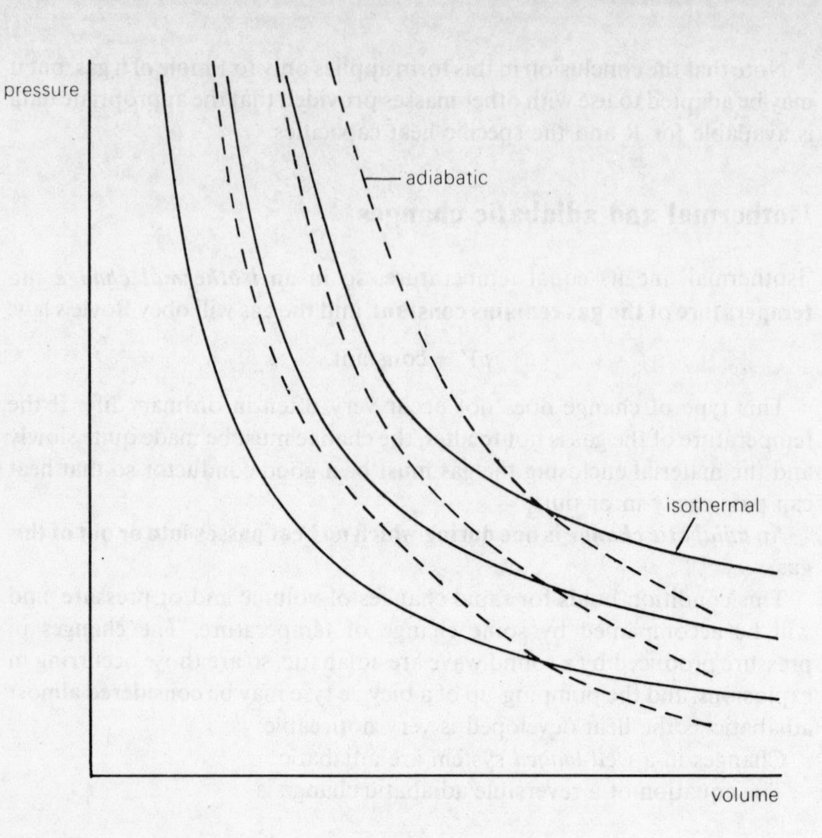

Fig. 154 Isothermal and adiabatic curves

The Joule–Kelvin effect

If a gas expands freely through a small hole into a low-pressure region, a drop in temperature is observed. This is interpreted as due to the work being done against the intermolecular attractions (the Van der Waals' forces). It would not occur for an ideal gas.

By increasing the pressure difference, and repeating the process cyclically, the effect has been used to liquefy many gases.

EXERCISES

4 How much work is done by a gas whose pressure is 1×10^5 Pa if its volume increases by 0.5 m^3?

5 In an adiabatic expansion, the volume of a quantity of helium increases from 0.6 m^3 to 0.75 m^3. If its initial pressure was 1.5×10^5 Pa, calculate the pressure of the helium after the expansion.

6 A gas is compressed rapidly into half its original volume. If its original temperature was 27°C, find the temperature of the gas after compression, taking $\gamma = 1.40$.

Examination questions 30

1 Distinguish between *isothermal* changes and *adiabatic* changes. Discuss the concept of an *ideal* gas.

Show that the difference between the principal molar heat capacities C_p and C_v of an ideal gas is given by the relationship $C_p - C_v = \mathscr{R}$.

One mole of a gas, whose behaviour may be assumed to be that of an ideal gas, is compressed isothermally at a temperature T_1 until its pressure is doubled. It is then allowed to expand adiabatically until its original volume is restored. In terms of the initial pressure find (i) the final pressure, (ii) the final temperature T_2. (The ratio of the principal molar heat capacities = 1.4.)

What energy must be given to the gas to produce the same temperature change, $T_1 - T_2$, at constant volume? Express your answer in terms of the original temperature T_1. [AEB, June 1976]

2 A large tank contains water at a uniform temperature to a depth of 20 m. The tank is open to the atmosphere and atmospheric pressure is equivalent to that of 10 m of water. An air bubble is released from the bottom of the tank and rises to the surface.

Assuming surface tension effects to be negligible, the volume of the air bubble

A halves before it reaches the surface.
B doubles before it reaches the surface.
C remains constant.
D doubles before it rises 10 m.
E halves before it rises 10 m. [C]

3 Explain what is meant by the statement that a gas expands adiabatically, and explain why there is a change in temperature.

A fixed mass of an ideal gas at a pressure of 2.0×10^5 N m^{-2} expands reversibly and adiabatically to twice its original volume. If the ratio of the principal specific heat capacities of the gas is 1.4, calculate the final pressure. [JMB]

4 Define the terms (a) *isothermal change*, and (b) *adiabatic change*, as applied to the expansion of a gas. Explain how these changes may be approximated to in practice. What would be the relationship between the *pressure* and *temperature* for each of them for an ideal gas?

Sketch, using the same axes, the *pressure–volume* curves for each of (a) and (b) for the expansion of a gas from a volume V_1 and pressure p_1 to a volume V_2. How, from these graphs, could you calculate the work done in each of the expansions?

Explain why the temperature falls during an adiabatic expansion and discuss whether or not the temperature fall would be the same for an ideal gas and a real gas. [L]

5 The first law of thermodynamics can be stated in the form $\Delta Q = \Delta U + \Delta W$. Which of the quantities ΔQ, ΔU, ΔW must necessarily be zero when a real gas undergoes a change which takes place at constant pressure?

A ΔQ only is necessarily zero.

B ΔU only is necessarily zero.
C ΔW only is necessarily zero.
D None of ΔQ, ΔU and ΔW is necessarily zero.
E All of ΔQ, ΔU and ΔW are necessarily zero. [O]

31 The kinetic theory of gases

If certain basic assumptions are made about the molecules of a gas, it is possible to set up a simplified mathematical theory to account for many of the observed experimental results given in the previous chapter.

Assumptions

(1) All the molecules of a particular element are identical.
(2) The volume of an individual molecule may be neglected in comparison with the total volume occupied by the gas.
(3) Collisions between molecules are perfectly elastic.
(4) The time involved in a collision is negligible.
(5) There is no force of attraction between the molecules.

The average speed of a molecule

Molecules move throughout the gas in all directions at random, hence the total momentum of the molecules is always zero.

The total kinetic energy of the gas is not necessarily zero, however, since kinetic energy is not a vector. Consequently the average speed used in these calculations is taken as the root-mean-square average (p. 175) so that this value, denoted by $\bar{c}$, gives a justifiably **average value $\frac{1}{2}m\bar{c}^2$ for the kinetic energy of a single molecule.**

Calculation of pressure

The pressure exerted by a gas is observed as the result of collisions between the molecules and the boundaries of the space in which they are contained. It is calculated as in Chap. 15 (p. 127) from the rate of change of momentum of the molecules of the gas, taken as a whole.

Consider a cube of side 1 m containing the gas, and suppose the number of molecules in the cube is n. Figure 155 shows an individual molecule inside this cube. Suppose the velocity of this molecule has components u, v and w parallel to the various edges of the cube. If this molecule strikes the shaded face of the cube, its momentum in the u direction will change from $+mu$ to $-mu$, since the collision is elastic.

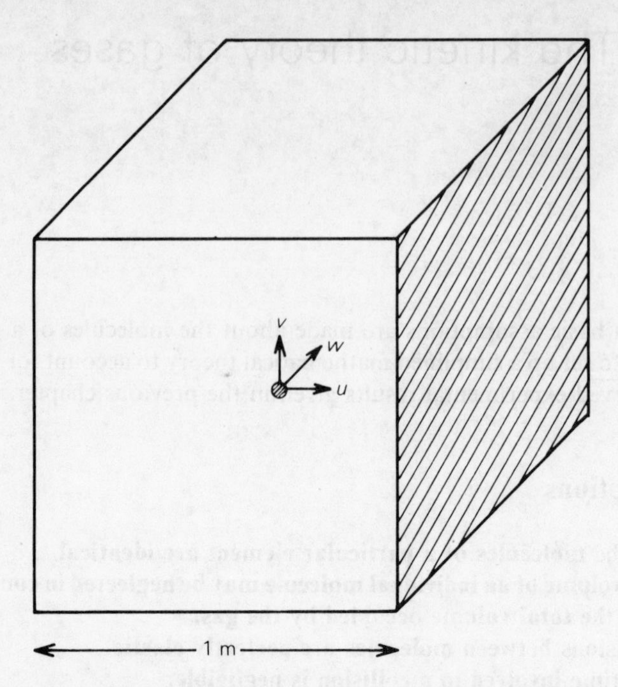

Fig. 155 Calculation of the pressure of a gas using the kinetic theory

The molecule will then move across the cube, rebound from the opposite face, and strike the shaded face again after a time interval of $2/u$ seconds (ignoring any possible collisions with other molecules on the way).

Hence the force exerted on the shaded face by this particular molecule is given by

$$\begin{aligned}\text{rate of change of momentum due to these collisions}\\ &= \frac{2\,mu}{2/u}\\ &= mu^2.\end{aligned}$$

Summing for all the n molecules contained in the cube,

$$\text{total force} = nm\bar{u}^2 \tag{1}$$

where $\bar{u}$ is the r.m.s. average speed of the molecules in the direction of u.

Now, since there are a large number of molecules involved, moving in all directions with various speeds, $\bar{u} = \bar{v} = \bar{w}$.

Also, for any particular molecule,

$$\text{velocity}^2 = u^2 + v^2 + w^2.$$

Hence
$$\begin{aligned}\bar{c}^2 &= \bar{u}^2 + \bar{v}^2 + \bar{w}^2\\ &= 3\bar{u}^2.\end{aligned}$$

Substituting in (1),

$$\text{force on face of cube} = \tfrac{1}{3}nm\bar{c}^2.$$

Since the face of the cube has an area of 1 m^2, this gives the pressure p on the face:

$$p = \frac{1}{3} nm\overline{c}^2 . \qquad (2)$$

Collisions between molecules during their movement across the cube may be ignored since they cause no loss of momentum from the system as a whole, but simply redistribute velocities among the molecules, the average $\overline{c}$ remaining the same. The assumptions (3) and (4), made at the beginning of the chapter, were to deal with this point.

It is generally more convenient to rewrite equation (2) to involve the density of the gas, ρ.

$$\text{Since density} = \frac{\text{total mass}}{\text{volume}},$$

$$\rho = nm$$

and equation (2) becomes

$$p = \frac{1}{3}\rho\overline{c}^2$$

that is,

$$\text{pressure} = \frac{1}{3} \times \text{density} \times (\text{r.m.s. speed of molecules})^2.$$

Dalton's law of partial pressures

This law states that **the pressure of a mixture of gases is equal to the sum of the pressures of the individual gases.**

It was first given as an experimental conclusion, but can be seen to agree with the calculation just performed, provided that the gases do not exert any influence upon each other (assumption (5)).

EXERCISES

1 2×10^{26} oxygen molecules, each of mass 5×10^{-26} kg, move at random inside a cubic box whose edge is 2 m long. If the average (r.m.s.) speed of the molecules is 400 m s^{-1}, calculate (a) the average change of momentum when a molecule is in head-on collision with a wall, and (b) the pressure of the gas.

2 If the density of hydrogen at s.t.p. (1.013×10^5 Pa) is 0.099 kg m^{-3}, what is the average speed of a hydrogen molecule under these conditions?

Temperature of the gas

Consider 1 mole of the gas, which by definition always contains Avogadro's number, N_A, of molecules, and suppose it occupies a volume V.

Then n, the number of molecules per m^3, $= \frac{N_A}{V}$.

Using equation (2), the pressure of the gas is given by

$$p = \frac{1}{3} \frac{N_A}{V} m\overline{c^2}$$

which may be rewritten as

$$pV = \tfrac{1}{3} N_A m\overline{c^2}. \tag{3}$$

Now the general results of experimental work with gases described in the previous chapter were combined into the Equation of State (p. 259):

$$\frac{pV}{T} = \mathcal{R}.$$

where $\mathcal{R}$ is the molar gas constant.

Writing this in the form

$$pV = \mathcal{R}T$$

and comparing it with the theoretical equation (3), it can be seen that the two are consistent if $\overline{c^2} \propto \boldsymbol{T}$.

The average kinetic energy of the molecule is then given by

$$\tfrac{1}{2}m\overline{c^2} = \frac{3}{2N_A} pV \text{ from (3)}$$

$$= \frac{3\mathcal{R}}{2N_A} T. \tag{4}$$

$\mathcal{R}$ and N_A are both universal constants, and the ratio $\mathcal{R}/N_A$ is denoted by k and called *Boltzmann's constant* (or the gas constant per molecule). Its value is 1.38×10^{-23} J K^{-1}.

Substituting k into equation (4) gives

$$\tfrac{1}{2}m\overline{c^2} = \tfrac{3}{2}kT \tag{5}$$

i.e. the average kinetic energy of a molecule of the gas is proportional to its absolute temperature.

Specific heat capacity

If the kinetic energy of a single molecule is given by equation (5), then the total internal energy of 1 *mole* of the gas will, correspondingly, be $\frac{3}{2}\mathcal{R}T$.

A rise in temperature of 1 K will mean an increase in internal energy of $\frac{3}{2}\mathcal{R}$ which must be supplied from outside, i.e. the heat capacity of 1 mole of gas at constant volume must be $\frac{3}{2}\mathcal{R}$.

$$C_V = \tfrac{3}{2}\mathcal{R}. \tag{6}$$

From equation (2) of Chap. 30 (p. 262)

$$C_p - C_v = \mathscr{R}$$

hence
$$C_p = \tfrac{5}{2}\mathscr{R}. \qquad (7)$$

The ratio C_p/C_v, γ, has been determined experimentally for many gases, and for the *monatomic* gases $\gamma = 1.67$, which is in agreement with the ratio 5/3 predicted by equations (6) and (7). However, for other gases, γ has values lower than this.

The theory has been extended in an attempt to account for the observed results for *diatomic* and *polyatomic* gases.

Molecules made up of two or more atoms may have quite considerable *rotational* kinetic energy, the individual atoms revolving about a common centre. This motion would be in addition to the linear velocity of the molecule as a whole, and so the total kinetic energy of such a molecule would be greater than that of a similar monatomic molecule.

Molecular models

Figure 156 shows two of the simple molecular models often used to illustrate theoretical ideas. These are constructed from small coloured plastic spheres which can be connected by short wooden rods fitting into holes in the spheres.

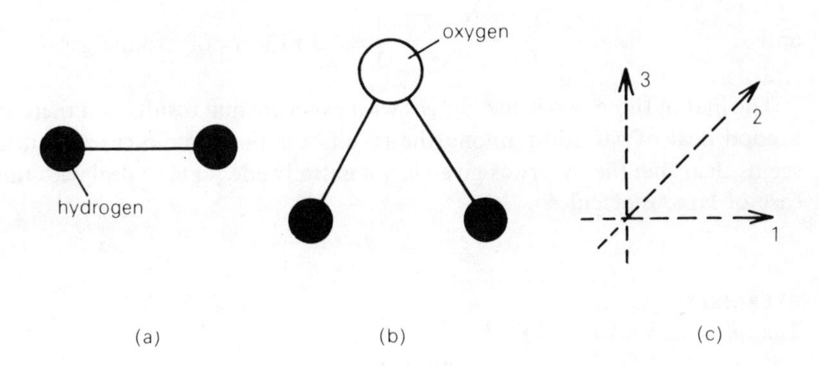

Fig. 156 Representation of molecules by constructed models

Figure 156(a) is a model of the hydrogen molecule, a diatomic gas. The wooden rod represents the electron bond between the two atoms which holds them together into a molecule. Figure 156(b) shows a molecule of water vapour, using two black spheres again as the two hydrogen atoms, and one larger white sphere for the oxygen atom. Again the electron bonding is represented by the rods which hold the molecule together.

Take three mutually perpendicular directions for reference, as shown in Fig. 156(c). Now if the diatomic model (a) rotates about an axis in the direction 2, it will possess a quantity of rotational kinetic energy since the two atoms will be moving at some distance from the axis. Similarly for a rotation about axis 3, but no energy will be involved in a rotation about axis 1 since this passes through the centres of the atoms. Such a model is said to have *two degrees of rotational freedom.*

The water molecule, however, will have *three* degrees of rotational freedom, since none of the axes passes through the centres of all the atoms at once. This is generally true of all polyatomic models.

The kinetic theory type of molecule also possesses *three degrees of translational freedom* since it is free to move linearly in any of the three given directions.

The principle of equipartition of energy assumes that **the total internal energy of a gas will be divided equally among translations and rotations**, since collisions between the molecules will be continually redistributing their individual energies.

The molecules of a monatomic gas, with only three degrees of freedom and average kinetic energy of $\frac{3}{2}kT$ (equation (5)), may be said to have $\frac{1}{2}kT$ **of energy per degree of freedom.**

Then a diatomic gas would have average kinetic energy $\frac{5}{2}kT$ per molecule, and a polyatomic gas would have $\frac{6}{2}kT$.

This would lead to $\dfrac{C_p}{C_v} = \dfrac{5+2}{5} = \dfrac{7}{5} = 1.4$ for a diatomic gas,

and $\dfrac{C_p}{C_v} = \dfrac{3+1}{3} = \dfrac{4}{3} = 1.33$ for a polyatomic gas.

The first of these two values agrees with experimental results, but there is a good deal of variation among the results for polyatomic gases, and it seems clear that the theory as given is not entirely adequate to deal with the case of large molecules.

EXERCISES

Take $k = 1.38 \times 10^{-23}\ \mathrm{J\,K^{-1}}$
and $\mathscr{R} = 8.31\ \mathrm{J\,K^{-1}}$ as appropriate.

3 What would be the absolute temperature of a gas in which the average molecule, of mass 8×10^{-26} kg, is moving with a speed of $500\ \mathrm{m\,s^{-1}}$?

4 Calculate the total internal energy of 1 mole of a monatomic gas at 100°C.

5 What is the theoretical value of the molar heat capacity, at constant volume, of oxygen?

6 Find the molar heat capacity, at constant pressure, of neon.

Examination questions 31

1 State the assumptions made in the simple kinetic theory of gases. (You are *not* expected to derive the relation $p = \frac{1}{3}\rho\overline{c^2}$.)

Explain how the kinetic theory attempts to account for the following:
(i) work is required to compress a gas;
(ii) the volume of a saturated vapour at constant temperature is independent of pressure;
(iii) when a gas expands rapidly against a moving piston its temperature falls;
(iv) the volume of a gas is proportional to its absolute temperature at constant pressure.

0.001 kg of helium gas is maintained at 300 K in a container of volume 0.30 m^3. Estimate the number of molecules in the container and the r.m.s. speed of the molecules.

(relative molecular mass of helium = 4,
Ideal gas constant $\mathscr{R} = 8.3$ J mol^{-1} K^{-1},
Avogadro constant $N_A = 6.0 \times 10^{23}$ mol^{-1}.) [AEB, June 1975]

2 The number of effective degrees of freedom of an N_2 molecule in gaseous nitrogen at room temperature is
A 1.4 B 2 C 3 D 5 E 7 [C]

3 Write down *four* assumptions about the properties and behaviour of molecules that are made in the kinetic theory in order to define an ideal (or perfect) gas. On the basis of this theory derive an expression for the pressure exerted by an ideal gas.

Show that this expression is consistent with the ideal gas equation.

Use the kinetic theory to explain why hydrogen molecules diffuse rapidly out of a porous container into the atmosphere even when the pressure of the hydrogen is equal to the atmospheric pressure outside the container.

Air at 273 K and 1.01×10^5 N m^{-2} pressure contains 2.70×10^{25} molecules per cubic metre. How many molecules per cubic metre will there be at a place where the temperature is 223 K and the pressure is 1.33×10^{-4} N m^{-2}? [L]

4 The pressure of a fixed mass of gas at constant volume is greater at a higher temperature. One reason for this is that
A the molecules collide with the container walls more frequently.
B the number of intermolecular collisions increases.
C the mean free path of the molecules increases.
D the size of each individual molecule increases.
E the energy transferred to the walls during collisions increases. [O]

5 What do you understand by the term *ideal gas*?

Describe a molecular model of an ideal gas and derive the expression $p = \frac{1}{3}\rho\overline{c^2}$ for such a gas. What is the reasoning which leads to the assertion that the temperature of an ideal monatomic gas is proportional to the mean kinetic energy of its molecules?

The Doppler broadening of a spectral line is proportional to the r.m.s. speed of the atoms emitting light. Which source would have less Doppler broadening, a mercury lamp at 300 K or a krypton lamp at 77 K? (Take the mass numbers of Hg and Kr to be 200 and 84 respectively.)

What causes the behaviour of real gases to differ from that of an ideal gas? Explain qualitatively why the behaviour of all gases at very low pressures approximates to that of an ideal gas. [OC]

32 Vapours

Molecules of a liquid which have escaped from its surface and move freely in the space above, form a *vapour.*

In an enclosed space, the vapour comes into *dynamic equilibrium* with the liquid; this means that molecules are continually passing into and out of the liquid surface, the overall density of the vapour remaining constant. When this state is reached, the vapour is said to be *saturated.*

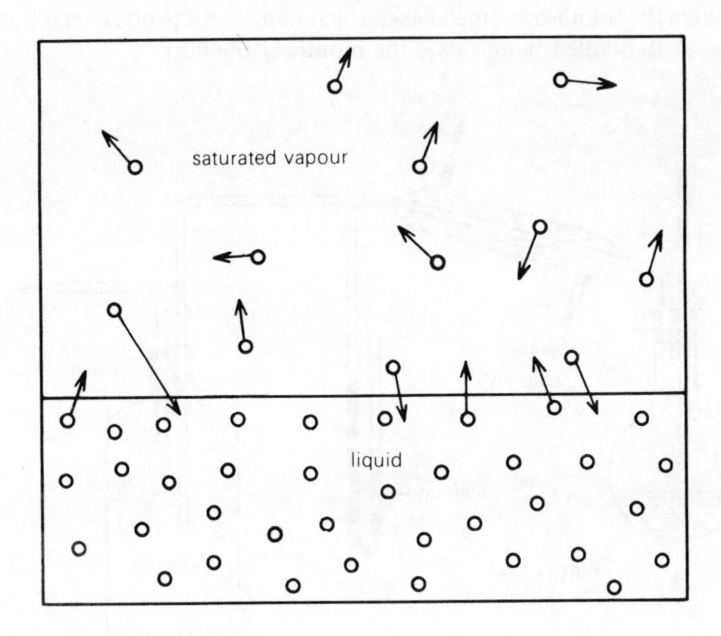

Fig. 157 Behaviour of molecules of a vapour

If heat energy is supplied to the liquid, the average speed of the liquid molecules increases and more of them will be able to escape into the vapour. The density of the vapour will thus increase until a new state of dynamic equilibrium is reached.

The pressure exerted by the molecules of a saturated vapour is called the *saturated vapour pressure* of the liquid. This will increase with temperature, but there is no simple mathematical relation between the two variables.

Boiling

A liquid is commonly accepted to be *boiling* when bubbles of vapour are forming rapidly throughout the liquid and breaking at its surface. Under these conditions, **the saturated vapour pressure of the liquid is equal to the external pressure on its surface**. The temperature at which this occurs is defined as the *boiling-point* of the liquid.

Variation of boiling-point with external pressure

A simple experiment can be set up to measure the boiling-point of water under various pressures, over a range of about half an atmosphere. Greater variation of pressure may cause difficulties under normal laboratory conditions.

The apparatus is shown in Fig. 158. About a litre of distilled water is placed in the round-bottomed flask, which stands on a tripod over a Bunsen flame. A two-holed bung closes the mouth of the flask.

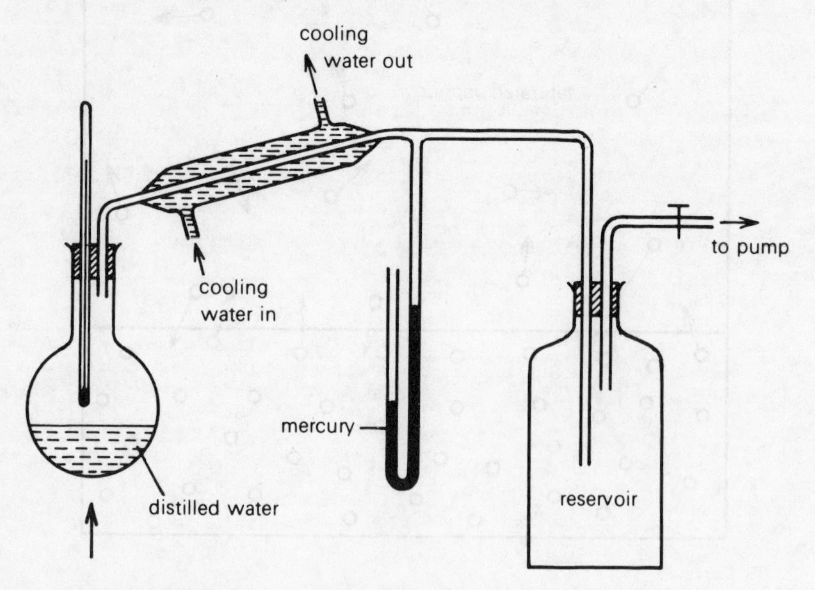

Fig. 158 Measurement of the boiling-point of water under different pressures

The thermometer which fits through one hole is placed with its bulb at the centre of the flask, well clear of the walls, and just above the level of the liquid. This thermometer should be graduated to 0.1°C.

A short glass tube through the second hole is connected to a Liebig condenser, which is clamped at an angle so that the condensed water vapour returns to the flask.

The exit from the far side of the condenser leads to a T-piece. One arm of the T-piece is connected to an open-ended mercury manometer, which needs to be at least 0.4 m high if the pressure is going to be reduced by half an atmosphere. The other arm of the T-piece is connected, via a large reservoir vessel, to a small pump. The reservoir is used not only to even out the pressure in the system, but also as a safety precaution to prevent water or mercury being sucked into the pump.

At the start of the experiment, the bung should be loosened in the reservoir vessel so that the pressure becomes atmospheric. The water in the flask is brought to the boil and the temperature of the vapour is recorded when conditions have become steady.

The Bunsen flame is then removed, and the thermometer reading allowed to fall about five degrees. The bung of the reservoir is tightened up and the pump turned on to suck for a few seconds, until the water is again seen to be boiling. Enough heat may be retained in the glass walls of the flask to keep the temperature conditions steady while both temperature and manometer readings are recorded; if not, a very small Bunsen flame should be returned under the flask.

The experiment is repeated to obtain a set of readings covering the whole desired range. It may be possible to take some readings at pressures greater than atmospheric, if the apparatus will remain airtight. The readings of the manometer must be converted into actual pressures by combining them with the reading of a barometer in the same room.

A graph of boiling-point against pressure will be a smooth curve. Since the saturated vapour pressure of a liquid is equal to the external pressure at the boiling-point, **this graph is also the relation between the saturated vapour pressure of water and temperature**. The experimental results may be compared with the standard values given in Table 10.

Table 10

Temperature/° C	Saturated vapour pressure of water/10^5 Pa.
100	1.013
90	0.701
80	0.473
70	0.312
60	0.199
50	0.123
40	0.073
30	0.042
20	0.023

Mixtures of gases and vapours

Consider an enclosed volume which contains both a pure gas and a small quantity of a liquid in equilibrium with its own vapour. The resulting pressure in the space will be the combination of the partial pressure due to the gas and the saturated vapour pressure of the liquid (Dalton's law). Both of these vary with temperature, and their behaviour must be considered independently.

Should the temperature of the system be raised sufficiently to evaporate all the liquid, the vapour will become unsaturated and can then be assumed to obey the gas laws (Fig. 159).

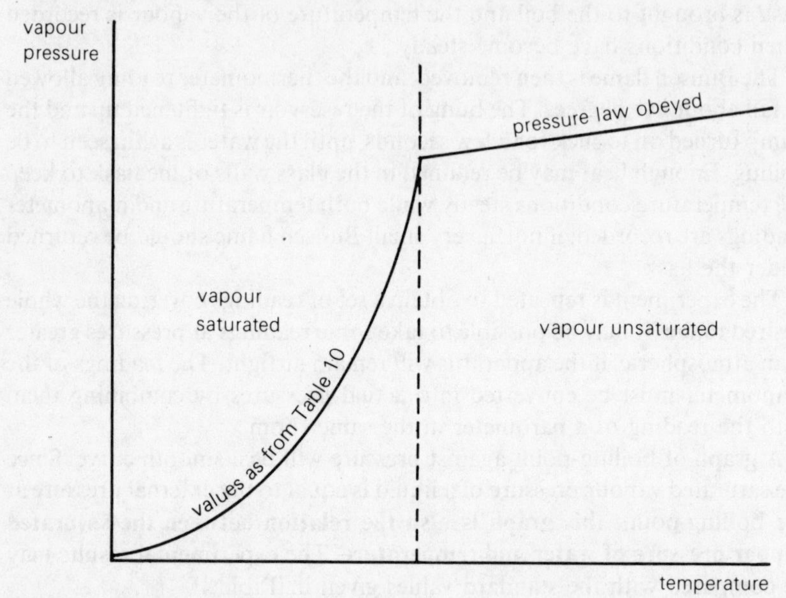

Fig. 159 Graph showing the variation of vapour pressure with temperature

EXERCISES

Take the appropriate values for the saturated vapour pressure of water from Table 10 when required.

1 A saturated vapour of density 1.3 kg m^{-3} occupies a volume of 2 $\times 10^{-2}$ m^3. If this volume is slowly reduced by half, keeping the temperature constant, what mass of vapour will condense into liquid?

2 A small quantity of water is placed in an enclosed vessel which already contains air at a pressure of 1×10^5 Pa. If the initial temperature of the vessel is 20°C, calculate the pressure that will exist inside it at 60°C, assuming that not all the liquid will evaporate.

3 0.001 m^3 of air saturated with water vapour at 100°C is cooled to 20°C while the total pressure is maintained constant at 1.2×10^5 Pa. What volume will the mixture now occupy?

The behaviour of carbon dioxide

An important study of the behaviour of carbon dioxide was carried out by Andrews in 1863. This is a very useful experimental substance since it can easily be studied in any of the three states, solid, liquid and gas, under normal laboratory conditions.

Andrews' apparatus is shown in Fig. 160. It is now kept in the Science Museum at South Kensington. The carbon dioxide was enclosed in a thick-walled glass capillary tube which was graduated to read the volume directly. Mercury sealed the lower end of the tube, which then widened out and was enclosed in a strong metal container filled with water. A screw plunger fitting tightly into the base of the container was used to vary the pressure inside the system.

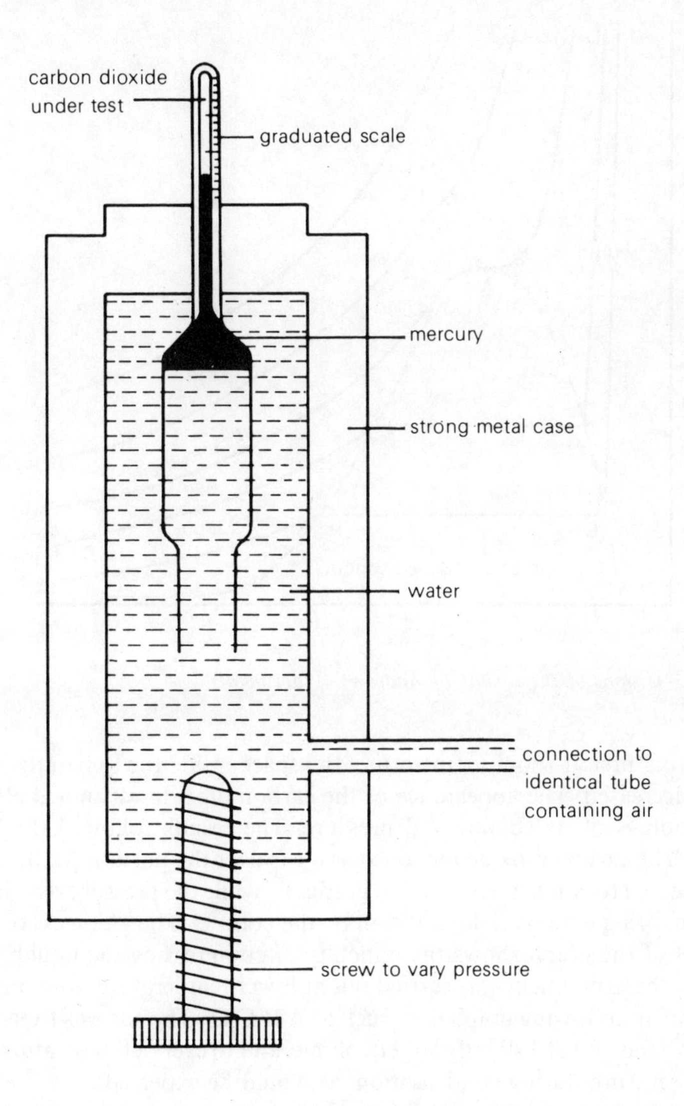

Fig. 160 Andrews' apparatus for investigating the behaviour of carbon dioxide

A second identical tube was connected to the first so that the pressure in the two was always equal. The second capillary tube contained air. Assuming that the air obeyed Boyle's law, its pressure could be calculated from its observed volume, and this would be also the pressure of the carbon dioxide.

The whole apparatus was placed in a water bath to control the temperature at which the experiments were carried out.

A sketch of the results is shown in Fig. 161 in the form of a graph of pressure against volume.

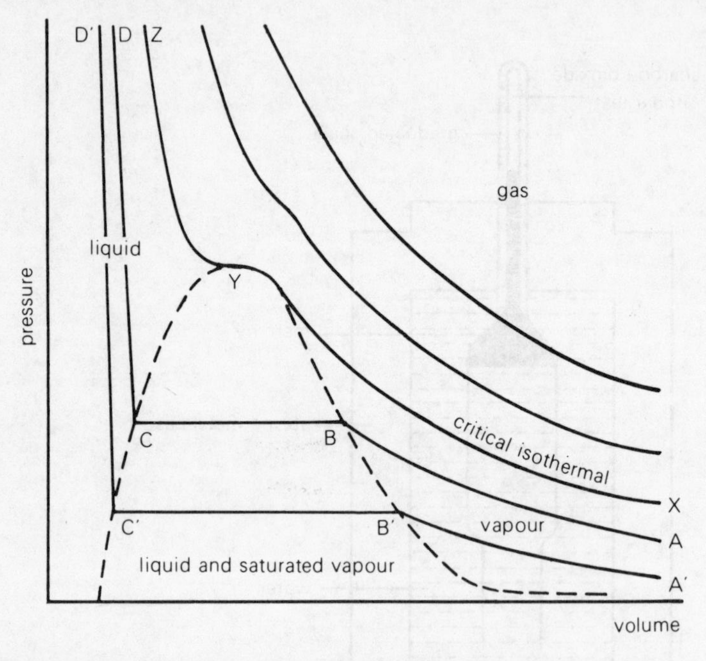

Fig. 161 Graphs of the results of Andrews' experiments

Working first at room temperature, the space inside the apparatus was slowly decreased. The appearance of the carbon dioxide was noted along with readings of its volume and pressure. The results followed the line ABCD. The carbon dioxide remained as a gas over the portion AB, then it was observed to condense into liquid gradually while the pressure remained constant, being entirely in liquid form by the point C. The steepness of the last part of the graph shows the difficulty of compressing the liquid.

When the experiment was carried out at lower temperatures, the results were similar, following graph lines such as A′B′C′D′. The gas was found to begin to condense at a slightly larger volume, and to exert a lower saturated vapour pressure during condensation, as would be expected.

The most important result of the whole experiment was found when readings were taken at certain higher temperatures. No condensation was observed even at high pressures, and the relation between pressure and volume agreed fairly well with Boyle's law over the whole range. The carbon dioxide did not liquefy.

Continuing the experiment carefully, Andrews was able to plot a set of readings which formed a critical boundary XYZ between the two types of behaviour of the gas. This occurred at a temperature of 31.1°C, which is called the *critical temperature* for carbon dioxide, while the point of inflection of the curve, Y, is called the *critical point*.

The complete set of curves, of which only some are shown in Fig. 161, are known as *isothermals*.

Critical temperature

The behaviour of carbon dioxide is typical of all other gases, which can be shown to demonstrate the same sort of isothermals. Some values of the critical temperature for other gases are given in Table 11. **Above this temperature, the gas cannot be liquefied.** It is common practice to distinguish between substances in this state, which are referred to as *gases*, and those below their critical temperatures, which are called *vapours*.

Table 11

Substance	Critical temperature/° C
Water	374
Ammonia	130
Carbon dioxide	31.1
Oxygen	−118
Nitrogen	−146
Hydrogen	−240 (33K)
Helium	−268 (5K)

The triple point

Vapours are also formed in the same way in the space around a solid which is below its melting-point; this process is called *sublimation*. The experimental results of measurement of the saturated vapour pressure in this region are, of course, much lower than those for the liquid, since it is correspondingly more difficult for a molecule to escape from a solid.

The line APB in Fig. 162 shows the general shape of the graphs of temperature and saturated vapour pressure. The portion A → P corresponds to liquid and vapour in dynamic equilibrium, and P → B corresponds similarly to solid and vapour. The common point P marks the change of state of the substance from liquid to solid, i.e. the liquid is freezing.

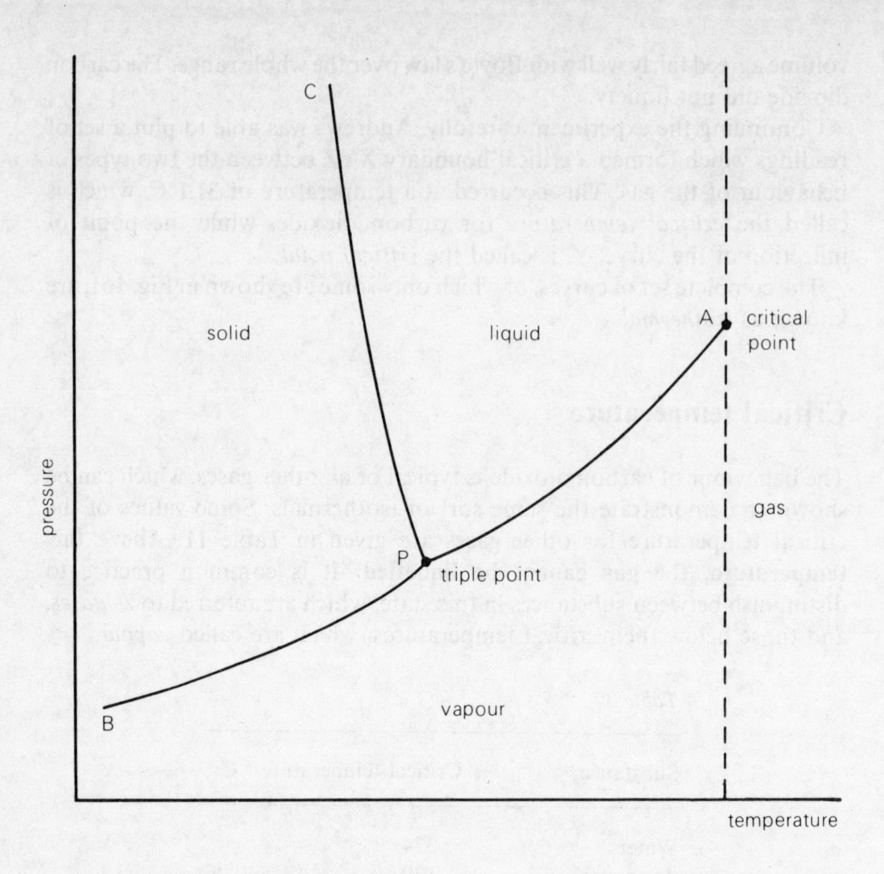

Fig. 162 Graphs of temperature and pressure relating to the changes of state of a substance

It is known that the freezing-point of a liquid varies with the external pressure, and the way in which it varies is represented in Fig. 162.by the line C → P. This corresponds to dynamic equilibrium between solid and liquid. CP may have either negative or positive slope according to whether the substance expands or contracts on solidifying.

These three graph lines form the boundaries between the three states of the substance. The common point P is called the *triple-point*; **this is the unique combination of temperature and pressure at which solid, liquid and vapour can exist together in dynamic equilibrium.**

The triple-point temperature for water is easy to determine accurately, and it has been chosen as the fundamental point of the thermodynamic temperature scale (p. 2) since it seems of greater universal significance than is the freezing-point of water at standard terrestrial atmospheric pressure. The value assigned to this temperature is 273.16 K.

Examination questions 32

1 (a) Describe an experiment to determine the saturation vapour pressure of water at various temperatures in the range 75°C to 110°C. Sketch a graph showing the results which would be obtained from the experiment. (S.v.p. of water at 75°C = 38 kPa; standard atmospheric pressure = 101 kPa.)
(b) By considering the effect of temperature on the s.v.p. of water vapour, explain why it is essential to have a safety valve on the water boiler of a central heating system.
(c) In a Boyle's law experiment using damp air, the following results were obtained:

Initial pressure (air unsaturated) = 8.5 kPa
Pressure when volume reduced to $\frac{1}{2}$ of initial volume = 16.0 kPa
Pressure when volume reduced to $\frac{1}{3}$ of initial volume = 23.0 kPa.

(i) Show that the vapour exerts its saturation pressure when the volume is reduced to half its initial value.
(ii) Calculate the saturation vapour pressure at the temperature of the experiment.
(iii) Calculate the initial pressure of the water vapour.

[AEB, Nov. 1979]

2 Explain what is meant by the saturated vapour pressure of a liquid. Give a simple molecular explanation of why saturated vapour pressure may be expected to increase rapidly with temperature.

A theory suggests that the saturated vapour pressure p of any liquid varies with temperature T approximately according to the relation

$$p = p_0 e^{-A/\mathscr{R}T},$$

where p_0 and A are constants characteristic of the liquid and $\mathscr{R}$ is the molar gas constant. The following experimental results were obtained for water:

p/Pa	3.6×10^3	1.4×10^4	4.2×10^4	2.4×10^5	9.2×10^5
T/K	300	325	350	400	450

(a) By plotting a graph of log (p/Pa) against $1/(T/\mathrm{K})$, investigate the degree to which these results support the theoretical relation. Estimate the value of A.
(b) The molar latent heat of vaporisation of water is approximately 4×10^4 J mol^{-1}. Compare this value with your estimate of A, and comment on the result.

(Molar gas constant $\mathscr{R} = 8.3$ J K^{-1} mol^{-1}; log e = 0.43) [C]

3 Use the kinetic theory of matter to explain how the saturation vapour pressure of water vapour might be expected to vary between 273 K and 383 K. Illustrate your answer with a sketch graph. [JMB]

4 What is meant by an isothermal process? Describe briefly how the isothermal behaviour of a real substance may be investigated in the region of its critical point. Sketch typical isothermal curves of pressure against volume for a real fluid in this region and indicate on your graph the critical point.

Discuss the state of the fluid in the various sections of such a curve below the critical temperature. [L]

5 Why has the triple-point of water, rather than the melting-point of ice, been chosen as the fixed point for the establishment of the kelvin?

A It is more precisely reproducible.

B It is closer to the defining temperature of 273.16.

C It gives a more convenient scale between 0°C and 100°C.

D Very accurate gas thermometers have shown that it is best.

E It ensures a more linear scale for gas thermometers. [O]

Light II

33 Lenses

Location of the image formed by a lens

Figure 163 represents light travelling from a point object O, on the principal axis of a lens, to form a real image at I. The dotted curves show the positions of two wavefronts, one (AA′) diverging from O which has just reached the lens surface at A, and the second (BB′) converging to I, having just left the second surface of the lens at B.

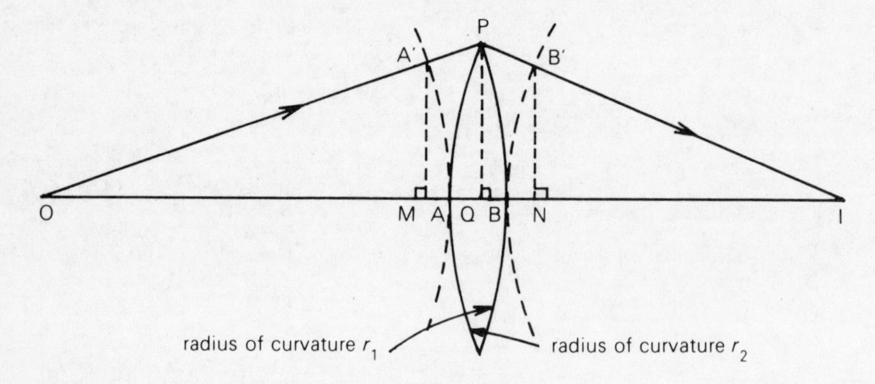

Fig. 163 Location of the image formed by a lens

A ray of light travelling from O to I via the point P at the extreme edge of the lens must take the same time as a ray which has passed straight through the lens.

Hence the time to travel A′P + PB′ in air must be the same as the time to travel AB in the material of the lens. Suppose the velocities of light in these two media are c and c'.
Then

$$\frac{\mathrm{A'P} + \mathrm{PB'}}{c} = \frac{\mathrm{AB}}{c'}.$$

Considering only a narrow beam of light close to the principal axis, A′P + PB′ is approximately equal to MN, where M and N are the feet of the perpendiculars from A′ and B′ to the principal axis.

$$\mathrm{MN} = \mathrm{MA} + \mathrm{AB} + \mathrm{BN}$$

so the equation becomes

$$\frac{(\text{MA}+\text{BN})}{c}+\frac{\text{AB}}{c}=\frac{\text{AB}}{c'}$$

or, rearranging,

$$\frac{\text{MA}+\text{BN}}{c}=\left(\frac{1}{c'}-\frac{1}{c}\right)\text{AB}$$

$$\Rightarrow \quad \text{MA}+\text{BN}=\left(\frac{c}{c'}-1\right)\text{AB}.$$

Since $c/c' = n$, the refractive index of the material of the lens,

$$\text{MA}+\text{BN}=(n-1)\text{AB}. \tag{1}$$

Now AA′ is the arc of a circle, centre O and radius u (using the customary symbol for the distance between object and lens). From the geometry of the circle

$$\text{MA}\,(2u-\text{MA})=\text{A'M}^2$$

Similarly
$$\text{BN}\,(2v-\text{BN})=\text{B'N}^2$$

and
$$\text{AB}=\text{AQ}+\text{QB}$$

where
$$\text{AQ}\,(2r_2-\text{AQ})=\text{PQ}^2$$

and
$$\text{QB}\,(2r_1-\text{QB})=\text{PQ}^2,$$

Q being the foot of the perpendicular from P to the principal axis.

If the lens is thin, then MA, BN, AQ and QB may be neglected in comparison with $2u$, $2v$, $2r_2$ and $2r_1$ inside the brackets, and A′M = B′N = PQ.

Hence $\text{MA}=\dfrac{\text{PQ}^2}{2u}$, $\text{BN}=\dfrac{\text{PQ}^2}{2v}$, $\text{AQ}=\dfrac{\text{PQ}^2}{2r_2}$ and $\text{QB}=\dfrac{\text{PQ}^2}{2r_1}$.

Substituting in equation (1) gives

$$\frac{\text{PQ}^2}{2u}+\frac{\text{PQ}^2}{2v}=(n-1)\left(\frac{\text{PQ}^2}{2r_2}+\frac{\text{PQ}^2}{2r_1}\right)$$

which simplifies to

$$\frac{1}{u}+\frac{1}{v}=(n-1)\left(\frac{1}{r_1}+\frac{1}{r_2}\right). \tag{2}$$

Note the two assumptions that have been made in deducing this equation.

If the beam of light incident on the lens is a parallel beam, it will form an image at the principal focus. $1/u$ will be zero, and $1/v = 1/f$, where f is the focal length of the lens.

Hence
$$\frac{1}{f}=(n-1)\left(\frac{1}{r_1}+\frac{1}{r_2}\right)$$

and in general,

$$\frac{1}{u}+\frac{1}{v}=\frac{1}{f}. \tag{3}$$

Although this proof has been applied specifically to the case of a biconvex lens with a real object and image, its results, equations (2) and (3), can be used for any shape of lens and either real or virtual objects and images provided that a suitable sign convention is adopted.

Sign conventions

Real-is-positive

Distances measured to real objects and images are positive.
Distances measured to virtual objects and images are negative.
A converging lens has a positive focal length.
A diverging lens has a negative focal length.
With these conventions, the lens formula applies in all cases in the form

$$\frac{1}{u}+\frac{1}{v}=\frac{1}{f}.$$

Cartesian

The centre of the lens is taken as the origin of coordinates.
The incident light travels towards the right.
Object and image distances measured to the right are positive.
Object and image distances measured to the left are negative.
A converging lens employs a principal focus on the positive side, hence its focal length is positive.
A diverging lens employs a principal focus on the negative side, hence its focal length is negative.
With these conventions, the lens formula applies in all cases in the form

$$\frac{1}{v}-\frac{1}{u}=\frac{1}{f}.$$

Either of these alternative sign conventions may be used for problem solving, though it should always be stated which one is in fact being used.

Conjugate foci

Since equation (3) is symmetrical for u and v, it follows that if the object distance is equal to v, then the image will be formed at a distance u from the lens. The pair of values, u and v, are said to be *conjugate*, and the object and image lie at *conjugate foci* of the particular lens.

Figure 164 shows the general relation between u and v for (a) a converging lens, and (b) a diverging lens. The graphs are hyperbolae, with asymptotes corresponding to the focal length of the lens.

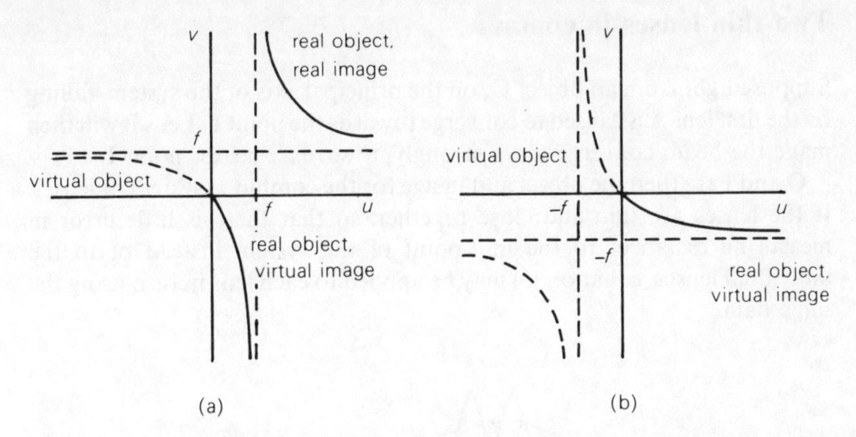

Fig. 164 Graph of u and v for (a) a converging lens, (b) a diverging lens

Experimental determination of the focal length of a lens is best checked by the use of a straight-line graph, which can be obtained by plotting $1/v$ against $1/u$. Figure 165 shows how such a graph would look, for each type of lens. If the lines are extrapolated (where necessary) to cut the axes, the intercepts will be at $1/f$.

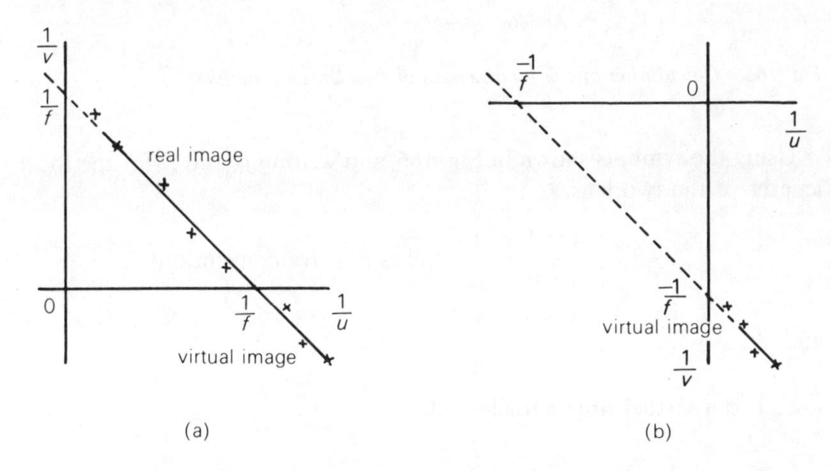

Fig. 165 Graph of $1/u$ and $1/v$ for (a) a converging lens, (b) a diverging lens

EXERCISES

1 Calculate the focal length of a bi-convex lens whose faces have radii of curvature 1.00 m and 0.67 m if the refractive index of the material of the lens is 1.50.

2 Find the required radius of curvature of the concave face of a plano-concave lens of focal length 300 mm made of glass of refractive index 1.51.

Two thin lenses in contact

Suppose light from an object O, on the principal axis of the system, falling on the first lens A is caused to converge towards the point I′. Lens B will then make the beam converge more strongly, towards a nearer point I.

O and I are then the object and image for the combined system of lenses. If the lenses are thin and close together, so that there is little error in measuring distances to the mid-point of the system instead of to the individual lenses, equation (3) may be applied to each lens in turn using the same data.

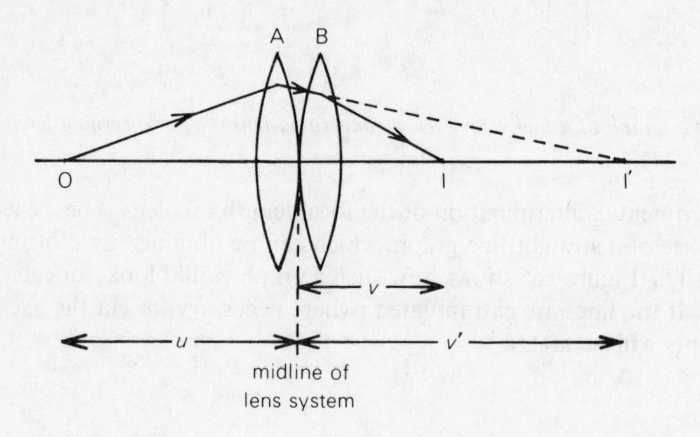

Fig. 166 Calculation of the focal length of two lenses combined

Using the symbols shown in Fig. 166, and writing f_A and f_B for the focal lengths of the two lenses,

$$\frac{1}{u}+\frac{1}{v'}=\frac{1}{f_A} \text{ (real-is-positive convention)}$$

and

$$-\frac{1}{v'}+\frac{1}{v}=\frac{1}{f_B}$$

since I′ is a virtual object for lens B.
Adding,

$$\frac{1}{u}+\frac{1}{v}=\frac{1}{f_A}+\frac{1}{f_B}.$$

But for the combined system, if its focal length is F,

$$\frac{1}{u}+\frac{1}{v}=\frac{1}{F}.$$

Hence $$\frac{1}{F}=\frac{1}{f_A}+\frac{1}{f_B}. \qquad (4)$$

A similar argument applies if one or both of the lenses is diverging, and equation (4) will still hold.

This result is only approximate because of the assumptions made regarding the measurement of distance. Experimental determination of the combined focal length of the system using the displacement method will generally yield a more accurate value.

The displacement method for measurement of the focal length of a thick converging lens

Measurements of the distances of objects and images from a lens generally involve some inaccuracy because of the thickness of the lens. In this experiment, the lens is moved from one position A to another position B and it is the *displacement* y of the lens which is used in subsequent calculation. This can be measured by reference to any convenient point of the lens, and in practice it is read from a marker on the lens holder which moves along a metre scale.

The thick lens, or combination of lenses, is set up between an illuminated object and a screen, as shown in Fig. 167. These are fixed a known distance x apart, which must be greater than four times the focal length of the lens.

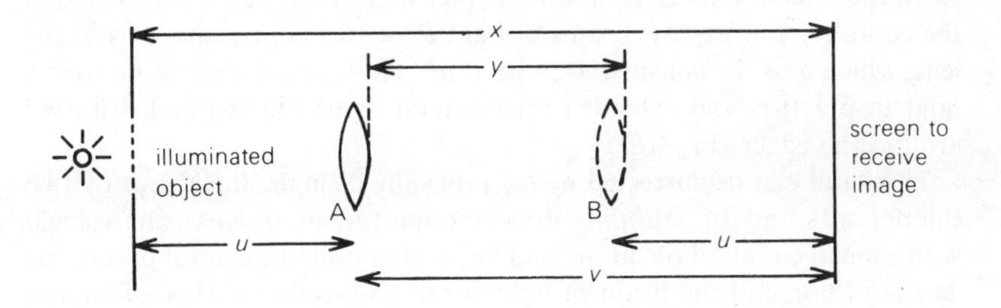

Fig. 167 Displacement method for measurement of the focal length of a thick lens

(The focal length of a converging lens can be found roughly by focusing light from a distant object, such as a window, on the palm of one's hand; it is advisable to do this automatically when picking up any lens, to see immediately how strong or weak it is.) If this condition is fulfilled, it will be possible to find two different positions, A and B, for the lens in which an image of the object is sharply in focus on the screen.

The values to be used for u and v in equation (3) can be deduced from the measured values of x and y, since $x = u + v$ and $y = v - u$, which gives

$$f = \frac{x^2 - y^2}{4x}.$$

This expression does not have a simple graphical representation, and it is best to repeat the experiment several times with different values for x, and calculate the results separately.

EXERCISES

3 Two converging lenses of focal lengths 500 mm and 400 mm are placed in contact, with a common axis. What is the focal length of the combination?

4 What lens should be used in combination with a converging lens of focal length 100 mm to form a converging system whose focal length is 200 mm?

5 A thick lens, moved between an illuminated object and a screen 1.50 m apart, forms a sharply focused image of the object in two positions which are 70 mm apart. Calculate the focal length of the lens.

6 A converging lens of focal length 200 mm produces a real diminished image of an object when the distance between object and image is 1.20 m. How far must the lens be moved to obtain a sharply focused magnified image on the screen, if screen and object remain in their original positions?

Spherical aberration

In all the calculations given in this chapter it has been necessary to restrict the conditions to narrow beams of light close to the principal axis of the lens, which must be considered to be 'thin'. These conditions do not really hold in practice, and extended images tend to be blurred and distorted around the edges (Fig. 168).

This fault can be corrected by (a) replacing a single 'thick' lens by two thinner ones, and (b) 'stopping' down the aperture of the lens using a shield with a small circular hole in the middle, so that only the central part of the lens is in use, and the beam of light is not so divergent. This, of course, decreases the brightness of the image.

Chromatic aberration

Different colours of light are refracted by different amounts, and hence the focal length of a lens varies with colour. This means that white light from an object will give a *real* image which appears to be surrounded by a coloured halo, not all the colours being brought to a focus at the same point.

This fault is not noticeable with *virtual* images viewed by the eye, since the actual location of these images is not important and the eye sees them superimposed.

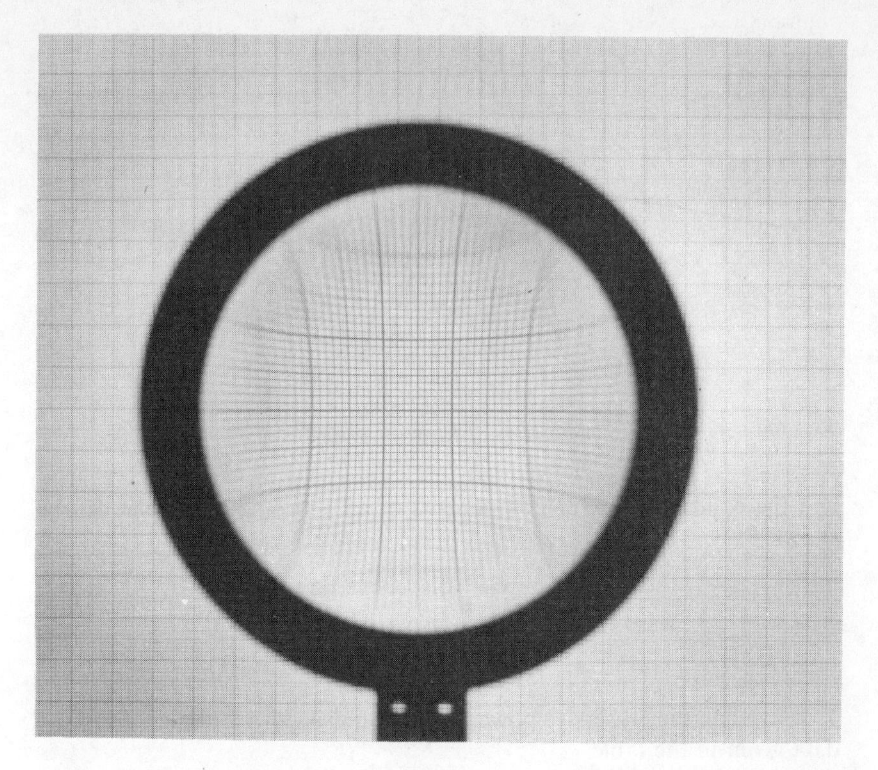

Fig. 168 Spherical aberration

High-quality optical instruments correct for chromatic aberration by the use of *achromatic combinations* of lenses whenever a real image is being formed.

Achromatic combinations

The dispersion produced in white light by a prism or section of a lens can be counteracted by placing a second prism, inverted, next to the first. If these are made of different types of glass, it is possible to find suitable shapes which will annul dispersion while still deviating the light.

As shown in Fig. 169, the two rays considered emerge parallel, and can be brought to the same focus. Dispersion of the intermediate colours may not be completely compensated but this will be only a second-order fault.

In practice, this result is achieved by using crown glass for the converging lens A and flint glass to make a plano-concave lens for B, the two fitting closely together over one pair of faces. Crown glass produces only about half the dispersion of a similar piece of flint glass so that the achromatic combination (or *doublet*) can still be a converging system.

The *dispersive power* of a small-angle prism or lens is defined as

$$\omega = \frac{\text{difference in refractive index for blue and red light}}{(\text{average refractive index}) - 1}$$

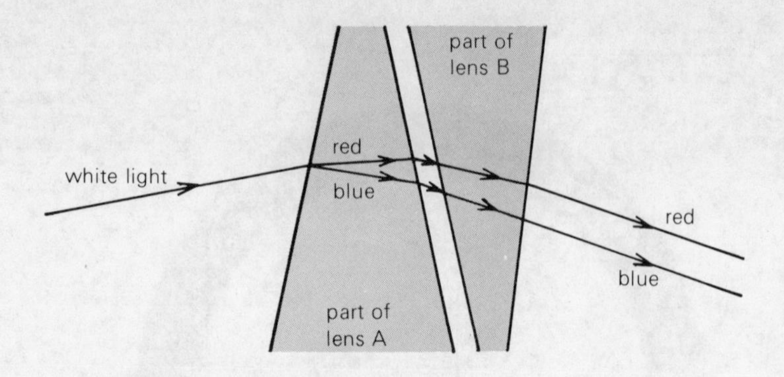

Fig. 169 Achromatic combination of lenses or prisms

and the condition for the components of an achromatic doublet can be shown to be

$$\frac{\text{focal length of A}}{\text{focal length of B}} = \frac{\text{dispersive power of A}}{\text{dispersive power of B}}.$$

EXERCISE

7 Calculate the dispersive powers of crown glass and flint glass from the data given in the table:

	Refractive index	
	Red light	Blue light
Crown glass	1.510	1.522
Flint glass	1.643	1.665

Hence find the focal length of a lens made of flint glass to be used with a converging crown glass lens of focal length 200 mm to form an achromatic doublet.

Examination questions 33

1 Deduce an expression for the focal length of a lens combination consisting of two thin lenses in contact.

Describe how you would verify this expression experimentally.

Two equiconvex lenses, each of focal length 150 mm, are placed in contact and the space between them is filled with liquid of refractive index 1.3. If the refractive index of the material of the lens is 1.5, calculate the focal length of the combination. [AEB, June 1976]

2 A lens of focal length 0.3 m is placed between an illuminated object and a screen which are 1 m apart. By varying the position of the lens, it is possible to produce on the screen

A 2 real inverted images of the object.
B 1 real inverted image of the object.
C 2 real erect images of the object.
D 1 real erect image of the object.
E no images of the object at all. [C]

3 (a) Explain with the aid of a ray diagram the action of a spectacle lens in correcting the far point of an eye from 2 m in front of the eye to infinity. Repeat the procedure for correcting the near point of an eye from 0.50 m to 0.25 m.
(b) In each case calculate the focal length of the spectacle lens, ignoring the separation of lens and eye. [JMB]

4 A plano-convex thin lens is to be used as a reading glass. If glass of refractive index 1.5 is used and a magnification of 10 is desired when the print is 50 mm from the lens, what should the radius of the curved face be? [L]

5 An object is placed 0.15 m in front of a converging lens of focal length 0.10 m. It is required to produce an image on a screen 0.40 m from the lens on the opposite side to the object. This is to be achieved by placing a second lens midway between the first lens and the screen. What type and of what focal length should this lens be? Sketch a diagram showing two rays from a non-axial point on the object to the final image. [L]

6 A bi-convex lens has surfaces with radii of curvature 1.0 m and 1.5 m, and the refractive indices of its material for red light and for blue light are respectively 1.535 and 1.545. Calculate the distance between the images of a distant white-light source formed by the lens in red light and in blue light. [O]

34 Optical instruments

The slide projector

Figure 170 shows the full optical system of a slide projector.

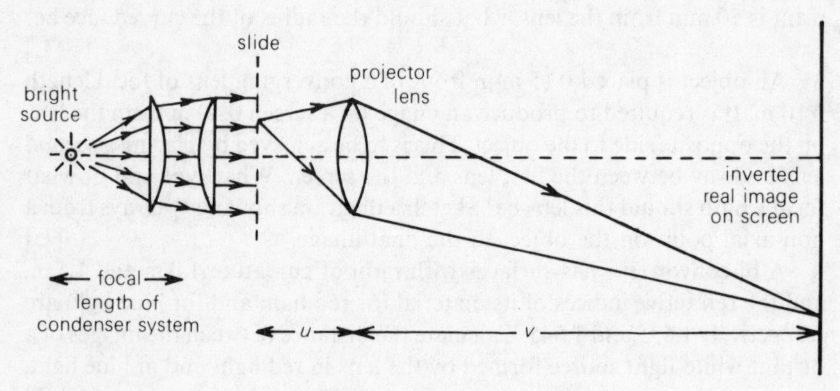

Fig. 170 Optical system of a slide projector

The source of light must be a small, high-power lamp so that enough light is emitted to ensure adequate intensity at the screen. It is placed at the principal focus of the condenser lens.

Parallel light emerges from the condenser and is used to illuminate the slide evenly. The diameter of the condenser lens must be large enough to ensure that the whole area of the slide is covered, and as it must also be of fairly short focal length (to receive the maximum emission from the source), the lens will inevitably be thick. To cut down spherical aberration (p. 292), the condenser system is, in practice, made out of two plano-convex portions, mounted as shown in the figure. Having the plane face of the lens towards the source reduces the danger of the glass cracking from the heat given off.

Each point of the slide can now be regarded as an object for the projector lens, which produces a magnified, inverted, real image on the screen, placed some distance away. Focusing of the image is carried out by a screw movement of the projector lens, which should be an achromatic doublet (p. 293) if colour is to be used.

The *linear magnification* of the system is defined as

$$\frac{\text{height of image}}{\text{corresponding height of object}} = \frac{\text{distance of image from lens}}{\text{distance of object from lens}} = \frac{v}{u}.$$

The camera

A simple camera uses an achromatic converging lens of short focal length, which produces a real image in or near the focal plane, for all objects more than 2 m or so from the camera. A focusing adjustment is usually provided so that the lens can be moved through a few millimetres. Professional photographers would change their lens to accommodate close-up objects.

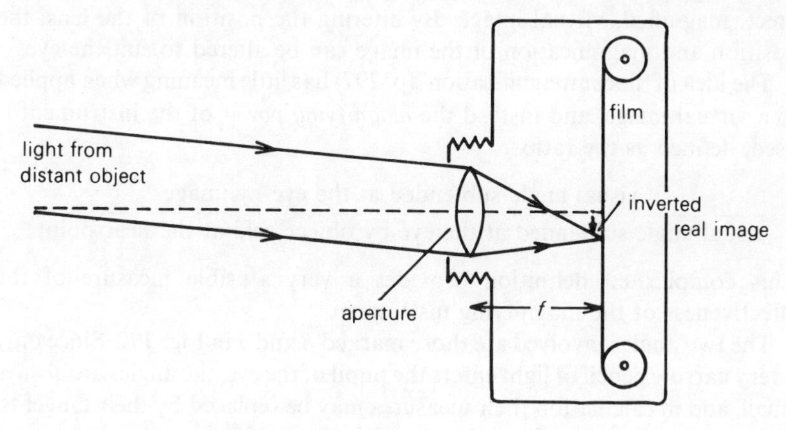

Fig. 171 Optical system of a simple camera

The exposure is made by opening a shutter in front of the lens for a specified time. As the chemical reaction which takes place in the emulsion on the film depends on the quantity of light energy reaching it, it follows that if the aperture is small, the exposure time must be comparatively long, and vice versa.

The size of the aperture can be varied by selecting one of a set quoted as $f/7.9$, $f/11$, $f/16$, $f/22$ etc. f refers to the focal length of the camera lens. Using this form gives a series of numbers which apply to all cameras whatever their focal distance may be. The numbers 7.9, 11, 16 . . . bear the relation that their squares 62, 121, 256 . . . are in the ratio 1:2:4 . . . and have been chosen so that in the series of apertures available, each allows twice as much light through as the next smaller one.

The use of magnifying instruments

When examining an object in detail, e.g. reading the date on a coin, it is found best to hold the object some distance away from the eye. This distance varies from person to person, and alters noticeably with age; it is called the *least distance of distinct vision* and given the symbol D. The point at which the object is held is called the *near point*.

Optical instruments which are designed to help the eye by magnifying the object frequently give an image at the near point. However, the eye muscles must be partially contracted when viewing at this distance, and it is easier, for prolonged viewing, to arrange for the final image to be at infinity, when the normal eye would be relaxed.

The magnifying glass

A converging lens, used with an object within its focal distance, produces an erect, magnified, virtual image. By altering the position of the lens, the position and magnification of the image can be altered to suit the eye.

The idea of 'linear magnification' (p. 297) has little meaning when applied to a virtual image, and instead the *magnifying power* of the instrument is used, defined as the ratio

$$\frac{\text{visual angle subtended at the eye by image}}{\text{visual angle subtended at the eye by object held at the near point}}.$$

This complicated definition provides a very sensible measure of the effectiveness of the magnifying instrument.

The two angles involved are those marked β and α in Fig. 172. Since only a very narrow pencil of light enters the pupil of the eye, the angles are always small, and in calculation their measures may be replaced by their tangents.

By applying the lens formula (equation (3), p. 288) it can be shown that for this instrument

$$\text{magnifying power } M = \frac{D}{f} - 1.$$

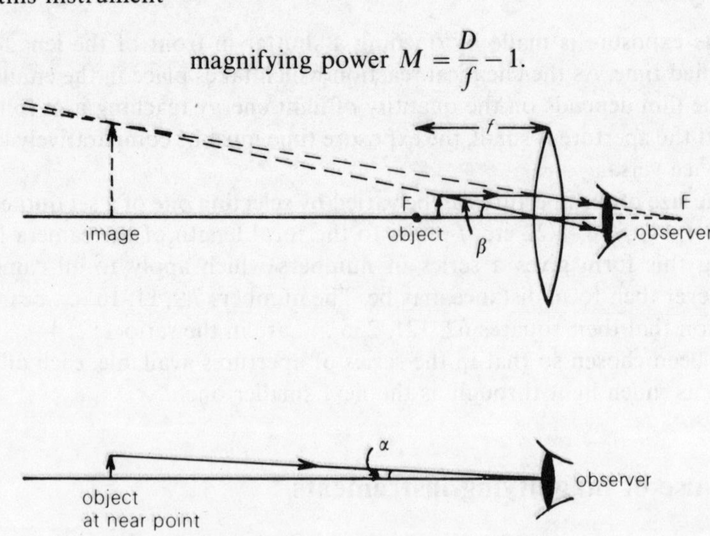

Fig. 172 Use of a magnifying glass

EXERCISES

1 An image of a slide which measures 25 mm × 35 mm is to be projected on to a screen 10 m from the lens of a slide projector. If the distance between the slide and the lens is 30 mm, calculate the size of the image produced.

2 The recommended exposure for a certain photographic film is $f/16$ at 1/50 s. Since the subject of the picture will be moving, it is considered better to shorten the exposure to 1/200 s. What aperture should then be used?

3 A magnifying glass of focal length 80 mm is being used by an observer whose near point is at a distance of 0.24 m. What magnifying power will the lens have?

The simple astronomical telescope

A simple telescope is designed to produce a magnified virtual image of an object a great way off. This is done by using an achromatic lens (the objective) to produce a real image of the object and then looking at this image through a magnifying glass (the eyepiece). Such an optical system gives an *inverted* image, but this is no problem in astronomy.

The two lenses are set up as in Fig. 173 so that their focal planes coincide (for normal adjustment). The real image is formed in this plane. To locate this image using a ray diagram, draw in the central ray of the incident beam from a point P on the object. Since this passes through the optical centre O of the objective lens, the ray is undeviated and can be produced to meet the focal plane of the objective at P′, the corresponding point of the image.

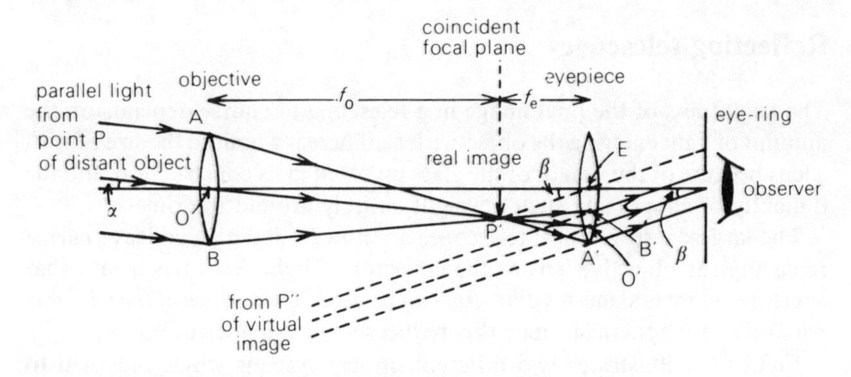

Fig. 173 Optical system of a simple astronomical telescope

The path of this ray can be continued up to point O′ on the eyepiece lens. The emergent beam will be of parallel light, since P′ lies also in the focal plane of the eyepiece. To find its direction, draw in the line from P′ which passes through the optical centre E of the eyepiece lens. A ray following this path would be undeviated, and hence all emergent rays will be parallel to it.

When viewed by the eye, these rays will appear to come from a magnified virtual image.

An *eye-ring* is incorporated at the end of the telescope to indicate where the eye should be placed for best use of the instrument. The image is 'seen' by light which has entered the telescope through the objective lens, a beam of light limited by the two points A and B marked in Fig. 173. The rays AP′ and BP′ reach the eyepiece at A′ and B′, and emerge parallel to P′E. The eye should be positioned to receive the whole of the beam between these limiting rays. OP′O′ is the central ray of the beam, and so the eye-ring is built into the tube of the telescope at the point at which the emergent ray from O′ crosses the central axis.

Magnifying power

For an instrument used to view an object at infinity, this is defined as

$$\frac{\text{angle subtended at eye by image}}{\text{angle subtended at eye by object}}$$

or $$M = \frac{\beta}{\alpha} \text{ in Fig. 173.}$$

Since α and β are small angles in all practical telescopes, their values may be replaced by their tangents, leading to

$$M = \frac{f_o}{f_e}$$

i.e. **the ratio of the focal lengths of the objective and eyepiece lenses.**

Reflecting telescopes

The brightness of the final image in a telescope of course depends on the amount of light entering the objective lens. There is a limit to the size of such a lens because of the weight of the glass involved in its construction, and the difficulty of supporting such a weight entirely around the rim.

The largest astronomical telescopes are now built with a concave *mirror* replacing the objective lens as the collector of light. This has meant that apertures of several metres diameter can be used. *The surface of the mirror is parabolic*, not spherical, since this reduces spherical aberration.

Figure 174 illustrates two different optical systems which are used to deflect the reflected light so that the real image is produced at a convenient place for observation. (In both diagrams, the curvature of the mirror has been greatly exaggerated, and as a result the rays do not always appear to be reflected at quite the right angle.) Figure 174 (a) represents the Newtonian system, which is used in the largest British telescope; (b) shows an alternative design due to Cassegrain.

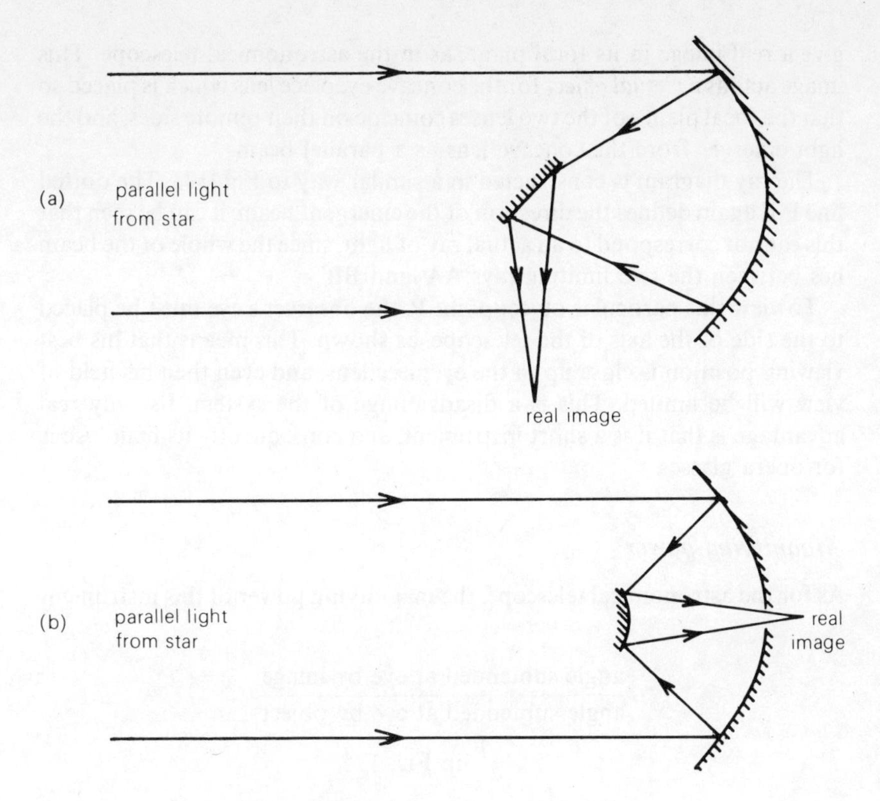

Fig. 174 Use of concave mirrors in astronomical telescopes

The Galilean telescope

This is a simple optical system which produces an *erect* magnified virtual image, and is therefore useful when viewing everyday objects.

Figure 175 shows the arrangement of the two lenses in normal adjustment. The convex achromatic objective lens converges the incident light to

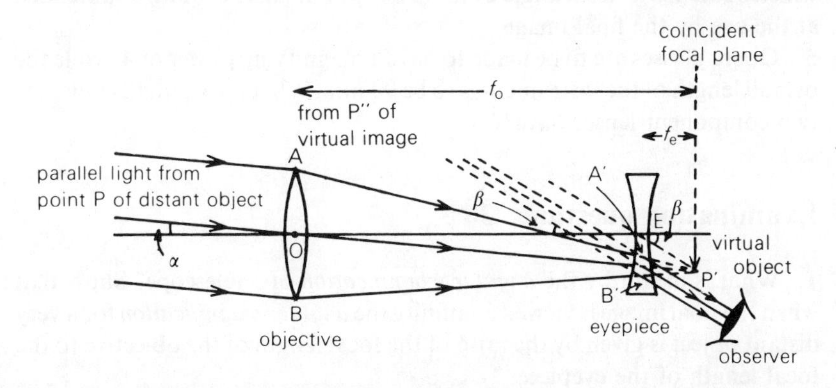

Fig. 175 Optical system of the Galilean telescope

give a real image in its focal plane, as in the astronomical telescope. This image acts as a *virtual object* for the concave eyepiece lens which is placed so that the focal planes of the two lenses coincide on their remote sides, and the light emerges from the concave lens as a parallel beam.

The ray diagram is constructed in a similar way to Fig. 173. The dotted line P′E again defines the direction of the emergent beam; it can be seen that this cannot correspond to an actual ray of light, since the whole of the beam lies between the two limiting rays AA′ and BB′.

To view this particular object point P, the observer's eye must be placed to the side of the axis of the telescope, as shown. This means that his best viewing position is close up to the eyepiece lens, and even then his field of view will be limited. This is a disadvantage of the system. Its only real advantage is that it is a short instrument, and consequently its main use is for opera glasses.

Magnifying power

As for the astronomical telescope, the magnifying power of this instrument is

$$\frac{\text{angle subtended at eye by image}}{\text{angle subtended at eye by object}}$$

$$= \frac{\beta}{\alpha} \text{ in Fig. 175}$$

$$= \frac{f_o}{f_e},$$

the ratio of the focal lengths of the objective and eyepiece lenses, as before.

EXERCISES

4 A telescope is constructed using two converging lenses whose focal lengths are 1.2 m and 0.04 m respectively. It is used to observe a planet which subtends a visual angle of $(2 \times 10^{-5})^c$. Calculate the angle subtended at the eye by the final image.

5 Opera glasses are to be made to have a magnifying power of 4, while the overall length of the instrument is to be 90 mm. What focal lengths must the two component lenses have?

Examination questions 34

1 What is meant by the *angular magnification* of a telescope? Show that when the final image is viewed at infinity the *angular magnification* for a very distant object is given by the ratio of the focal length of the objective to the focal length of the eyepiece.

How is the angular magnification affected if the final image of the very distant object is formed at the near point of an eye positioned very close to the eyepiece?

An astronomical telescope has an objective of focal length 0.25 m and an eyepiece of focal length 0.04 m. Calculate the angular magnification if the final image of a very distant object is formed 0.25 m from the eye, the latter being positioned very close to the eyepiece. [AEB, June 1975]

2 An astronomical telescope in normal adjustment has a converging eyepiece of focal length 0.05 m separated by 0.85 m from the objective of focal length f_0. Which one of the following statements is *correct*?

A The image is formed at the least distance of distinct vision and $f_0 = 0.90$ m.

B The image is formed at infinity and $f_o = 0.80$ m.

C The image is formed 0.90 m from the eye and $f_0 = 0.80$ m.

D The image is formed at infinity and $f_o = 0.90$ m.

E The image is formed at the least distance of distinct vision and $f_o = 0.80$ m. [C].

3 Draw a ray diagram to illustrate the action of a thin lens used as a magnifying glass, the image being at the near point of the eye and the eye being close to the lens.

Distinguish clearly between the *magnification* of the image and the *magnifying power* of the lens. Explain why, in the case shown in your ray diagram, these quantities are numerically equal and derive an expression giving the magnifying power of the lens in terms of its focal length f and the nearest distance of distinct vision D. [JMB]

4 Describe the optical system of a slide projector. Your answer should include ray diagrams showing (a) how the slide is illuminated, and (b) how the image is produced on the screen.

What is the purpose of the condenser lens?

A projector is designed to produce an image of a 35 mm × 24 mm transparency on a screen 2.0 m square placed 15 m from the projection lens. What focal length lens would produce the largest possible image of the entire transparency on the screen? When the projector is used to produce the same size image on a screen 30 m from the projection lens, a different focal length lens is required. How far from this lens must the transparency be placed? [L]

5 Under certain light conditions a suitable setting for a camera is: exposure time 1/125 s, aperture $f/5.6$.

If the aperture is changed to $f/16$ what would the new exposure time be in order to achieve the same film image density? What other effect would this change in f-number produce? [L]

6 (a) In terms of the wave theory of light, explain qualitatively (without using any formulae) how a thin converging lens brings a beam of light from a distant source to a focus, and why the focal length depends on the refractive index of the glass and the radius of curvature of each face.

(b) A thin converging lens (focal length 0.50 m) and a thin diverging lens (focal length 0.10 m) are mounted so that their axes coincide and their separation can be varied, and light from a distant object falls on the converging lens. Explain how the system serves as a Galilean telescope when the separation of the lenses is about 0.40 m. Define *magnifying power* for this telescope and obtain an expression for its value with the telescope in normal adjustment for parallel incident light.

(c) Give a short account of the advantages and disadvantages of the Galilean telescope as a practical instrument. [O]

35 Photoelectricity and the line spectrum

Photoelectricity is the name given to the emission of electrons from metal surfaces under the action of electromagnetic radiation. All metals can be made to show photoelectric effects for suitable frequencies of radiation.

The phenomenon was discovered towards the end of the last century by experimenters working with zinc exposed to the light from a mercury vapour lamp. The photoemission was not obtained when other sources of light were used. Further experimental work led to three conclusions:
(1) **Only radiation above a certain frequency produces photoemission,** the frequency varying for different metals.
(2) When using radiation of suitable frequency, **the emission of electrons is proportional to the intensity of the radiation** falling on the metal surface.
(3) **There is no variation of photoemission with the frequency of the radiation if its intensity remains constant.**

The photocell

The construction of a photoemissive cell is shown in Fig. 176. Caesium, or a combination of caesium and another metal, is generally used for the photoemitter, since this responds to light in the visible and near ultraviolet regions. The electrons are drawn across the cell and received by an anode made in the form of a rod or loop so that it does not block the incoming radiation which should fall normally on the emitting surface.

The current produced by the photoelectrons is very small, only about 1 μA, and must be amplified before it can be measured. In a *vacuum* photocell, the current may be taken as directly proportional to the intensity of the radiation. Other types of cell contain a small amount of an inert gas, which is ionised by the passage of the electrons through it. A larger current is obtained in this way, but it is no longer directly proportional to the intensity of the radiation.

The work function

If a low *negative* voltage is applied to the anode, it is possible for this to repel the electrons so that no current leaves the cell. A very sensitive circuit is needed to find the cut-off point accurately. Figure 177 shows how the

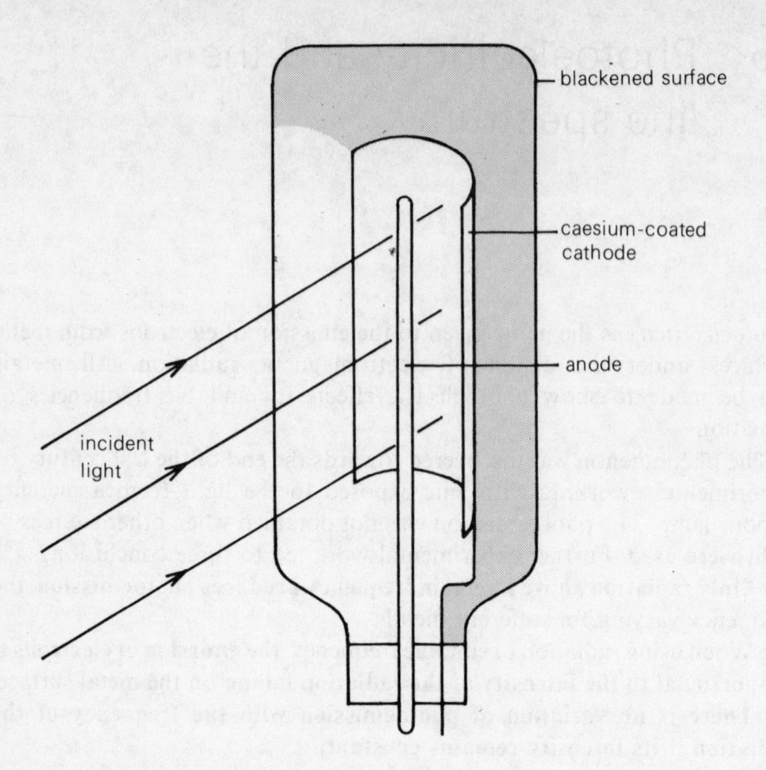

Fig. 176 Construction of a photoemissive cell

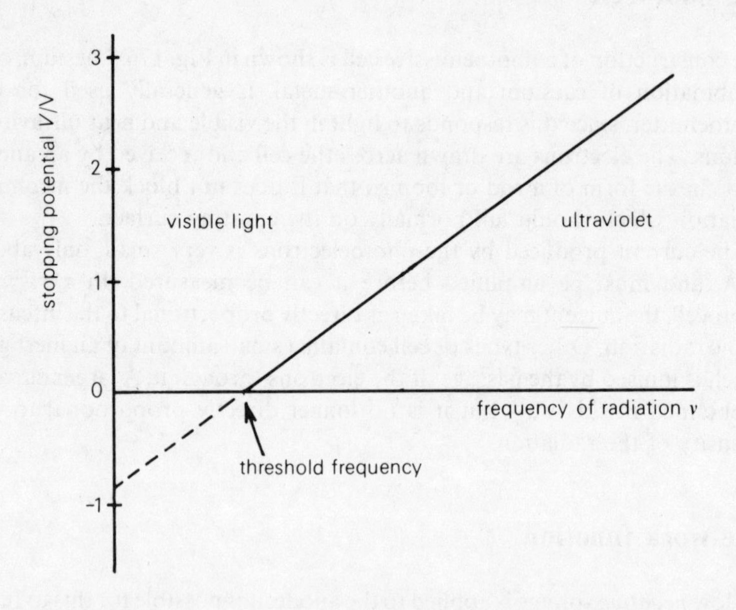

Fig. 177 Graph showing the stopping potential required for different frequencies of radiation falling on a photocell

stopping potential V (anode relative to photocathode) varies with the frequency ν of the radiation used, for a particular photocell.

The graph is a straight line. The point at which it cuts the frequency axis is known as the *threshold frequency*, below which no photoemission occurs. The intercept on the potential axis is interpreted as meaning that a small amount of energy is needed to draw electrons out of the surface, away from the attraction of the ions in the lattice. This energy is called the *work function* Φ of the particular surface. It is calculated by multiplying the value of the intercept by the charge on an electron, and is stated in electron-volts.

The same work function is found in connection with *thermionic emission*, i.e. the emission of electrons from a heated metal. This is interpreted as showing that the necessary energy to withdraw an electron from the metal lattice can alternatively be supplied as thermal energy. The picture is then similar to the kinetic theory of evaporation, in which only some of the most energetic particles are able to escape from a surface. Thermionic emission is used to produce the electron supply in cathode-ray tubes and X-ray tubes.

The quantum theory

The nineteenth-century wave theory of light can offer no explanation for the existence of a threshold frequency. Emission of an electron would be expected as soon as sufficient wave energy had been absorbed by the surface.

Einstein, in 1905, suggested that the explanation could be found if electromagnetic radiation were to be considered as made up of particles (photons) whose energy was related to the frequency of the radiation. This was a development of an earlier proposal by Planck (1902) who was attempting to find a mathematical basis for the curves of the continuous spectrum (p. 87).

Einstein's photoelectric theory gives

$$E = h\nu \tag{1}$$

for the energy of a *quantum* of radiation, i.e. a single photon of frequency ν.

$h = 6.63 \times 10^{-34}$ J s, and is known as *Planck's constant*. This value agrees with the results of experimental work on both photo emission and black body radiation.

If a photon of suitable frequency strikes a metal surface, it can only cause photoemission if its energy is greater than the work function of the surface. This is why low-frequency photons, towards the red end of the spectrum, do not produce photoelectric effects.

Energy of photoelectrons

The difference between the energy of the photon and the work function of the surface appears as kinetic energy of the photoelectron.

$$h\nu - \Phi = \tfrac{1}{2}m_e v^2.$$

The symbol v in this equation gives the maximum velocity possible for an emitted photoelectron (of mass m_e). Slower electrons, from below the surface, may also be observed.

Referring back to the idea of a 'stopping potential' (p. 307) this will be the potential difference across a photocell sufficient to prevent the fastest photoelectrons getting across. The kinetic energy of emission of these electrons must be just equal to the work done by the electrons as they move across the cell against the electric field. Using the previous symbols, this can be written as

$$\tfrac{1}{2}m_e v^2 = eV$$
$$\Rightarrow \quad eV = h\nu - \Phi$$

and this will be the equation of the straight-line graph of Fig. 177.

The gradient of this line is h/e. Since the value of e is known from Millikan's experiment (p. 204), h may be determined.

Momentum of photons: de Broglie's equation

Einstein's quantum theory was extended by Louis de Broglie in 1924, as follows.

A photon of frequency ν travelling at speed c (the velocity of light) is considered to have energy $E = h\nu$.

By Einstein's mass–energy equation (equation (4) of Chap. 28, p. 240), $E = mc^2$, so that the photon may also be considered to have a mass $h\nu/c^2$, and hence it will have momentum $= \dfrac{h\nu}{c} = \dfrac{h}{\lambda}$, where λ is the wavelength of the photon.

This is the momentum concerned in the impact of a photon with an electron, which may lead to photoemission.

De Broglie suggested that the same theory might be extended to other particles which have momentum, and that equation (2) would give a wavelength which could be associated with them. This was the basis of the wave-mechanics mathematics later developed.

Support for de Broglie's ideas was given by experimental work on electron diffraction in gold films. This suggested a wavelength of about 10^{-11} m for an electron beam travelling at about 10^8 m s^{-1}.

EXERCISES

Take $h = 6.6 \times 10^{-34}$ J s
and 1 eV $= 1.6 \times 10^{-19}$ J where required.

1 What is the energy in joules of a single photon of yellow light of frequency 5×10^{14} Hz?

2 The work function of a zinc surface is 4.2 eV. What is the lowest frequency of radiation that can cause photoemission from this surface?

3 Ultraviolet light of frequency 4×10^{16} Hz falls on a metal surface whose threshold frequency is 5×10^{14} Hz. Find the maximum kinetic energy of the photoelectrons emitted.
4 What wavelength could be associated (by de Broglie's theory) with an alpha-particle of mass 6.64×10^{-27} kg travelling at 4×10^{7} m s^{-1}?

Energy states of an atom

The model of atomic structure consisting of a small nucleus surrounded by 'shells' of electrons, was put forward in 1913 by Niels Bohr as a development of Planck's quantum theory of radiation.

Generally the electrons occupy the shells nearest to the nucleus, where they have least potential energy. This is called the *ground state*. However it is possible, if energy is supplied to the atom, for a single electron to move temporarily into an outer shell; the atom is then said to be in an *excited* state. For this to happen, exactly the right amount of energy must be supplied to the electron. After a short time, the atom will de-excite, the electron returning to its original position, and the energy released as it does so will appear as an emitted quantum of radiation.

This means that **the only radiation which can be emitted or absorbed by an atom are those quanta corresponding to the possible energy changes of the electrons in its shells.**

Now, the atoms of a substance can be made to emit radiation in large quantities by heating the substance until it becomes incandescent. The atoms are then colliding with each other with sufficient force to become excited, and the radiation emitted as they de-excite forms a spectrum as described in Chap. 10.

The lines in the line spectrum emitted by a hot gas correspond to separate distinct frequencies of radiation, i.e. to the particular quanta corresponding to the possible energy changes of the atoms of that gas.

Figure 178 shows the customary way of illustrating these changes. The atom as a whole can exist only in one of the states shown by the horizontal lines. Excitation of the atom lifts it to one of the higher energy levels; de-excitation allows it to fall back with the emission of radiation, as indicated by the vertical arrow.

Careful analysis of the line spectra of gases has shown that there is a mathematical relation between the frequencies associated with the various lines. This relation has been used to deduce the *permitted* energy levels of the atom.

The values given in Fig. 178 apply only to the hydrogen atom. The lowest level, called the ground state of the atom, is labelled -13.6 eV. This means that to remove the single electron completely from a hydrogen atom, i.e. to ionise it, requires 13.6 eV of energy from some external source.

To raise the atom into its lowest excited state, labelled -3.4 eV, only 10.2 eV of energy is required. This may come from one of three possible sources:

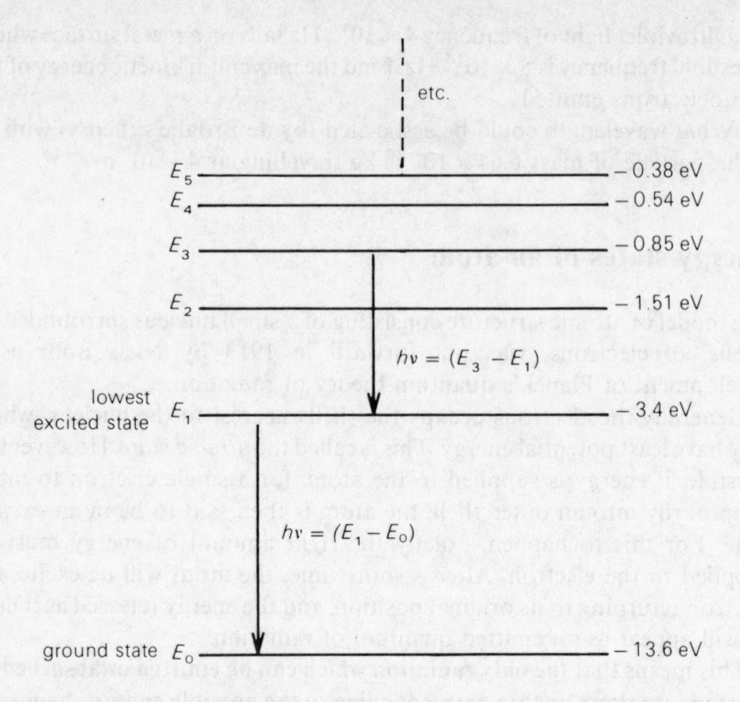

Fig. 178 Illustration of the theoretical permitted energy levels of the hydrogen atom

(a) thermal collisions at high temperature, as mentioned above,
(b) collision with a high-energy particle, either a radioactive emission or a product of some artificial accelerating device, or
(c) absorption of a suitable quantum of radiation. The frequency ν of this radiation may be calculated by applying Einstein's equation (1):

$$\text{(difference between energy levels)} = h\nu$$
$$10.2 \times 1.6 \times 10^{-19} = 6.63 \times 10^{-34}\ \nu$$
$$\Rightarrow \nu = 2.5 \times 10^{15}\ \text{Hz.}$$

This lies just inside the ultraviolet region of the spectrum.

Higher energy states are also possible; only a few of these are shown in Fig. 178.

As the atom de-excites, radiation is emitted at frequencies characteristic of the particular element. These correspond to transitions from one energy level to another lower level. The frequency calculated above is the lowest frequency in the Lyman series of the line spectrum of hydrogen, which contains no visible light.

Similar calculations should apply to the atoms of other elements, but the mathematics required to deal with even two orbital electrons (the case of helium) is very complicated.

EXERCISE

5 The Balmer series of lines in the hydrogen spectrum is formed by electron transitions from higher levels to the energy level −3.4 eV. Show these on a diagram similar to that of Fig. 178, and calculate the lowest two frequencies of the series.

Energy bands

The electrons in the atomic lattice of a solid are considered, as are the electrons in the shells of a single atom, to occupy only certain permitted energy states. However, since in the solid state each electron is affected by many neighbouring atoms, there are a very large number of permitted energy levels which together form what is called an *energy band*. The band normally occupied by the electrons is called the *valency band*, and the next higher group of energy levels form the *conduction band*.

(a) *Metals*. The atoms of metals possess only one or two electrons in their outer shells, which implies that there are many vacant energy states easily available. The small amount of energy supplied by applying an external potential difference is sufficient to cause conduction.

(b) *Insulators*. These are elements and compounds whose outermost shells are completely filled. A comparatively large amount of energy would be needed to raise electrons into a state where conduction would occur.

(c) *Semiconductors*. These are supposed to have only a small gap, less than 1 eV, between the valence band and the conduction band. The thermal energy of the atoms at everyday temperature is enough to allow some electrons into the conduction band.

This theory is supported by the existence of *band spectra*, which are obtained from incandescent diatomic gases (Fig. 179). Whereas a monatomic gas such as hydrogen gives a sharp line spectrum (Fig. 46, p. 83), the greater number of possible transitions between the energy levels of a *molecule* results in the large number of closely spaced lines in a band spectrum.

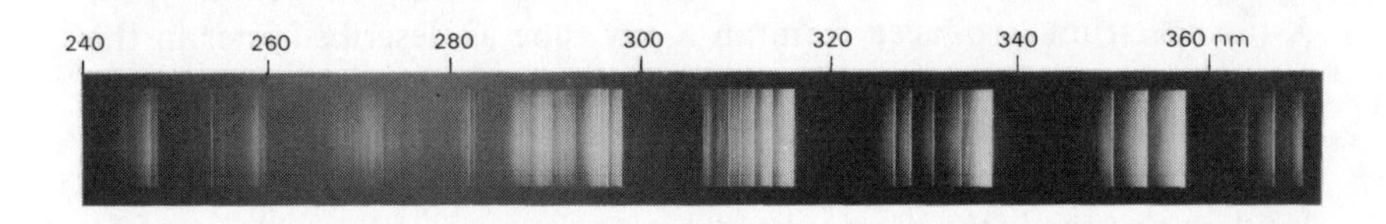

Fig. 179 Band spectrum of nitrogen

In the case of an incandescent solid, the bands broaden out further to form a continuous spectrum.

X-ray spectra

X-rays are a form of electromagnetic radiation having short wavelengths and high frequency, about 10^{18}–10^{19} Hz.

Their production can be explained using the energy levels theory, since in everyday life X-rays are produced by streams of high-energy electrons in collision with atoms of high atomic number, such as tungsten or molybdenum. Such complex atoms will have many possible energy levels. If, by collision, an electron is removed from one of the innermost shells, one of the outer electrons may jump into the vacant place, and a quantum of radiation will be emitted to balance the energy equation. The energy difference may be as much as 125 eV, which corresponds to a photon of frequency $2 \times 10^{-17}/h = 3 \times 10^{18}$ Hz by Einstein's equation, i.e. an X-ray.

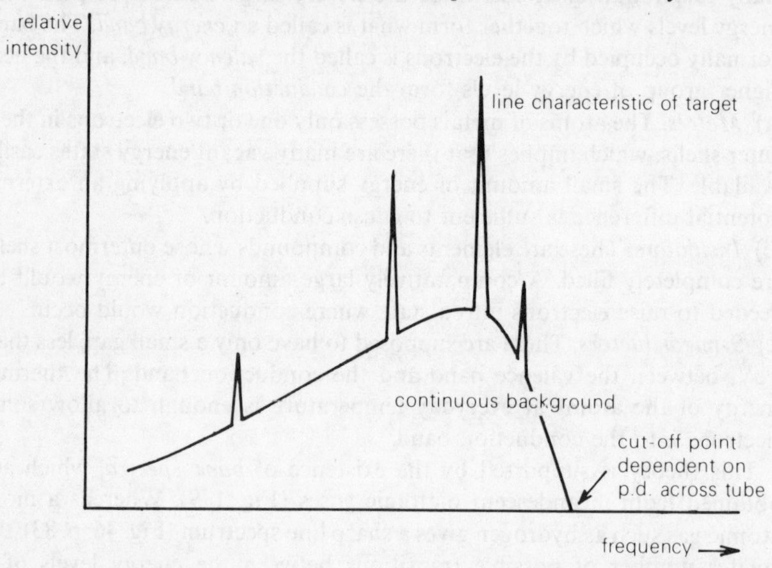

Fig. 180 Graph showing the variation of intensity with frequency in a typical X-ray spectrum

Figure 180 shows the variation of intensity with frequency for a typical X-ray spectrum, produced from an X-ray tube as described later in this chapter. Comparing this graph with Fig. 49 on p. 87, it can be seen that the main curve is the same shape in both cases. This continuous background radiation arises in the same way as the continuous heat/light spectrum. The incident electron in the X-ray tube makes one or more *elastic* collisions with atoms within the target, losing some energy in the form of heat, before causing an inelastic collision which gives rise to an X-ray.

The peaks of the X-ray graph show the line spectrum characteristic of the target metal, superimposed on this continuous background.

The distinct cut-off point marks the highest frequency of the spectrum, corresponding to an inelastic collision by an electron of maximum energy, equal to (electronic charge) × (p.d. across the X-ray tube).

The X-ray tube

Figure 181 shows the essentials of a commercial X-ray tube. An indirectly heated cathode emits electrons by thermionic emission. These are attracted towards an anode which is at a very high voltage, perhaps 500 kV. The anode carries a target area, generally of tungsten, and when the high-speed electrons strike the target, X-rays are emitted.

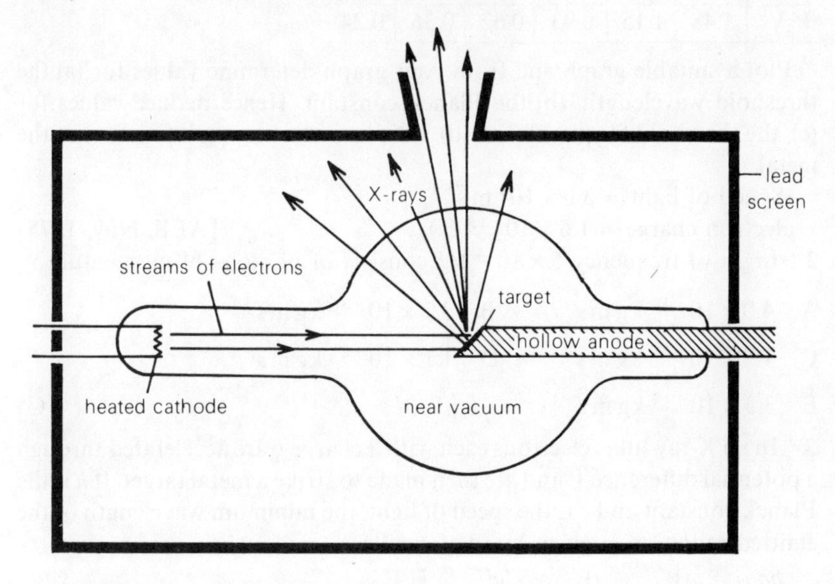

Fig. 181 Construction of an X-ray tube

The tube is made of strong glass and is almost completely evacuated, so that the electrons lose no energy through collisions on their way across the tube. The target and anode are thick pieces of metal so that the heat produced may be conducted away quickly; the anode is hollow, and cooled by oil flowing through it. The target is set at 45° to the electron stream to direct the X-rays out of the tube, though of course many are emitted at unsuitable angles and a screening of lead must be used to absorb them.

The high voltage needed is supplied from a transformer. The anode is earthed, since so many external connections are made here, and the cathode must then be negative. The tube can operate only during alternate half-cycles since it acts as its own rectifier. Tube current is of the order of 200 mA.

Examination questions 35

1 When ultraviolet radiation falls on to an insulated metal plate it is found that the plate becomes positively charged. The potential reached by the plate, measured relative to earth, is found to be independent of the intensity of the incident radiation. When infrared radiation falls on to the same plate, the plate remains uncharged. Explain these observations and give the relevant theory.

When such a metal plate was illuminated with radiation of different wavelengths λ the following values of maximum potential V were observed:

λ/nm	366	405	436	492	546	579
V/V	1.48	1.15	0.93	0.62	0.36	0.24

Plot a suitable graph and from your graph determine values for (a) the threshold wavelength, (b) the Planck constant. Hence deduce values for (c) the threshold frequency and (d) the photoelectric work function of the metal.

(Speed of light = 3.0×10^8 m s^{-1},
electron charge = 1.6×10^{-19} C.) [AEB, Nov. 1975]

2 Light of frequency 5×10^{14} Hz consists of photons of momentum

A 4.0×10^{-40} kg m s^{-1} B 3.7×10^{-36} kg m s^{-1}

C 1.7×10^{-28} kg m s^{-1} D 1.1×10^{-27} kg m s^{-1}

E 3.3×10^{-19} kg m s^{-1} [C]

3 In an X-ray tube, electrons each with a charge q are accelerated through a potential difference V and are then made to strike a metal target. If h is the Planck constant and c is the speed of light, the minimum wavelength of the emitted radiation is given by the formula

A $\dfrac{hq}{cV}$ B $\dfrac{qV}{hc}$ C $\dfrac{qV}{h}$ D $\dfrac{hc}{qV}$ E $\dfrac{hcV}{q}$ [C]

4 (a) Explain what is meant by the duality of electrons, and state the relation between the momentum and the wavelength for a monoenergetic beam of electrons.

(b) Electrons are accelerated from rest through a potential difference V. Derive an expression for the wavelength of the electrons. [JMB]

5 When a metallic surface is exposed to monochromatic electromagnetic radiation, electrons may be emitted. Apparatus is arranged so that (a) the intensity (energy per unit time per unit area) and (b) the frequency of the radiation may be varied. If each of these is varied in turn whilst the other is kept constant, what is the effect on (i) the number of electrons emitted per second, and (ii) their maximum speed? Explain how these results give support to the quantum theory of electromagnetic radiation.

The photoelectric work function of potassium is 2.0 eV. What potential difference would have to be applied between a potassium surface and the collecting electrode in order just to prevent the collection of electrons when the surface is illuminated with radiation of wavelength 350 nm? What would be (iii) the kinetic energy, and (iv) the speed, of the most energetic electrons emitted in this case?

(Speed of electromagnetic radiation in vacuo = 3.0×10^8 m s^{-1},
electronic charge = -1.6×10^{-19} C,

mass of an electron = 9.1×10^{-31} kg,
Planck's constant = 6.6×10^{-34} J s.) [L]

6 Describe one simple experiment in each case to demonstrate (a) thermionic emission of electrons and (b) photoelectric emission of electrons, from a clean metal surface in a good vacuum.

Explain what is meant by the term *work function.* The values of the thermionic work function and the photoelectric work function agree fairly closely for a given metal—each about 1.9 eV for caesium and each about 4.5 eV for tungsten. Why would you expect this agreement?

Describe how you would measure the photoelectric work function of caesium, and explain why a metal with such a low work function is very useful. [O]

7 An element in the vapour or gaseous state may be caused to emit a characteristic line spectrum. It may also yield a line absorption spectrum which is related to, but not identical with, the line emission spectrum.

(a) Explain the terms 'line spectrum', 'characteristic' and 'line absorption spectrum'.

(b) Describe, in outline, the experiments you would make (i) to test whether the source of an unwanted line in the spectrum of a mercury arc lamp is a particular impurity; (ii) to demonstrate a line absorption spectrum.

(c) Explain, qualitatively and in outline only, how line emission and line absorption spectra are interpreted in terms of atomic structure. Formulae may be quoted without proof. [OC]

Wave Theory

36 Interference

The phenomenon called *interference* is caused by the overlapping of two sets of waves from the same or similar coherent sources, which results in the setting up of some form of a stationary state of oscillation in the medium.

This is simply demonstrated by using a ripple tank, and Fig. 182 may be interpreted either as an instantaneous picture of water waves in a ripple tank, or alternatively as a Huygens' construction (p. 73) of the wavelets from two radiating sources.

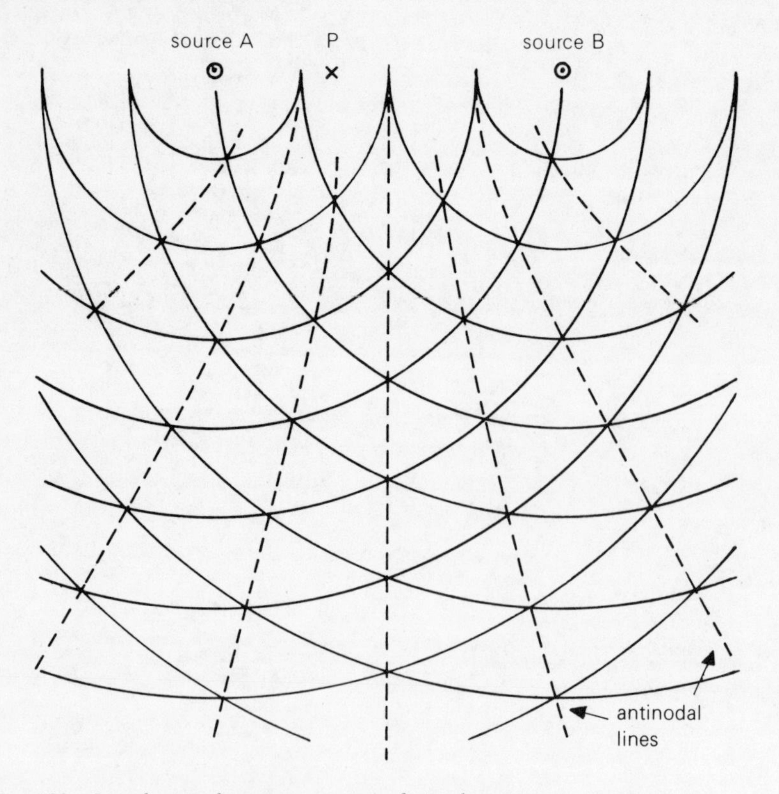

Fig. 182 Interference between two sets of circular waves

The stationary state is seen most clearly along the line AB. If P is a point on this line, then the wave from A has reached it by travelling a distance AP while the wave from B travels a distance equal to BP, i.e. the path difference of the two waves is (BP – AP).

If this path difference is zero, or a whole number $n \times$ the wavelength λ, the two waves will reach P in phase and will reinforce. P will be an antinode of the stationary state.

Similarly **a node will occur at P if the path difference is half a wavelength, or an odd number of half-wavelengths.**

This may be expressed mathematically by writing

$$(BP - AP) = n\lambda \text{ for an antinode,}$$

and

$$(BP - AP) = (n \pm \tfrac{1}{2})\lambda \text{ for a node.}$$

The separation between consecutive nodes (or consecutive antinodes) is $\frac{1}{2}\lambda$.

General calculation of the path differences leading to the formation of the other antinodes and nodes shown in Fig. 182 is not so simple unless the point considered is a long way from the sources, relative to the wavelength of the system, in which case certain approximations can be made. This condition applies to the optical experiments discussed in the remainder of this chapter.

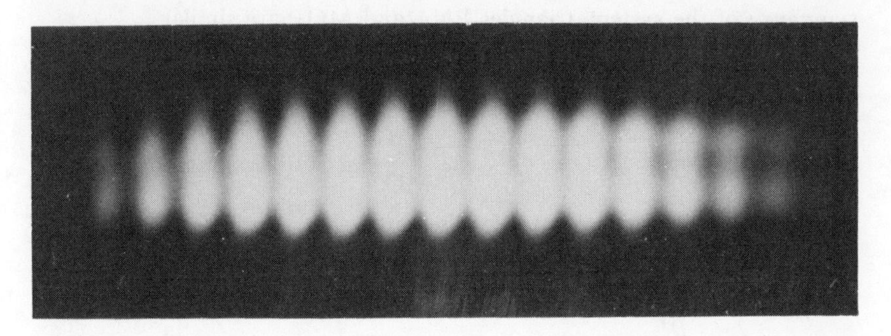

Fig. 183 Interference pattern produced by two coherent sources of light

Young's slits

This experiment uses two slits A and B which must be close together and very narrow, so that they can be considered as line sources of equal intensity.

As the light waves radiate from these slits, interference occurs where they overlap. A screen placed in the overlapping beam will show a pattern of dark and bright bands. The naked eye, looking towards the slits, will also see the interference pattern, since the eye lens will focus the waves on to the retina. If accurate measurements of the bands are to be made, a microscope and scale must be used.

Referring to Fig. 184, the path difference between the two waves reaching a point P can be found by constructing the arc of a circle centre P and radius PB; if this arc cuts AP at N, then AN is the path difference.

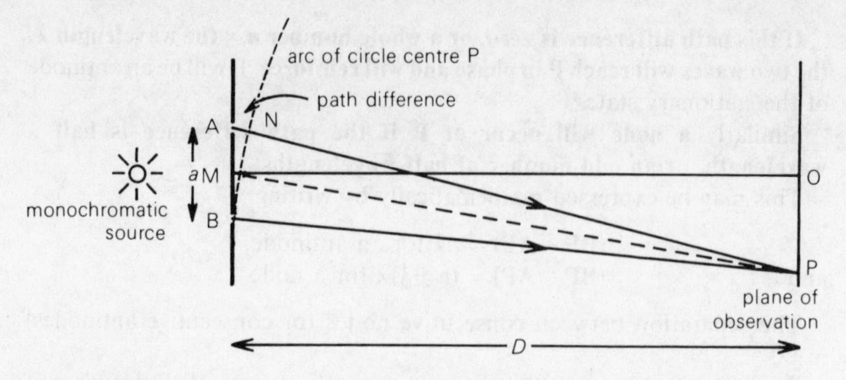

Fig. 184 Calculation of the separation between interference bands produced by Young's slits

If the plane of observation containing P is a long way from the slits, it may be taken as a good approximation that BN is straight and perpendicular to AP. Then if M is the midpoint of AB, and MO is the line of symmetry of the system, triangles BNA and MOP are similar.

Hence
$$\frac{\mathrm{AN}}{\mathrm{AB}} = \frac{\mathrm{OP}}{\mathrm{MP}}$$
$$= \frac{\mathrm{OP}}{\mathrm{MO}} \text{ since OP is a long way from M.}$$

Writing AB $= a$, the separation of the slits,

and MO $= D$, the distance of the plane of observation from the slits,

$$\mathbf{OP} = \frac{D}{a} \times \textbf{(path difference).}$$

The mid-point O of the interference pattern will correspond to zero path difference, so it will be the centre of a bright band.

The centre of the next bright band will be at the point where the path difference is λ, at a distance OP $= (D/a)\lambda$, and successive bright bands will follow at the same separation, with dark bands spaced in between. A travelling microscope focused on the plane OP can measure the distance between bands quite accurately.

EXERCISES

1 Sound waves of wavelength 0.3 m are being reflected normally from a wall. What is the separation of the maxima and minima of intensity observed by a small microphone moved along the normal from the source to the wall?

2 Two narrow slits, 0.2 mm apart, are illuminated by a beam of sodium light of wavelength 6×10^{-7} m. Find the separation of the interference bands that would appear on a screen placed parallel to the slits and 1 m from them.

3 Two radio aerials are situated 2 km apart. If they are transmitting simultaneously, and the signals received along a line parallel to that joining the two aerials, but 10 km away, show a variation of intensity with maxima 1.25 km apart, calculate the wavelength of the transmission.

Phase change on reflection

Reflection of a progressive wave at a solid barrier was considered in Chap. 20 (p. 166). The conclusion then given, that a compression is reflected back as a compression, is equivalent to saying that **a phase change of π occurs in the wave,** which may alternatively be interpreted as **adding half a wavelength to the optical path of the reflected ray**.

This phase change does not occur in partial reflection on entering a less dense medium.

Interference between a wave and its reflection

If the waves from a single source are reflected in a plane mirror, an interference pattern similar to that of Fig. 182 will be produced.

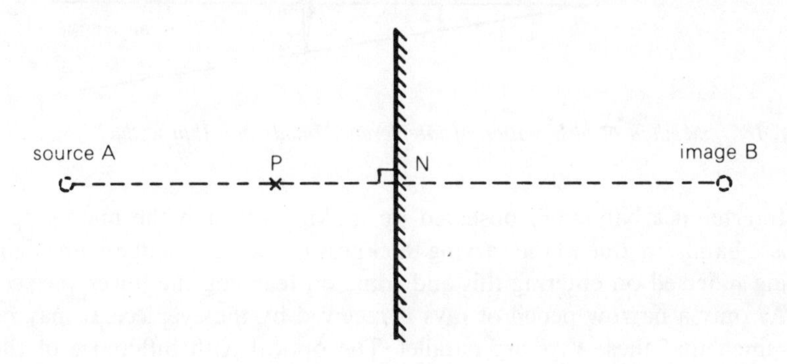

Fig. 185 The image of a wave-source produced by reflection in a plane mirror

However, because of the phase change occurring on reflection, the reflected wave arriving at a point P on the line AB will in this case have travelled an optical path equal to $(\mathrm{AN}+\mathrm{NP}+\frac{1}{2}\lambda)$, where N is the point at which AB intersects the mirror.

The condition for an antinode to be set up at P becomes

$$(\mathrm{AN}+\mathrm{NP}-\mathrm{AP}) = (n\pm\tfrac{1}{2})\lambda$$

or

$$2\mathrm{NP} = (n\pm\tfrac{1}{2})\lambda,$$

and for a node,

$$2\mathrm{NP} = n\lambda.$$

The separation between consecutive nodes or antinodes along AN will still be $\frac{1}{2}\lambda$.

A thin wedge

Figure 186 shows a very narrow wedge of air formed between two glass plates. These are in contact along one edge, and separated at the opposite edge only by a thickness of paper. Light from a monochromatic source is directed by partial reflection at a large glass plate so that it falls normally on the wedge. The reflected rays are collected by a microscope eyepiece focused on the air layer.

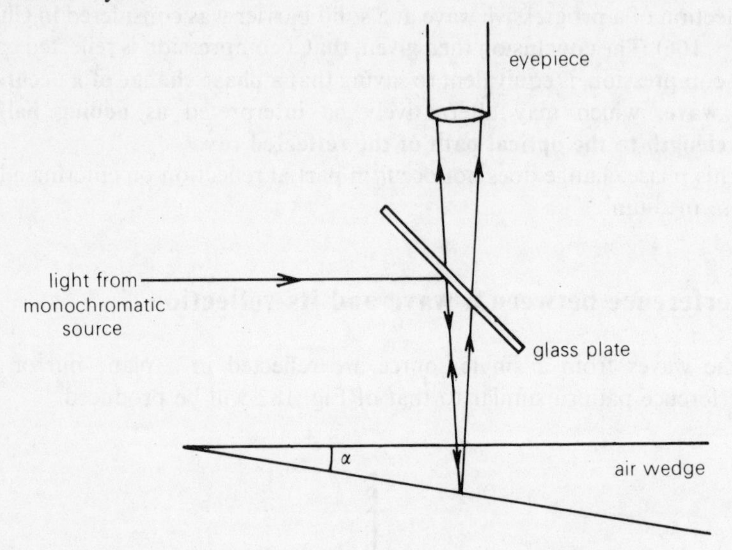

Fig. 186 Method of observation of interference bands in a thin wedge

Interference bands are observed on looking through the microscope. These bands are due to the varying thickness t of the air-wedge, some light being reflected on entering this and some on reaching the lower surface.

As only a narrow pencil of rays is received by the eyepiece, it may be assumed that these rays are parallel. The optical path difference of the interfering rays is therefore $(2t + \frac{1}{2}\lambda)$, a phase shift of π being introduced by reflection at the lower surface of the air layer.

At the narrow end of the wedge, $t = 0$ so that a dark band is seen there. The next dark band occurs when $t = \frac{1}{2}\lambda$. If the angle of the wedge is α^c, this band will be at a distance $t/\alpha = \lambda/2\alpha$ from the end of the wedge, and throughout the interference pattern, this expression will give the separation of consecutive dark or bright bands.

The appearance of the bands gives an indication of the size of the angle of the wedge, bands being closer together for larger angles and separating more widely as the glass plates are more nearly parallel. This result is often used as a test of parallelism when setting up apparatus.

Interference bands of the same kind can often be seen by looking obliquely at the surface of a glass block which is illuminated by monochromatic light. The existence of such bands indicates that the opposite faces of the block are not truly parallel, and the small angle between them can be calculated from the observed separation of the bands, as for an air wedge.

The phase change of π due to reflection at a denser medium would, in this case, occur at the upper surface and not at the lower one.

Newton's rings

Newton's rings are an interference effect seen when a lens of large radius of curvature is placed on a reflecting surface and observed in the same way as the air wedge.

The air between the lens and the plane surface forms a circular small-angle wedge, and the interference bands observed are a set of circles around the point of contact of the lens with the surface. Using a lens of about 1.0 m focal length, the set of rings can be seen by the naked eye as a patch a few millimetres across.

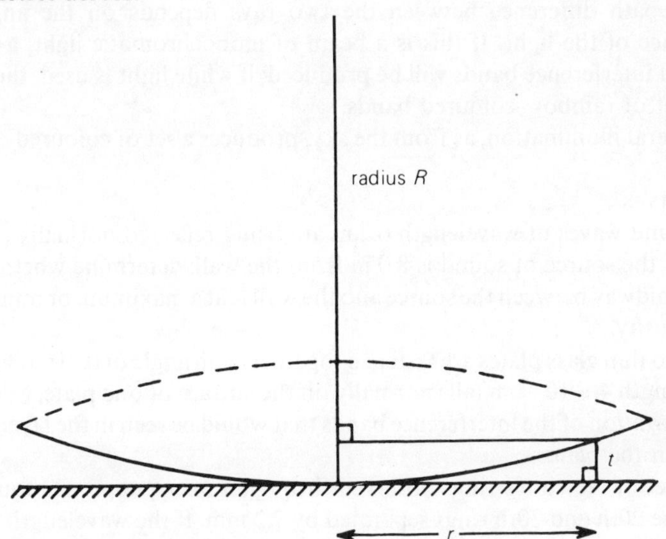

Fig. 187 Calculation of path difference for Newton's rings

The path difference involved in the production of a circular band of radius r can be deduced from the geometry of the circle as in Fig. 187, in which the curvature of the lens is grossly exaggerated. If the radius of the curvature of the lens is R, then

$$r^2 = t(2R - t) \qquad \text{(intersecting chords)}$$

Since R is large and t is small, this can be taken as

$$r^2 = t \,.\, 2R$$

$$\Rightarrow \quad t = \frac{r^2}{2R}.$$

Remembering the phase change introduced on reflection at the plane mirror, the optical path difference between the interfering rays is $(2t + \frac{1}{2}\lambda)$ as before.

A bright ring will occur when $2t + \frac{1}{2}\lambda = n\lambda$,

$$\Rightarrow \quad \frac{r^2}{R} = (n - \tfrac{1}{2})\lambda. \qquad (1)$$

The radii of Newton's rings can be measured accurately using a travelling microscope, and equation (1) then applied to give either the radius of curvature of the lens or the wavelength of the light.

Thin parallel-sided films

Interference can occur when light is obliquely reflected at the two surfaces of a thin transparent film, such as a soap bubble, thin sheet of glass, or film of oil.

The path difference between the two rays depends on the angle of incidence of the light. If this is a beam of monochromatic light, a set of parallel interference bands will be produced; if white light is used, they will be a set of rainbow-coloured bands.

General illumination, as from the sky, produces a set of coloured circles.

EXERCISES

4 Sound waves of wavelength 0.2 m are being reflected normally from a wall. If the source of sound is 8.0 m from the wall, determine whether the point midway between the source and the wall is at a maximum or minimum of intensity.

5 Two thin glass plates are fastened together at an angle of 0.01^c. If light of wavelength 4×10^{-7} m falls normally on the surface of one plate, calculate the separation of the interference bands that would be seen in the layer of air between the plates.

6 A set of Newton's rings observed through a microscope was found to have the 20th and 30th rings separated by 7.2 mm. If the wavelength of the light used in the experiment was 6×10^{-7} m, find the radius of curvature of the lens involved.

Examination questions 36

1 Explain why 30 mm radio waves rather than light waves are often used to demonstrate interference phenomena. Name two other wave phenomena which could be demonstrated using radio waves.

Describe, with the aid of a diagram, how you would measure the wavelength of visible monochromatic radiation using an interference method. Derive an expression to be used in calculating the result from the observations.

A plane soap film is formed in a vertical wire loop and viewed by reflected light incident normally on the film. Describe and explain what would be seen if (a) a sodium lamp and (b) a white light source were used.

[AEB, Nov. 1974]

2 Two radio transmitters emit electromagnetic waves of frequency 9×10^7 Hz. The speed of the waves is 3×10^8 m s^{-1}.

(a) Calculate the internodal distance in the standing wave set up along the line joining the transmitters.

A mobile receiver moves along the straight line joining the transmitters at a speed of 6×10^2 m s^{-1}.

(b) Calculate the rate at which nodes in this standing wave are passed by the moving receiver. [C]

3 The only practical way to produce visible interference patterns with light is, in effect, to derive two sources from a single source. Explain why this is so.

Describe an experimental arrangement to observe interference in a wedge of air. How would you use this to determine a value for the wavelength of the light used?

A piece of wire of diameter 0.050 mm and two thin glass strips are available to produce the air wedge. If a total of 200 fringes are produced, what is the wavelength of the light used? [L]

4 Describe how you would use Newton's rings to determine the wavelength of yellow sodium light.

Monochromatic light is incident normally on the thin air film contained between the lower face of a convex lens (which has a radius of curvature of 1.50 m) and a plane glass surface. Newton's rings are formed in the air film and are observed in reflected light. According to simple theory, if there is perfect contact between the lens and the plate at the centre of the ring system, the radius r_n of the nth bright ring is given by the equation

$$\frac{r_n^{\,2}}{\rho} = (n + \tfrac{1}{2})\lambda$$

where ρ is the radius of curvature of the lower lens surface and λ the wavelength of the light. The following values are observed:

n	5	10	15	20
r_n/mm	2.14	2.98	3.60	4.16

Plot a graph of $r_n^{\,2}$ against n, and estimate the value of λ with the help of your graph. [O]

5 A thin, vertical rod is partially immersed in a large deep pool of water. It moves vertically with simple harmonic motion of small amplitude. Describe the waves produced on the water surrounding the rod. State and explain how (i) the wavelength, and (ii) the amplitude of the waves depend on the distance from the rod of the point at which they are measured.

Describe and briefly explain what happens when a second rod, similar to the first and vibrating with the same frequency and amplitude, and in phase with it, is placed in the water at a point distant d from the first rod.

Discuss the difficulties encountered in attempting to demonstrate similar behaviour for two sources of visible light and describe an experiment you would perform to achieve this. [OC]

37 Diffraction

The spreading out of waves through slits and around obstacles is known as *diffraction*. It is a familiar effect as far as water waves and sound waves are concerned. For electromagnetic waves, which have much shorter wavelength, the phenomenon is harder to observe, but examples do exist in everyday life.

The yellow bar of a sodium street lamp can be seen to be accompanied by parallel bright fringes. One's eyelashes, viewed in sunlight through slitted eyes, are framed in coloured fringes. An ordinary electric street light seen through the condensation on a bathroom window is surrounded by haloes, sometimes white, sometimes in colours. All these are produced by diffraction.

Only a wave theory of light can account for these phenomena.

Diffraction at a single slit

If a beam of parallel light falls on a narrow slit, a pattern of interference fringes as in Fig. 188 can be seen, either with the naked eye or on a suitably placed screen.

Fig. 188 Diffraction pattern produced by a single slit

A simple explanation of how these fringes are formed follows from consideration of Fig. 189, which shows a section of the slit illuminated by a beam of light incident from the left side of the diagram. Sets of Huygens' wavelets are shown spreading out from two points A and B, one at the extreme edge of the slit and the other at its centre. The wavelets reinforce in the straight-through direction, in which most of the light is propagated.

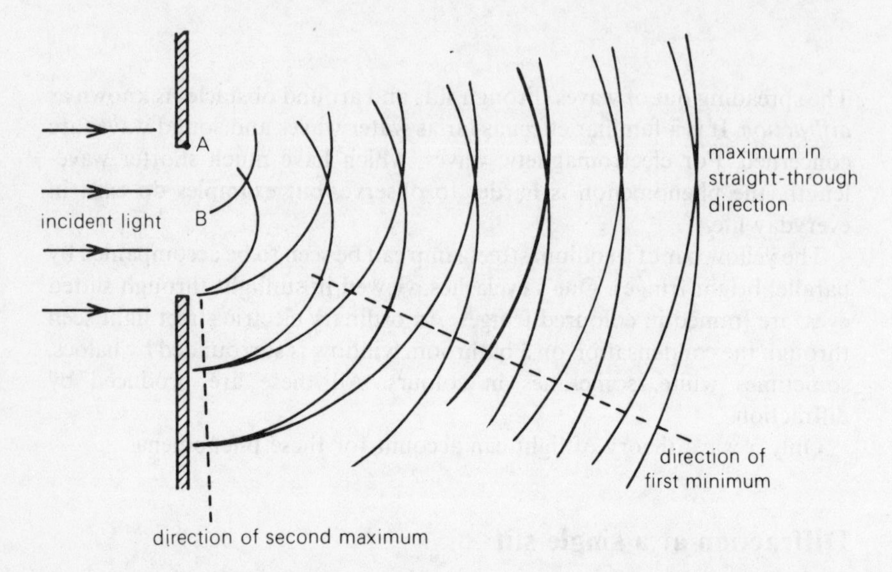

Fig. 189 Interference between waves from the two halves of a slit

However, in the particular direction marked by the dotted line, the two sets of wavelets can be seen to be half a wavelength out of step, so that their effects will cancel out. Now for every point between A and B, a corresponding point exists in the other half of the slit for which the secondary waves will similarly be 180° out of phase in this direction, and in fact *all the waves from the one half of the slit will be cancelled by the waves from the other half.* No light will actually travel in this direction, and a black fringe will be observed.

Between the maximum of the straight-through position and the minimum just considered, the intensity of the beam gradually falls off as shown in Fig. 190. A second maximum, of much lower intensity, occurs at a greater angle of diffraction, and other weaker fringes may also be observed as in Fig. 188.

Angular width of the central band

The first minimum of the diffraction pattern is formed when the path difference of the rays from corresponding points in the two halves of the slit is half a wavelength. From Fig. 191, it can be seen that this will be in the

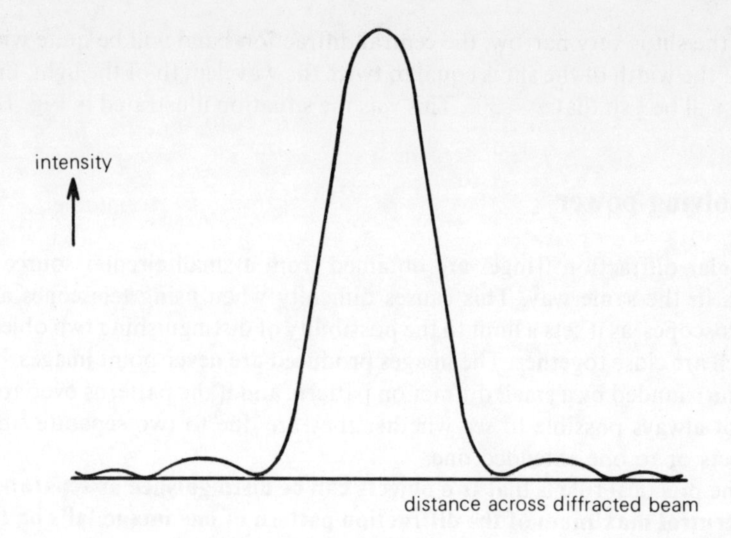

Fig. 190 Graph showing the variation of intensity across the diffracted beam produced by a slit

direction which makes an angle α with the straight-through position, where

$$\sin\alpha = \frac{\frac{1}{2}\lambda}{\frac{1}{2}a} = \frac{\lambda}{a} = \frac{\text{wavelength of light}}{\text{width of slit}}.$$

If the slit is wide compared with the wavelength of the incident light, this angle will be very small. Since the wavelength of visible light is about 5×10^{-7} m, diffraction effects will only be noticeable when using a slit about 10^{-6} m wide. This explains why Newton believed that diffraction of light did not occur.

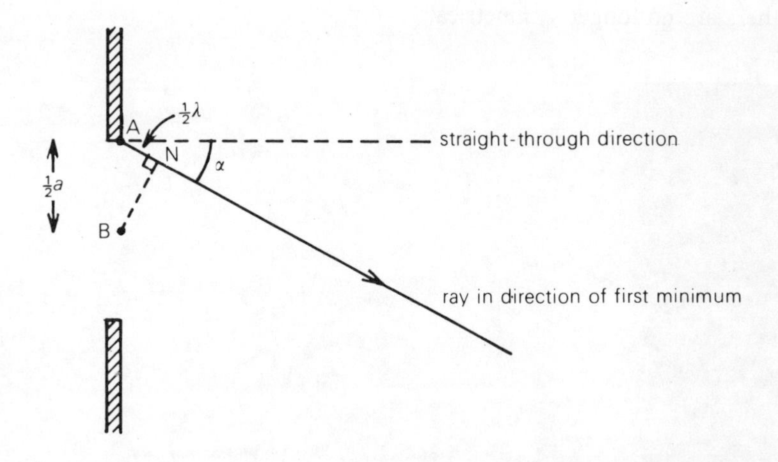

Fig. 191 Calculation of the direction of the first minimum in the diffraction pattern of a slit

If the slit is very narrow, the central diffraction band will be quite wide; e.g. if the width of the slit is equal to twice the wavelength of the light, then $\sin\alpha$ will be $\frac{1}{2}$ so that $\alpha = 30°$. This was the situation illustrated in Fig. 189.

Resolving power

Circular diffraction fringes are obtained from a small circular source of light, in the same way. This causes difficulty when using telescopes and microscopes, as it sets a limit to the possibility of distinguishing two objects which are close together. The images produced are never point images, but are surrounded by a small diffraction pattern, and if the patterns overlap, it is not always possible to see whether they are due to two separate small objects or to one extended one.

The practical rule is that **two objects can be distinguished as separate if the central maximum of the diffraction pattern of one image falls on the first minimum of the pattern of the second.**

For a telescope, it can be shown that the angle between two stars whose images can just be resolved according to this criterion is

$$\frac{1.22 \times (\text{mean wavelength of light})}{\text{diameter of objective lens}}.$$

This expression gives the *resolving power* of the telescope.

Diffraction by an obstacle

Diffraction fringes occur close to the edge of an obstacle, as shown in the photograph of Fig. 192. This is because many of the secondary wavelets are cut off by the obstacle and the interference effects produced among the others are no longer symmetrical.

Fig. 192 Diffraction fringes produced by an obstacle (a razor blade)

There is also a slight spreading out of the waves into the geometrical shadow so that the intensity of the light does not fall to zero for a short distance.

The diffraction grating

A diffraction grating consists of a set of accurately ruled parallel opaque lines, on a sheet of glass or plastic. The space between the rulings which forms a slit is about 10^{-6} m wide, and so for visible light a considerable amount of energy will be diffracted out of the main beam. Interference then occurs between the diffracted waves from the various slits in the same way that it does for Young's slits (p. 319).

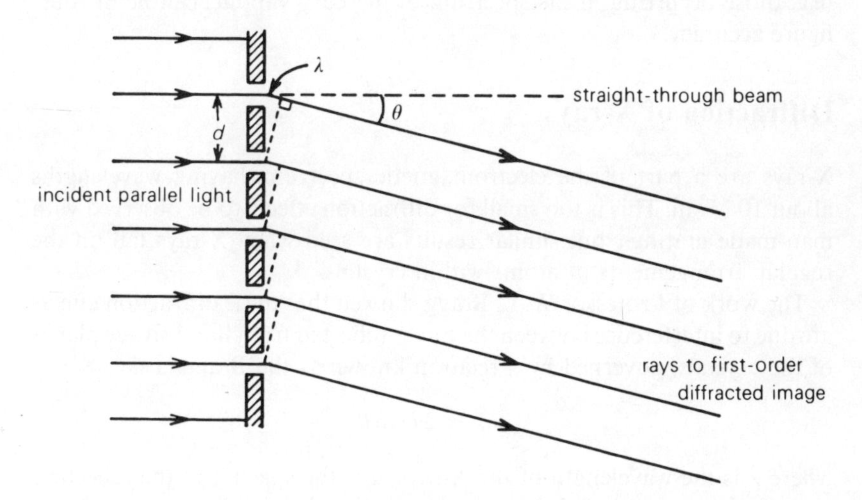

Fig. 193 Calculation of the direction of the first-order diffracted image produced by a grating

Figure 193 shows parallel monochromatic light falling on a diffraction grating. In the direction shown at an angle θ to the main beam, the diffracted waves are one complete wavelength λ out of step across the whole grating so that they reinforce each other. If this beam of light is brought to a focus, a bright image of the source of light will be obtained. The required angle θ is given by

$$\sin\theta = \frac{\lambda}{d} = \frac{\text{wavelength of light}}{\text{distance between centres of adjacent slits}}$$

A second-order image would be formed in a direction θ_2 given by $\sin\theta_2 = 2\lambda/d$, and in general **the *n*th-order image occurs where $\sin\theta_n = n\lambda/d$.** Several images can be seen if d is sufficiently large, but they will be progressively less bright as the angle of diffraction increases.

Measurement of wavelength using the diffraction grating

The optical system of the spectrometer (Chap. 10, p. 80) is used to provide a beam of parallel light which illuminates the grating. This is secured to the spectrometer turntable using a special attachment, and the diffracted light is received in the telescope. Several images of the slit of the collimator are to be seen.

The positions of the various orders of diffracted images are noted on both sides of the straight-through position, and the difference between two corresponding readings is then halved to give the most accurate value of the angle of diffraction θ.

The value of the grating constant d is supplied by the manufacturers, and the results obtainable for the wavelengths of various colours of visible light (e.g. those occurring in the spectrum of mercury vapour) can be of four-figure accuracy.

Diffraction of X-rays

X-rays are a part of the electromagnetic spectrum, having wavelengths about 10^{-10} m. This is too small for diffraction effects to be observed with man-made gratings, but similar results are seen when X-rays fall on the regular arrangements of atoms within crystals.

The work of Professor W. L. Bragg showed that these diffraction effects are due to interference between the waves reflected from the cleavage planes of the crystals, governed by a relation known as the Bragg law:

$$n\lambda = 2d \sin \theta$$

where λ is the wavelength of the X-rays, d is the spacing of the reflecting planes in the crystal, and θ is the observed angle of diffraction.

The science of X-ray crystallography is based on this relation which has made possible detailed analysis of the geometry of many different types of crystal lattice.

EXERCISES

1 A diffraction grating is ruled with 400 lines/mm. If it is used with light of wavelength 6.0×10^{-7} m, at what angle would the first-order image be seen?

2 A diffraction grating whose constant is 2×10^{-6} m is used with light of wavelength 5.4×10^{-7} m. How many orders of image would be produced?

3 X-rays of wavelengths 10^{-10} m are used to investigate the structure of a crystal, and an intense diffracted beam is observed at an angle of $12° 12'$. What value does this suggest for the distance between two adjacent layers of atoms in the crystal?

Examination questions 37

1 (a) Draw a labelled diagram showing the optical components of a spectrometer.
(b) Describe the adjustments that you would make when using the spectrometer with a diffraction grating to ensure that
(i) a parallel beam of light is incident on the grating,
(ii) the grating is normal to the beam of light.
(c) Prove that, for a diffraction grating of which the line spacing is d, and which is placed normally to light of wavelength λ, the angle θ, between the normal and the nth order image is given by the formula $n\lambda = d \sin\theta$.
(d) Sodium light of wavelength 5.98×10^{-7} m falls normally on a diffraction grating and the angle between the normal and the first order image is $17° 24'$. Calculate (i) the number of lines per mm ruled on the grating, (ii) the angle which the largest order image makes with the normal.
[AEB, Nov. 1969]

2 The spacing between a certain set of planes of atoms in a crystal is 0.300 nm. The crystal is used in an X-ray spectrometer. How many orders of diffraction could be observed from these planes if the incident X-rays were of wavelength 0.154 nm? [C]

3 A sodium flame is viewed through a grating held close to the eye. Two images of the flame are seen, one on either side and each about 0.8 m from the flame. If the flame is 3.0 m from the grating, which has 450 lines/mm, what is the approximate wavelength of the sodium light? [L]

4 A parallel beam of white light (range of wavelengths 4.5×10^{-7} m to 7.5×10^{-7} m) is incident normally on a diffraction grating which is set up on a spectrometer table. When the telescope is moved to one side away from the normal through an angle of 60° it passes through two complete spectra. How many lines per metre are there on the grating?
A 5.8×10^5 B 9.6×10^5 C 11.6×10^5 D 19.2×10^5 E 23.0×10^5
[O]

5 Explain the term *resolving power* as applied to a telescope objective, and show how it is related to the diameter of the objective and the wavelength of the light used.

The distance between the components of a double star system subtends an angle of $1''$ of arc ($= 5 \times 10^{-6}$ rad) at the earth. Calculate the minimum diameter of an objective which will separate the images of the two stars at a wavelength of 550 mm.

Under ideal conditions the eye can just resolve the images of two objects whose distance apart subtends an angle of $1'$ of arc ($= 3 \times 10^{-4}$ rad) at the eye. If the focal length of the above objective is 1 m what is the maximum focal length of the eyepiece which can be used if the eye is to resolve the double star system?

Outline the laboratory experiment you would perform to test the formula for resolving power using a small telescope such as the telescope from a spectrometer. [OC]

38 The mathematical treatment of waves

The simplest waveform is a sine curve. Further (Fourier) analysis shows that it is possible to break down more complex waveforms into a summation of sine curves, so that it is sufficient to consider only this simple form as a first approach.

Figure 194 can be used to represent such a wave, either (a) as an instantaneous picture showing the displacement of the particles along the direction in which the wave is travelling, or alternatively (b) as the variation of displacement with time for a particular particle.

(a) The axis across the page represents x, the distance of any particular particle from an origin O at which the displacement y is zero. The distance x at which the curve begins to repeat itself is defined as its wavelength λ.

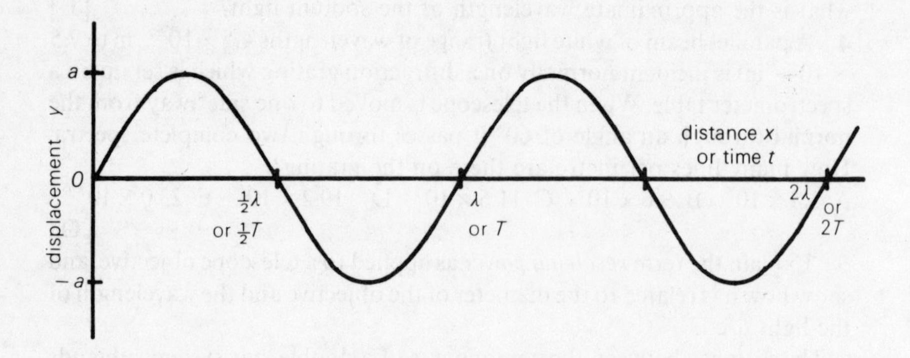

Fig. 194 Graph showing the displacement of a particle produced by a simple wave

The equation of the curve is then

$$y = a \sin 2\pi \left(\frac{x}{\lambda}\right) \qquad (1)$$

where a is the maximum value of the displacement, called the *amplitude* of the wave.

(b) Alternatively, the axis across the page can be taken to represent t, the time counting from a moment at which the displacement y of the particular particle considered is zero. Then the time t at which the curve begins to repeat itself is defined as the period T.

The equation of this curve is

$$y = a \sin 2\pi \left(\frac{t}{T}\right) \tag{2}$$

where a has the same meaning as before.

These two equations may be combined as follows. Every particle affected by the wave will be displaced according to equation (2), but the displacements will vary with distance according to equation (1). The term $2\pi(x/\lambda)$ is called the *phase difference*, and shows how far out of step a particular particle is in relation to the origin. Since the wave is supposed to be travelling in the direction of x, the motion of a particle at distance x will lag behind that of the origin by this term $2\pi(x/\lambda)$.

The combined equation is therefore

$$y = a \sin 2\pi \left(\frac{t}{T} - \frac{x}{\lambda}\right). \tag{3}$$

Formation of stationary waves

A stationary wave can be set up by the superposition of two waves of equal amplitude and frequency, travelling in opposite directions.

If the equation of the first wave is equation (3) above

$$y = a \sin 2\pi \left(\frac{t}{T} - \frac{x}{\lambda}\right)$$

then the equation of a similar wave travelling in the opposite direction will be

$$y = a \sin 2\pi \left(\frac{t}{T} + \frac{x}{\lambda}\right).$$

Adding the two equations, the displacement of a particle resulting from the combination of the two waves will be

$$y = a \sin 2\pi \left(\frac{t}{T} - \frac{x}{\lambda}\right) + a \sin 2\pi \left(\frac{t}{T} + \frac{x}{\lambda}\right).$$

This is of the form

$$\begin{aligned} & y = a \sin(A - B) + a \sin(A + B) \\ \Rightarrow\ & y = a[\sin A \cos B - \cos A \sin B] + a[\sin A \cos B + \cos A \sin B] \\ & \quad = 2a \sin A \cos B. \end{aligned}$$

Hence the combined wave equation simplifies to

$$y = 2a \sin 2\pi \frac{t}{T} \cos 2\pi \frac{x}{\lambda}.$$

For a particular particle, x is constant, so that the motion of the particle will be simple harmonic (referring back to Chap. 12 p. 97). The period of the motion is T and its amplitude is $2a \cos 2\pi(x/\lambda)$.

All particles move with the same period, but the amplitude of the motion varies with x.

The maximum amplitude occurs when $2\pi(x/\lambda) = 0, \pi, 2\pi \ldots$, i.e. when $x = 0, \frac{1}{2}\lambda, \lambda \ldots$, and is equal to $2a$. Such points are antinodes of the stationary wave.

Zero amplitude occurs when $2\pi(x/\lambda) = \pi/2, 3\pi/2 \ldots$, i.e. when $x = \frac{1}{4}\lambda$, $\frac{3}{4}\lambda \ldots$. Such points are nodes, and they are formed at a separation of $\frac{1}{2}\lambda$ (Chap. 20 p. 166).

Measurement of the speed of sound in air

Since the speed of sound is relatively low (about 330 m s^{-1}), it is quite simple to obtain an experimental value for it directly. Sound and light signals can be sent out simultaneously, perhaps by the firing of a cannon, and the time interval between their arrivals at some distant point is then measured. Such experiments are not able to allow for any wind effects or temperature variation over the distance travelled by the sound waves. All accurate experimental work must be carried out under controlled laboratory conditions.

The resonance-tube method of determining v, the speed of sound in air, was described in Chap. 20. A certain degree of inaccuracy is difficult to avoid in this experiment because of the end-correction associated with the tube.

The method described below, first carried out by Hebb, gives accurate results for the speed of sound in free air.

A small source of sound, of known frequency f, is placed at the principal focus A of a parabolic reflector. A similar reflector is placed at the other end of a long laboratory; this collects the waves from the first mirror and directs them to its own principal focus B.

Small microphones are placed at A and B. The currents from these are

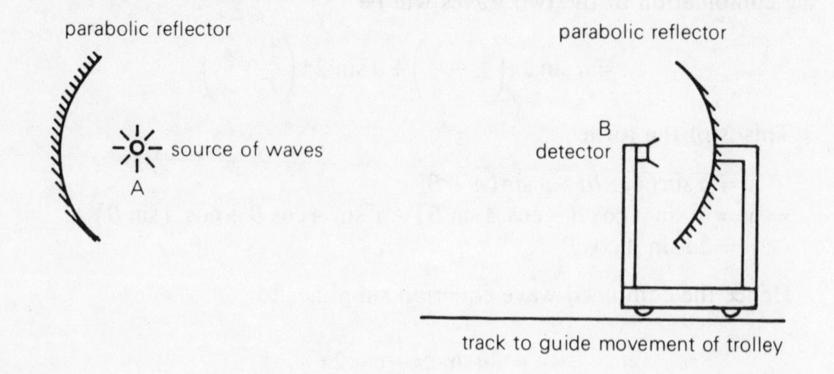

Fig. 195 Hebb's method for measurement of the speed of waves in free air

superimposed, and the combined current is studied using an oscilloscope.

The second reflector, together with its microphone, is mounted on a trolley so that it can be moved along a track towards the first reflector.

When the two signals arrive in phase, the path followed by the reflected signal must be a whole number of wavelengths. If the trolley is moved through a certain distance between successive positions of maximum (or minimum) intensity of the combined signal, this distance must be equal to one complete wavelength.

The speed of the wave can then be calculated using the relation $v = f\lambda$.

By controlling the temperature within the laboratory, it is possible to see how the speed of sound varies with temperature. The theoretical relation is discussed later in this chapter.

The experiment may be adapted to measure the speed of microwaves, using suitable reflectors and detectors. With the aid of modern electronic methods, this has given satisfactory results; since microwaves are a form of electromagnetic radiation, the value obtained in this way applies to all similar radiation, including visible light.

Theoretical expressions for the speed of pressure waves

Pressure waves are transmitted by the movement of particles, vibrating under the influence of interatomic restraining forces. It is therefore reasonable to suppose that the speed of such waves will be dependent on (a) the elastic modulus of the medium, and (b) the density of the medium. Although actual proof of the relation involves advanced mathematics, it may be accepted that

$$\text{velocity of wave} = \sqrt{\frac{\text{modulus of elasticity}}{\text{density of medium}}}. \tag{4}$$

For *longitudinal* waves travelling along a rod or stretched wire of density ρ the modulus of elasticity is Young's modulus E, and equation (4) may be written as

$$V = \sqrt{\frac{E}{\rho}}. \tag{5}$$

For *tranverse* waves travelling along a wire, the situation is more complicated. The relation

$$V = \sqrt{\frac{T}{m}}$$

can be deduced by the method of dimensions (Chap. 40), where T is the tension within the wire and m is its mass per metre length. This expression was used in Chap. 19 (p. 159), when considering a wave frequency f_0 and wavelength $2l$.

For waves travelling through a gas of density ρ, the appropriate modulus of elasticity is the adiabatic bulk modulus. Applying calculus methods to the equation of an adiabatic change (p. 263) shows that

$$\text{adiabatic bulk modulus} = \gamma p$$

where p is the pressure of the gas and γ is the ratio of the principal specific heats of the gas. Then equation (4) becomes

$$V = \sqrt{\frac{\gamma p}{\rho}}. \qquad (6)$$

Effects of pressure and temperature changes

Since sound is transmitted through a gas as a pressure wave, equation (6) may be used to consider how the speed of sound will vary under different atmospheric conditions.

Temperature constant

If the pressure of the gas changes, its density will change in the same way, and the two effects will cancel out.

Hence **the velocity of sound in a gas is independent of pressure.**

Pressure constant

The density of a gas is inversely proportional to its temperature, therefore **the velocity of sound through the gas will be proportional to the square root of its absolute temperature,**

$$V \propto \sqrt{T}.$$

EXERCISES

1 Sound waves of frequency 680 Hz are used in an experiment following Hebb's method. Taking the speed of sound in air as 340 m s^{-1}, find the distance through which one of the reflectors must be moved in order to change the strength of the combined signal from one minimum to the next.

2 What is the predicted speed of sound through a steel rail whose density is 8000 kg m^{-3} and Young's modulus is 2×10^{11} N m^{-2}?

3 Calculate the tension that must exist in a violin string of mass 10^{-3} kg m^{-1} if a transverse wave travels along it at 220 m s^{-1}.

4 Using the following data:

density of air at s.t.p. = 1.30 kg m^{-3}

γ for air = 1.40

standard atmospheric pressure = 1.0×10^5 Pa,

calculate the speed of sound in air at (i) 0°C (ii) 20°C.

The velocity of light

The first serious determination of the speed of light was made by Rømer in 1676, working from astronomical observations. Before this it was not thought that light took any time at all to travel, no matter how great a distance was involved.

As mentioned in Chap. 9, the conflict between the particle theory and the wave theory of the nature of light turned on the question of whether it travelled faster or slower in a more dense medium. Accordingly many different experimenters during the nineteenth and early twentieth centuries attempted to devise accurate methods of measuring the velocity of light under controllable conditions.

All these methods involved correlating the length of time taken by the light on its journey with some mechanical change in the apparatus, i.e. one tooth of a rotating wheel exactly replacing the next (Fizeau), or the movement of a mirror through a measurable angle (Foucault, also Michelson). The accuracy achieved was quite high, about 0.01 %, and the results not only established the wave theory of light but also provided a basis for Einstein's theory of relativity.

However, the whole problem has now been simplified by the invention of the 'atomic clock'. This measures time by the natural periodic vibration of a crystal incorporated into an electronic circuit, providing a digital display which can measure down to a few nanoseconds (10^{-9} s). Light moving at 3×10^8 m s^{-1} will travel less than one metre in a nanosecond, so that the distance over which it must travel during an experiment can now be greatly reduced and consequently can be measured more accurately.

The value at present accepted for the velocity of light in vacuo is 2.9979250×10^8 m s^{-1}, to about 0.0001 % accuracy.

Mathematical theory concludes that the velocity of an electromagnetic wave should be given by

$$\frac{1}{\sqrt{(\text{permeability} \times \text{permittivity})}} = \frac{1}{\sqrt{(\mu\varepsilon)}}.$$

The values of these two constants for air were given as

$$\mu_0 = 4\pi \times 10^{-7} \text{ henry m}^{-1} \text{ (p. 222)}$$
$$\text{and } \varepsilon_0 = 8.854 \times 10^{-12} \text{ farad m}^{-1} \text{ (p. 200),}$$

so that
$$\frac{1}{\sqrt{(\mu_0\varepsilon_0)}} = \frac{1}{\sqrt{(4\pi \times 8.854 \times 10^{-19})}}$$
$$= 2.99795 \times 10^8 \text{ m s}^{-1}$$

which agrees very well with the experimental result.

EXERCISES

5 In Fizeau's experiment, a narrow beam of light was sent out through the gap between two teeth of a rotating cog wheel having 720 teeth, and on its return after reflection at a mirror 8.6 km away, the passage of the light was blocked by the next tooth which had just moved into the gap. This occurred when the wheel was rotating at 12.6 rev/s. Calculate a value for the velocity of light.

6 In an experiment similar to that of Foucault, a ray of light was reflected from a rotating plane mirror, travelled a distance of 20 m to a fixed mirror, and returned to be reflected again from the rotating mirror. The ray was found to be deflected through an angle of 1.65×10^{-4} rad from its original path. If the mirror was rotating at 200 rev/s, calculate a value for the velocity of light.

7 In one of Michelson's experiments, light was sent out after reflection at one face of a rotating octagonal prism, and returned at the moment when the next face had moved exactly into the position of the first. The distance travelled by the light was 31 km in each direction. If the prism was rotating at 600 rev/s, calculate a value for the velocity of light.

8 At the McDonald observatory at the University of Texas, light traverses a path of 134 m (including a reflection) in a time of 447 nanosecond. Calculate a value for the velocity of light.

Examination questions 38

1 (a) Describe a laboratory experiment to determine the speed of sound in air. Show how you would calculate the speed from the readings which you would take.

(b) State the equation relating the speed of sound v to the pressure p and density ρ of the gas through which it is passing. Use this equation to show how the speed of sound depends upon

(i) the pressure of the air at a given temperature,

(ii) the presence of water vapour in the air. [AEB, Nov. 1979]

2 The displacement y at time t produced by a plane wave travelling in the negative x-direction may be represented by

$$y = a \sin \frac{2\pi}{\lambda}(ct + x),$$

where λ is the wavelength and c is the velocity of the wave. What does a represent?

Write the equation for the displacement y' produced by a similar wave travelling in the positive x-direction.

If $a = 20$ mm, $c = 300\ \text{m s}^{-1}$ and the wave has a frequency of 2500 Hz, *draw to scale* for the first equation

(i) a graph of displacement y against distance from origin x at time $t = 0$, and

(ii) a graph of displacement y against time t for the point $x = 30$ mm.

[L]

3 Show that the equation $\zeta = a \sin 2\pi f\left(t - \frac{x}{v}\right)$ represents a simple harmonic wave of amplitude a and frequency f travelling with velocity v in the positive direction of the x-axis.

When the displacement ζ is parallel to the x-axis, the longitudinal wave represented by the equation travels as a succession of compressions and rarefactions of the medium. Explain as fully as you can the connection between the variation with x of the particle displacement ζ and the travelling pattern of compressions and rarefactions in the wave. [O]

39 Polarisation

This usage of the word 'polarisation' refers to a phenomenon observed only in connection with electromagnetic waves, although it is often illustrated by mechanical demonstrations.

A *plane-polarised* wave is one in which the energy vibrations are restricted to a single plane.

Light emitted from an incandescent solid or gas is made up of many separate photons, each produced by the de-excitation of an individual atom as described in Chap. 35. The whole wave therefore consists of energy vibrations in all possible directions and it is said to be *unpolarised.*

If this unpolarised light falls on a natural transparent crystal, in many cases it is found that the transmitted light becomes split into two rays, each of which is plane-polarised, the planes of vibration of the two rays being mutually perpendicular.

Double refraction

Some crystals such as Iceland spar (a form of calcite) transmit the two refracted rays in slightly different directions. One ray obeys Snell's law of refraction and is called the *ordinary* ray; in this, the vibrations of the light are perpendicular to the natural axis of the crystal. The path of the other ray, called the *extraordinary* ray, depends on its direction relative to this axis; it is polarised in a plane parallel to the axis.

The two rays give rise to a double image of any object viewed through the crystal (Fig. 196).

Selective absorption

Other crystals such as tourmaline absorb the ordinary ray and transmit only the extraordinary ray.

If two such crystals are set with their axes at right angles, the second will absorb all the light transmitted by the first, and none will get through (Fig. 197). This fact shows that light must be a *transverse* wave, having its energy vibrations in a plane perpendicular to the direction in which the wave is travelling.

A *longitudinal* wave, whose vibrations are in its direction of propagation, could not be affected by displacing a crystal in a plane perpendicular to this

Fig. 196 Double refraction (a crystal of calcite on top of a single washer)

direction. This is why sound waves, which are carried by the longitudinal movements of particles, cannot be polarised by any method.

Polaroid material is made from small crystals which are selective absorbers, all set with their axes aligned in the same direction, so that the light transmitted by a sheet of Polaroid is plane-polarised. Such a sheet may be used to produce polarised light, or alternatively to investigate the state of polarisation of an incident beam.

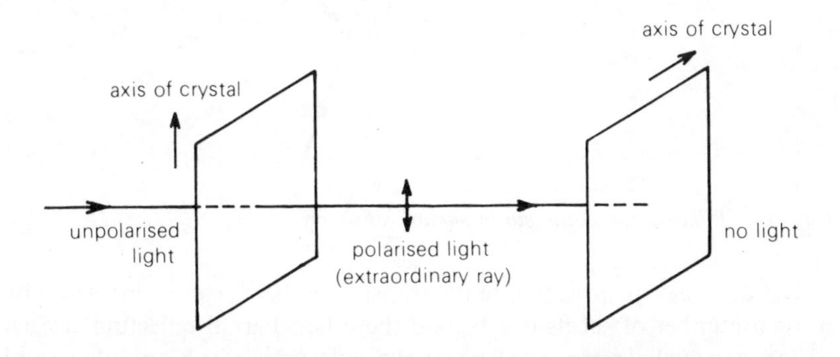

Fig. 197 Selective absorption of light by two crystals with crossed axes

Partial reflection

The light reflected from the surface of a plane sheet of glass or crystal is found to be partly polarised. The degree of polarisation may be studied by viewing the reflected rays through a Polaroid sheet set at the suitable angle.

Almost complete polarisation occurs for a particular value of the angle of incidence of the beam on the surface; this is known as the Brewster angle. For glass, it is about 57°.

Brewster's law states that this angle is $\tan^{-1} n$, where n is the refractive index of the material. This is an experimental conclusion. It leads to the result that the ray refracted into the medium must be at right angles to the ray reflected at the surface in this particular situation.

The plane of polarisation of the reflected ray is found to be at right angles to the plane of incidence (which is defined as containing the incident ray and the normal at the point of incidence). The refracted ray is of course also partially polarised, in the plane perpendicular to the polarisation of the reflected ray (Fig. 198).

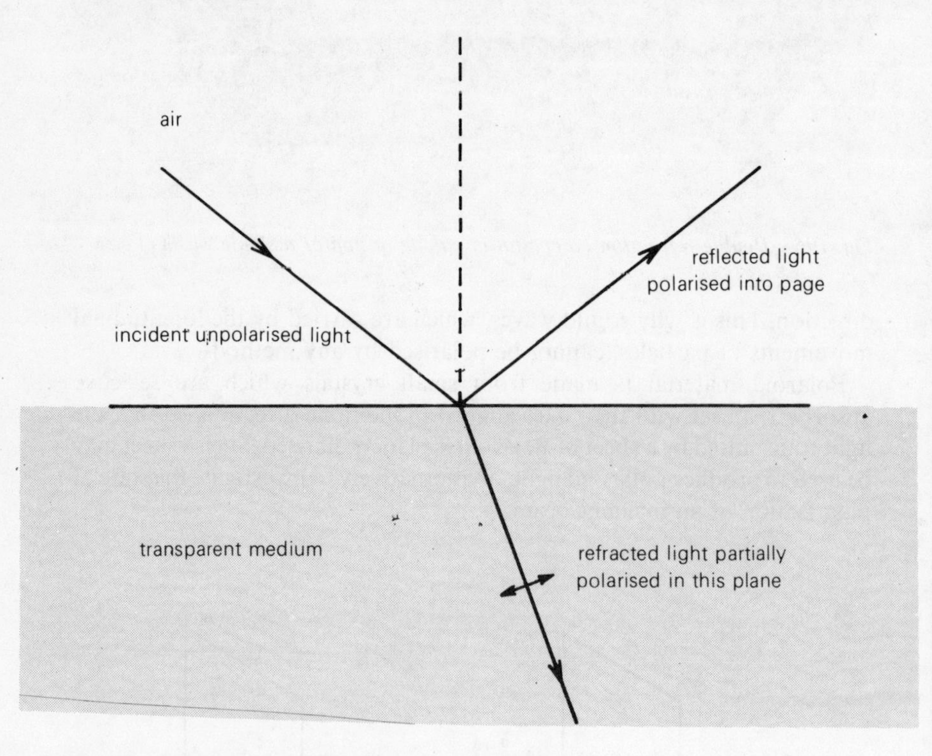

Fig. 198 Polarisation occurring in partial reflection

The degree of polarisation in the transmitted beam can be increased by using a number of sheets one behind the other. Partial reflection at each surface gradually removes all the light polarised in the one plane and transmits only the remainder.

Radio transmission

Polarisation is utilised in connection with the broadcasting of radio waves and microwaves.

These are transmitted in polarised form from a half-wave dipole aerial as in Fig. 199. High-frequency a.c. is supplied to the aerial, and electrons move forwards and backwards in the antennae BA and BC. By correctly choosing the lengths of these, a stationary state can be set up, with A and C as nodes (since the current is zero there) and B as an antinode.

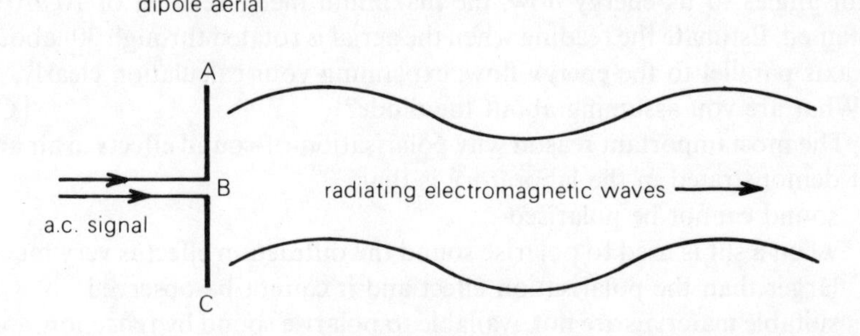

Fig. 199 Polarised radio waves emitted from a dipole aerial

The simple harmonic motion of the electrons causes changes in the surrounding electric and magnetic fields, and these changes radiate away through space with the speed of light. The waves are polarised since the electron movement is always in the one straight line AC, and the wavelength of the transmission is equal to 2.AC, since this is the distance between consecutive nodes.

To receive the waves, the aerial of the receiving circuit is best set in the same plane as that of the transmitter. The appropriate signal is selected by tuning the circuit to resonance (p. 194).

EXERCISE

1 Calculate the correct height for a transmitting aerial to be used at a frequency of 1 MHz, assuming that radio waves travel at a speed of 3×10^8 m s^{-1}.

Examination questions 39

1 Distinguish between *plane-polarised* light, *partially polarised* light and *unpolarised* light.

How could you use a single disc of Polaroid to detect the presence of each of these three types of light in an ordinary room in the house?

Two Polaroid discs are mounted on two stands on an optical bench so that their common axis is parallel to the bench. A source of light is placed on this common axis and an observer views it through the two Polaroid discs. Keeping one disc fixed, he rotates the other until he cannot see the light. Starting from this position he now rotates the same disc slowly through 360°. Describe what he should observe. [AEB, Nov. 1978]

2 What do you understand by a *plane-polarised* wave?

Describe how dipole radio aerials may be used to investigate plane-polarised radio waves. How might a d.c. meter and a diode be used to detect the signal in such an aerial?

When such an aerial is in the plane of polarisation of the signal, and at right angles to its energy flow, the maximum meter reading of 10 μA is obtained. Estimate the reading when the aerial is rotated through 30° about an axis parallel to the energy flow, explaining your calculation clearly.

What are you assuming about the diode? [C]

3 The most important reason why polarisation-of-sound effects in air are not demonstrated in the laboratory is that

A sound cannot be polarised.

B when a slit is used to polarise sound the diffraction effect is very much larger than the polarisation effect and it cannot be observed.

C suitable materials are not available to polarise sound by reflection, and this is the only practical possibility.

D the reflections from walls, benches etc. disrupt any experiments which attempt this demonstration.

E the width of any slits which would be suitable is comparable to the size of the laboratory, so the experiment is ineffective. [O]

4 Explain the terms *linearly polarised light* and *unpolarised light*.

A parallel beam of light is incident on the surface of a transparent medium of refractive index n at an angle of incidence i such that the reflected and refracted beams are at right angles to each other. Assuming the laws of reflection and refraction, find the relation between n and i.

At this value of i, the Brewster angle, the reflected beam is found to be linearly polarised. Describe the apparatus and procedure you would use to find the refractive index of a material, using the Brewster angle. Has the method any advantage over other methods for measurement of n?

[OC]

Extra Topics

40 Dimensions

All physical quantities are now measured in the units prescribed by the Système International d'Unités (SI).

In this system, the following four fundamental units are stated:

Mass .... kilogram (kg)
Length ... metre (m)
Time .... second (s)
Temperature .... kelvin (K)

All other units are derived in terms of these four.

Because of the way in which a particular quantity is defined, it is possible to consider it as being a combination of the fundamental units, and this results in a formula called the *dimensions* of the quantity. e.g. *force* is defined by Newton's second law of motion, and measured by the acceleration which it gives to a body,

$$\text{force} = \text{mass} \times \text{acceleration}$$

so that the dimensions of force are MLT^{-2}, abbreviating mass, length, and time to M, L, and T.

The dimension of temperature is written as K; this is only used in the special field of Heat.

In the fields of Electricity and Magnetism, it is necessary to introduce an additional dimension, that of current (I). Although the ampere is defined (p. 219) in terms of the fundamental units, it involves a new concept which is part of the definitions of all other electromagnetic quantities.

Consideration of the dimensions of a quantity can be helpful when stating the units in which it is measured, if these have no special name.

EXERCISES

Give the definitions of the following physical quantities, and hence deduce their dimensions.

1 Momentum
2 Angular momentum
3 Pressure
4 Torque
5 The universal gas constant $\mathscr{R}$.

Consistency of equations

During complicated calculations, one way to check that no errors have crept into the working is by considering the dimensions of the terms throughout an equation. These should be the same in each case.

For example, in the equation for the fundamental frequency of vibration of a stretched wire (p. 159),

$$f_0 = \frac{1}{2l}\sqrt{\frac{T}{m}},$$

the dimensions of the quantities involved are

f_0 (frequency) . . . . T^{-1}
l (length of wire) L
T (tension) . . . MLT^{-2}
m (mass per metre) . . . ML^{-1}.

Using these formulae in the equation gives

$$T^{-1} = \frac{1}{L}\sqrt{\left(\frac{MLT^{-2}}{ML^{-1}}\right)} \quad \text{(numerical constants have no dimensions)}$$

$$= \frac{1}{L}\sqrt{(L^2T^{-2})}$$

$$= \frac{1}{L}LT^{-1}$$

$$= T^{-1}.$$

This agreement shows that the equation is consistent as far as the physical quantities appearing in it are concerned. It does not, however, check the correctness of any numerical constants involved.

EXERCISES

Give the dimensions of both sides of the following equations, and thus show that they are consistent.

6 Newton's law of gravitation (p. 114), taking the dimensions of G from its value stated on p. 115.

7 The velocity of sound waves through a medium (p. 337), taking the dimensions of Young's modulus E from its definition on p. 132.

8 The equation for the conduction of heat through a solid (p. 23), taking the dimensions of thermal conductivity from the table on p. 23.

9 The resonant frequency of an a.c. circuit (p. 192), using the definitions of self-inductance L on p. 66, capacitance C on p. 209 and potential V on p. 199.

Examination questions 40

1 Explain the principle of the method of checking an equation by dimensions.

One of the following equations shows how the speed v of ripples on a deep liquid depends on the wavelength λ, the surface tension γ and the density ρ:

(i) $v = A\dfrac{\gamma}{\rho}\sqrt{\dfrac{1}{g\lambda^3}}$,

(ii) $v = B\dfrac{\gamma}{\rho}\sqrt{\dfrac{g}{\lambda}}$,

(iii) $v = C\sqrt{\dfrac{\gamma}{\lambda\rho}}$,

where g is the acceleration of free fall, and A, B and C are dimensionless constants. Identify any equation that is wrong dimensionally. [C]

2 Each of the following is an expression for the speed of a wave:

$$v = \sqrt{\frac{T}{m}}; \quad v = \sqrt{\frac{\gamma p}{\rho}}.$$

State a situation to which each expression applies, giving in each case the meanings of the symbols employed. Show that the right hand side of each expression has the dimensions of speed. [L]

3 A sphere of radius a and density ρ falling through a fluid of density σ reaches a terminal velocity v given by the expression $v = ka^2(\rho - \sigma)$, the value of k being constant for a given fluid. The dimensions of k are

A $ML^{-1}T^{-1}$ B $M^{-1}L^2T^{-1}$

C $ML^{-2}T^{-1}$ D $M^{-1}LT^0$

E none of the above [O]

41 Experimental errors

In a school laboratory, the degree of error of experimental work is generally about 1–2 %. If the limitations of the apparatus and the design of the experiment result in lower accuracy than this, then it cannot really be considered as more than a demonstration.

Such a degree of accuracy means that numerical results should be quoted only to three significant figures, although more may have been used during calculations, e.g. an answer given by an electronic calculator as 1.0392542 must be corrected to 1.04. To quote the value as 1.039 is a definite statement that it is not 1.038; such a statement implies that the probable error is less than 1 part in 1000.

The use of certain carefully made apparatus may justify results being given to four significant figures, but higher accuracy than this can be guaranteed only by painstaking, time-consuming attention to detail.

Errors of observations

In any experiment, there may have been readings of length or angle, weighings, thermometer readings, meter readings, measurements of time intervals, etc. None of these observations can be absolutely accurate, though some will be more accurate than others.

Unavoidable inaccuracy has three main sources:

(a) *Judgement involving the use of the human eye.* Example: as a general rule, the unaided eye cannot reasonably measure a length to better than 0.1 mm.
(b) *Limitations of the apparatus.* Example: the use of a given set of weights may mean that a mass cannot be determined to better than 0.01 g.
(c) *The personal factor.* An experimenter's performance improves with practice, but his reaction time has a definite biological minimum.

The maximum possible error of an observation must be assessed by considering the method which was involved. It is then expressed **as a percentage of the actual reading**.

Looking in this way at all the different observations involved in an experiment will show which one is the greatest source of possible error, and hence how the experiment might perhaps be improved. Example: the hand of a stopwatch moves in steps of $\frac{1}{5}$ s. If a time interval of about 2 min is to be measured, the use of this stopwatch limits the accuracy to 0.2 s in 120 s, or 0.17 %.

EXERCISES

Estimate the probable percentage errors of the following readings:

1 The 'height' of the barometer on an average day, read by eye to the nearest millimetre.

2 The 'height' of the barometer on an average day, read to the nearest 0.2 mm using a vernier scale.

3 A temperature difference of about 5°C, involving readings of two mercury thermometers graduated in $\frac{1}{5}$°C.

4 A current of about 8 mA, read on a meter whose scale is 80 mm long, graduated in 0.1 mA divisions.

5 A potential difference of about 100 mV, measured on a 1 m potentiometer scale calibrated directly against a Weston cadmium standard cell (see p. 45).

Probable maximum error of an experiment

The values obtained in a set of different observations will ultimately be combined in some way to give the final experimental results.

If the formula used to do this involves only multiplication and division operations, then it can be shown (by the Binomial Theorem) that the total error of the final result should not be greater than **the sum of the maximum possible errors of all the individual observations.** This is accepted as a working rule.

In the case of formulae involving addition and subtraction operations, such as a difference of temperature, these must be considered separately before the rule can be applied (see exercise 3, above).

Example: In an experiment to determine the density of a cubic object, the possible error involved in weighing is reckoned as 0.17%, and that in measuring the length of a side as 0.3%.
The formula used is:

$$\text{density} = \frac{\text{mass}}{(\text{length of side})^3},$$

so that the probable maximum error of the final result is

$$(0.17 + 3 \times 0.3) = 1.07\%.$$

Examination questions 41

1 A student takes the following readings of the diameter of a wire: 1.52 mm, 1.48 mm, 1.49 mm, 1.51 mm, 1.49 mm. Which of the following would be the best way to express the diameter of the wire in the student's report?

A between 1.48 mm and 1.52 mm

B 1.5 mm

C 1.498 mm
D (1.498 ± 0.012) mm
E (1.50 ± 0.01) mm [C]

2 The equation governing the volume rate of flow, v/t, of a fluid under streamline conditions through a horizontal pipe of length l and radius r is

$$\frac{v}{t} = \frac{\pi p r^2}{8 l \eta},$$

where p is the pressure difference across the pipe and η is the viscosity of the fluid.

In an experiment to find η for water, a student quotes his result as 1.137×10^{-3} kg m^{-1} s^{-1} and estimates the percentage uncertainties in his measurements of v/t, p, l and r as $\pm 3\%$, $\pm 2\%$, $\pm 0.5\%$ and $\pm 5\%$ respectively. How should he have written the value? [C]

3 The formula for the period of a simple pendulum is $T = 2\pi \sqrt{\frac{l}{g}}$. Such a pendulum is used to determine g.

The fractional error in the measurement of the period T is $\pm x$ and that in the measurement of the length l is $\pm y$. The fractional error in the calculated value of g is no greater than

A $x + y$ B $x - y$
C $2x - y$ D $2x + y$
E xy [C]

4 The errors in measurements made on a piece of wire are $\pm 0.1\%$ for the length, $\pm 2\%$ for the diameter and $\pm 0.5\%$ for the resistance. The value of the resistivity as calculated from these measurements has a maximum error of

A $\pm 0.1\%$ B $\pm 2.0\%$ C $\pm 2.4\%$ D $\pm 4.4\%$ E $\pm 4.6\%$ [O]

5 Explain the terms *random error* and *systematic error.*

To reveal systematic error, the measurement of a physical quantity is repeated under varying experimental conditions; to reduce the effect of random error, the readings are repeated under the same experimental conditions and their mean value is taken. Discuss the reasons for each of these procedures, and describe carefully how you would apply them when using a micrometer screw gauge to measure the diameter (about 15 mm) of a steel ball-bearing.

If after considering the incidence of errors the result is expressed as (15.05 ± 0.02) mm, what are the corresponding expressions for (a) the surface area of the sphere and (b) the volume of the sphere? [O]

6 The dimensions of a rectangular block are measured as (100 ± 1) mm $\times$ (80 ± 1) mm $\times$ (50 ± 1) mm. The value of the volume of the block calculated from these readings will have an error of at most about

A $\frac{1}{2}\%$ B 1% C 3% D 4% E 6% [O]

Answers

The value of $g = 10\ \mathrm{m\,s^{-2}}$ has been used throughout the exercises except where a different value has been stated.

Chapter 1

EXERCISES			EXAMINATION QUESTIONS		
p. 4	1	1.002 m	p. 9	1	$3.7 \times 10^{-3}\ \mathrm{K^{-1}}$
	2	(i) 40.067 ml		2	C
		(ii) 1.21 ml		3	C
	3	313 mm higher than A		4	49.05° C
p. 8	4	61.7° C			
	5	19.0° C			

Chapter 2

EXERCISES			EXAMINATION QUESTIONS		
p. 15	1	$376\ \mathrm{J\,kg^{-1}\,K^{-1}}$	p. 19	1	9.22 g
	2	$900\ \mathrm{J\,kg^{-1}\,K^{-1}}$		2	$1.64 \times 10^{3}\ \mathrm{J\,kg^{-1}\,K^{-1}}$
	3	(i) 18° C		3	(c) $1.67 \times 10^{3}\ \mathrm{J\,kg^{-1}\,K^{-1}}$
		(ii) approx. 31° C per min		4	(b) $2.38 \times 10^{6}\ \mathrm{J\,kg^{-1}}$
	4	approx. 40° C		5	E
p. 19	5	$4520\ \mathrm{J\,kg^{-1}\,K^{-1}}$			
	6	(i) $4100\ \mathrm{J\,kg^{-1}\,K^{-1}}$			
		(ii) $2.9\ \mathrm{J\,s^{-1}}$			

Chapter 3

EXERCISES			EXAMINATION QUESTIONS		
p. 28	1	50.4 J	p. 29	1	44 kW
	2	0.17		2	C
	3	3.7 kW		3	(d) 102.5° C
	4	261 kJ		4	6400 W, 66:1

Chapter 4

EXERCISES			EXAMINATION QUESTIONS		
p. 35	1	(i) 0.4 A	p. 39	1	D
		(ii) 1.6 W		2	(i) 1.0 Ω (ii) 3.75 J
		(iii) 32 J		3	0.367 m
	2	(i) 11.20 Ω		4	B
		(ii) 11.15 Ω			
	3	(i) 2×10^{19}			
		(ii) $2 \times 10^{-6}\ \mathrm{m\,s^{-1}}$			
	4	4.03 Ω			

EXERCISES			EXAMINATION QUESTIONS		
p. 39	5	0.71 Ω, 0.10 V			
	6	(i) 17 Ω (ii) 0.54 Ω 0.118 A, 2 A, 1 A, 0.5 A, 0.2 A			

Chapter 5

p. 43	1	2.67 Ω	p. 47	1	(b) 6.33 mV
	2	Near the 450 mm mark		2	(c) 294.7 Ω
p. 46	3	Leclanché: Daniell = 1.38:1		3	D
	4	0.60 A		4	D
	5	4×10^6 Ω			

Chapter 6

p. 52	1	0.152 Ω	p. 53	2	B
	2	5 mm mA^{-1}		3	55° C
	3	(i) Add 980 Ω in series		4	E
		(ii) Add 480 Ω in series			

Chapter 7

		No exercises	p. 59	2	B

Chapter 8

p. 64	1	0.05 weber	p. 68	1	2.62 mV
	2	115 T		2	B
	3	0.04 V			
	4	14.6 μV			
	5	1.57 V			
p. 67	6	0.8 V			

Chapter 9

p. 76	1	1.33	p. 76	1	39.6°
	2	2.04×10^8 m s^{-1}		2	A
	3	35.7°		5	A
	4	28.8°			
	5	61.1°			

Chapter 10

p. 83	1	32°	p. 85	1	C
	2	39.8°		2	B
	3	1.532			

Chapter 11

EXERCISES	EXAMINATION QUESTIONS
p. 88 **1** 7.3×10^{-7} m, red	p. 90 **1** 1.61×10^{3} W m^{-2}
2 5.85 kW	**2** C
3 115 K	**3** C
4 (i) 3.56 kW (ii) 0.46 kW	**4** (i) 251×10^{3} photons s^{-1} (ii) 19 %
p. 90 **5** 1.99 W m^{-2}	**5** 378 K
6 39.6×10^{22} kW	

Chapter 12

EXERCISES	EXAMINATION QUESTIONS
p. 102 **1** 1.99 s	p. 103 **2** 2.05 p.m.
2 0.34 s	**5** E
3 1.62 s	**6** initial acceleration $= 0.15$ m s^{-2}

Chapter 13

EXERCISES	EXAMINATION QUESTIONS
p. 109 **1** (i) 4π rad s^{-1} (ii) π m s^{-1}	p. 112 **1** (i) 5.62 rad s^{-2} (ii) 3.34 s (iii) 0.0204:1
2 0.2 N m	**2** B
3 2.4 kg m^{2} s^{-1}	**3** (b) 186 rad s^{-1} (c) 2.23×10^{-3} N m
4 137 J	**4** A
p. 112 **5** 505 m s^{-2}	**5** B
6 0.93 m s^{-1}	
7 26.1°	

Chapter 14

EXERCISES	EXAMINATION QUESTIONS
p. 119 **1** 1.334×10^{-10} N	p. 121 **1** (c) 41.4 mm
2 4.0×10^{-10} N m	**2** 3.4×10^{8} m
3 5.93×10^{24} kg	**4** A
4 9.802 m s^{-2}	**5** C
p. 121 **5** (i) 3.06 km s^{-1} (ii) 3.58×10^{4} km	**6** 1.95 hr
6 5.96×10^{24} kg	

Chapter 15

EXERCISES	EXAMINATION QUESTIONS
p. 123 **1** 1.47×10^{4} kg m s^{-1}	p. 128 **1** (c) (i) 20 kg m s^{-1} (ii) 2520 N (iii) 0.008 s
2 In the direction that the 2 kg ball was moving.	
p. 127 **4** 5.6 N s	**2** 1.85×10^{4} N
5 $500\sqrt{10}$ N	**3** 2.5×10^{5} N, 1.54×10^{5} N
6 10^{4} N	**4** 319 m s^{-1}
7 126 N	**6** $7\frac{1}{3} E$

EXERCISES — EXAMINATION QUESTIONS

Chapter 16

Exercises

p. 134 **1** (i) $1.25 \times 10^{8}\ \mathrm{N\,m^{-2}}$
(ii) 1.25×10^{-3}
(iii) $1.00 \times 10^{11}\ \mathrm{N\,m^{-2}}$
2 0.60 mm
p. 135 **3** 0.05 J
4 0.254 J
5 (i) 0.06 J
(ii) 0.16 J

Examination questions

p. 139 **1** 2.8×10^{5} N
3 (a) 8 : 1 (b) $\frac{1}{9}D$
4 3.5×10^{-2} m
5 (a) 2.02 J, 0.86 J, 1.16 J

Chapter 17

Exercises

p. 145 **1** 3.77×10^{-3} J
2 $6.8 \times 10^{-2}\ \mathrm{J\,m^{-2}}$
3 55.2 mm
p. 148 **4** $2.67\ \mathrm{N\,m^{-2}}$
5 $1.28 \times 10^{5}\ \mathrm{N\,m^{-2}}$
6 $340\ \mathrm{N\,m^{-2}}$, 34 mm depth

Examination questions

p. 149 **2** C
3 C
5 133 mm

Chapter 18

Exercises

p. 155 **1** 512 Hz
2 16 Hz
3 426.7 Hz
p. 158 **4** 515.1 Hz
5 125 Hz

Examination questions

p. 158 **2** E
3 about 2% accuracy

Chapter 19

Exercises

p. 163 **1** 50 Hz
2 339 Hz
3 768 Hz, 1280 Hz
4 800 Hz

Examination questions

p. 164 **1** 1.41 l
2 E
4 0.38 m, 4 N

Chapter 20

Exercises

p. 170 **1** 0.25 m
2 15 mm
3 0.50 m
4 0.367 m

Examination questions

p. 171 **2** B
3 E
5 D

Chapter 21

Exercises

p. 176 **1** 8.5 V
2 163 V, 0.0167 s

Examination questions

p. 177 **1** C
2 C

EXERCISES — EXAMINATION QUESTIONS

Chapter 22

Exercises

p. 180 **1** (i) $\frac{1}{240}$ s, $\frac{3}{240}$ s, $\frac{5}{240}$ s . . .
(ii) 0, $\frac{1}{120}$ s, $\frac{1}{60}$ s . . .

2 (i) 400π V
(ii) $200\sqrt{3}\pi$ V

3 $\frac{1}{8}$, $\frac{3}{8}$, $\frac{5}{8}$ and $\frac{7}{8}$ of each cycle
(ii) 900π W

4 (i) $\frac{\pi}{10}\,\Omega$
(ii) $2000\pi\,\Omega$

Examination questions

p. 183 **1** (b) $1000\pi\,\Omega$, 19.9 W

2 (d) (ii) 389 k W
(iii) 1480 V
(iv) 27.0%
(vi) 0.675 k W
(vii) 10 045 V
(viii) 95.6%

3 $\frac{\pi}{2\omega}$, 0.005 s *or* $\frac{3\pi}{2\omega}$, 0.015 s depending on the definition of inductance

Chapter 23

p. 188 **1** (i) $2\pi \times 10^{-2}$ A
(ii) $50\sqrt{3}$ V

2 (i) $\frac{1}{8}$, $\frac{3}{8}$, $\frac{5}{8}$ and $\frac{7}{8}$ of each cycle
(ii) 2.4π mA

3 (i) $\frac{1}{\pi} \times 10^{6}\,\Omega$
(ii) $\frac{50}{\pi}\,\Omega$

Chapter 24

Exercises

p. 192 **1** 330 Ω
2 332 Ω
3 74 Ω
4 (i) 0.73 A
(ii) 73.2°

p. 195 **5** 796 Hz
6 1.27 pF

Examination questions

p. 195 **1** (i) 15.9 mH
(ii) $\cos^{-1}\frac{5}{6}$

2 (i) $5 \times 10^{3}\,\Omega$
(ii) $5 \times 10^{3}\,\Omega$
(iii) $5\sqrt{2} \times 10^{3}\,\Omega$
(iv) $\pi/4$

4 A

5 P . . . 300 Ω, $\frac{4}{\pi}$ H
Q . . . $\frac{1}{4\pi} \times 10^{-4}$ F

Chapter 25

Exercises

p. 202 **1** 1.67 μC
2 0.159 MV m^{-1}
3 13.3 μC

p. 203 **4** 89,900 V

p. 206 **5** 2.05×10^{6} V

Examination questions

p. 207 **2** 1.6×10^{-19} C
3 D
4 6.4×10^{-19} C
6 D

6 1.92×10^{-14} N, 2.11×10^{16} m s^{-2}, 9.74×10^{-10} s
7 14.4 mm

Chapter 26

Exercises

p. 212
1 20 V
2 0.02 μC
3 9.56×10^{-9} F
4 8.92×10^{-12} F m^{-1}

p. 215
5 2.5×10^{-3} J
6 3.2×10^{-2} J
7 8.19×10^{-4} C

Examination questions

p. 215
1 (b) 7.0 (c) 8 : 1
2 D
3 (a) 20×10^{-6} C, 20×10^{-6} J
(b) 7.4×10^{-6} C
4 1.35×10^{-3} C, three times, 84 V
5 (a) 1 MV m^{-1}
(b) 32.5 pF, 2.6×10^{-4} J, 8.67×10^{-12} F m^{-1}
6 B

Chapter 27

Exercises

p. 221
1 2×10^{-5} N
2 0.283 N m
3 5×10^{-2} J T^{-1}

p. 226
4 3.93×10^{-5} T
5 2×10^{-6} T
6 3.77×10^{-3} T
7 2.51×10^{-3} T

p. 230
8 0.17 m
9 5×10^{7} m s^{-1}

Examination questions

p. 231
1 18.3 mm
2 (a) 3.6×10^{-9} s
(b) 3.6×10^{-9} s
3 (b) (i) 1.6×10^{-4} T
(ii) zero
4 B
5 D
6 0.8 T

Chapter 28

Exercises

p. 238
1 Polonium is element 84 in the periodic table; it has 84 protons + 126 neutrons in the nucleus
2 $^{238}_{92}\text{U} \rightarrow {}^{234}_{90}\text{Th} + {}^{4}_{2}\text{He}$
3 $^{210}_{82}\text{Pb} \rightarrow {}^{210}_{83}\text{Bi} + {}^{0}_{-1}\text{e}$
4 (i) 9.45 days
(ii) 4.0×10^{4} disintegrations/s
5 7.8×10^{-10} s^{-1}

p. 243
6 (i) 1.62×10^{-13} J
(ii) 1.01 MeV
7 0.0304 u
8 (i) 8.82 MeV
(ii) 7.85 MeV

Examination questions

p. 243
1 (b) A.D. 790
2 E
3 (d) (i) 5.40 MeV
(iii) 5.30 MeV
4 3.01602 u
5 A
6 $A = 216$, $Z = 88$; 1.16×10^{-19} N s; 3.37×10^{5} m s^{-1}

Chapter 29

EXERCISES			EXAMINATION QUESTIONS		
p. 248	1	26 mV	p. 255	2	C
	2	1.04×10^{29} m^{-3}			

Chapter 30

EXERCISES			EXAMINATION QUESTIONS		
p. 261	1	0.059 m^3	p. 265	2	B
	2	0.025 m^3		3	0.758×10^5 N m^{-2}
	3	22.7		5	D
p. 264	4	5×10^4 J			
	5	1.03×10^5 Pa			
	6	123° C			

Chapter 31

EXERCISES			EXAMINATION QUESTIONS		
p. 269	1	(i) 4×10^{-23} kg m s^{-1} (ii) $\frac{2}{3} \times 10^5$ N m^{-2}	p. 273	1	1.5×10^{23}, 43.2 m s^{-1}
	2	1.75×10^3 m s^{-1}		2	D
p. 272	3	483 K		3	4.35×10^{16} molecules/m^3
	4	4.65×10^3 J		4	A
	5	20.8 J K^{-1}			
	6	20.8 J K^{-1}			

Chapter 32

EXERCISES			EXAMINATION QUESTIONS		
p. 278	1	13 g	p. 283	1	(c) (i) 2.0 k Pa (ii) 1.5 k Pa
	2	1.34×10^5 Pa		2	(a) 4.1×10^4 J mol^{-1}
	3	0.000125 m^3		5	A

Chapter 33

EXERCISES			EXAMINATION QUESTIONS		
p. 289	1	800 mm	p. 294	1	107 mm
	2	153 mm		2	E
p. 292	3	222 mm		3	(b) −2.0 m, 0.5 m
	4	−200 mm		4	27.8 mm
	5	374 mm		5	−0.2 m
	6	693 mm		6	20 mm approx.
p. 294	7	0.0233, 0.0336, −288 mm			

Chapter 34

EXERCISES			EXAMINATION QUESTIONS		
p. 299	1	8.3 m × 11.7 m	p. 302	1	7.25
	2	*f*/7.9		2	B
	3	2		4	0.26 m, 0.53 m
p. 302	4	6×10^{-4c}		5	0.065 s
	5	+120 mm, −30 mm			

Chapter 35

EXERCISES	EXAMINATION QUESTIONS
p. 308 **1** 3.3×10^{-19} J	p. 313 **1** (a) 652 nm
2 1.02×10^{15} Hz	(b) 6.61×10^{-34} J s
3 2.61×10^{-17} J	(c) 4.6×10^{14} Hz
4 2.48×10^{-13} m	(d) 1.9 eV
p. 311 **5** 4.58×10^{14} Hz,	**2** D
6.18×10^{14} Hz	**3** D
	5 1.5 V
	(iii) 2.4×10^{-19} J
	(iv) 7.3×10^{5} m s^{-1}

Chapter 36

EXERCISES	EXAMINATION QUESTIONS
p. 320 **1** 175 mm	p. 324 **2** (a) 1.67 m
2 3 mm	(b) 360 s^{-1}
3 250 mm	**3** 500 nm
p. 324 **4** A minimum	**4** 56.4×10^{-7} m
5 0.02 mm	
6 83.8 mm	

Chapter 37

EXERCISES	EXAMINATION QUESTIONS
p. 332 **1** 13.9°	p. 333 **1** (d) (i) 500 (ii) 63.8°
2 Three	**2** Three
3 2.37×10^{-10} m	**3** 6×10^{-7} m
	4 A

Chapter 38

EXERCISES	EXAMINATION QUESTIONS
p. 338 **1** 0.5 m	
2 5 km s^{-1}	
3 48.4 N	
4 (i) 328 m s^{-1}	
(ii) 340 m s^{-1}	
p. 339 **5** 3.12×10^{8} m s^{-1}	
6 3.05×10^{8} m s^{-1}	
7 2.93×10^{8} m s^{-1}	
8 2.998×10^{8} m s^{-1}	

Chapter 39

EXERCISES	EXAMINATION QUESTIONS
p. 345 **1** 150 m	p. 345 **2** 8.66 μA if diode is 'linear', 7.5 μA if diode is 'square law'
	3 A

EXERCISES

Chapter 40

p. 348 **1** MLT^{-1}
2 ML^2T^{-1}
3 $ML^{-1}T^{-2}$
4 ML^2T^{-2}
5 $ML^2T^{-2}K^{-1}$
p. 349 **6** MLT^{-2}
7 LT^{-1}
8 ML^2T^{-3}
9 T^{-1}

EXAMINATION QUESTIONS

p. 349 **1** Equation (ii) is wrong
3 B

Chapter 41

EXERCISES

p. 352 **1** 0.5 mm in 760 mm = 0.07%
2 0.1 mm in 760 mm = 0.013%
3 $\frac{1}{2} \times \frac{1}{5}^\circ$ in 5° = 2%
4 $\frac{1}{4} \times$ 0.1 mA in 8 mA = 0.3%
5 1 mm in 1000 mm = 0.1%

These answers are all open to discussion.

EXAMINATION QUESTIONS

p. 352 **1** E
2 $(1.1 \pm 0.3) \times 10^{-3}$ kg m^{-1} s^{-1}
3 D
4 E
5 (a) 711.5 ± 1.9 mm^2
(b) 1784 ± 7.1 mm^3
6 D

Index